EQUIPMENT IN THE HOME

EQUIPMENT IN THE HOME
THIRD EDITION

Florence Ehrenkranz
Professor Emeritus, College of Home Economics
University of Minnesota

Lydia L. Inman
Coordinator of Resident Instruction
College of Home Economics
Iowa State University

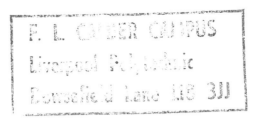
HARPER & ROW, PUBLISHERS
New York, Evanston, San Francisco, London

EQUIPMENT IN THE HOME, Third Edition

Copyright © 1958 1966, 1973 by Florence Ehrenkranz and Lydia L. Inman

Printed in the United States of America. All rights reserved. No part of this book may be used or reproduced in any manner whatsoever without written permission except in the case of brief quotations embodied in critical articles and reviews. For information address Harper & Row, Publishers, Inc., 10 East 53rd Street, New York, N.Y. 10022.

Standard Book Number: 06-041872-9

Library of Congress Catalog Card Number: 72-86371

CONTENTS

PREFACE

Equipment in the Home is planned as a textbook for college courses and a reference source for homemakers, home economists, and business people in the home appliance industries.

Principles and practical considerations involved in kitchen and laundry design and in selection, use, and care of appliances are one field of home economics. Our teaching experience, as well as the information obtained from the use of the first two editions of this book, especially from university students, suggests that the material in this new edition can be useful for an introductory course of one quarter, one semester, or two quarters, and as a reference text for advanced courses in refrigeration equipment, cooking equipment, laundry equipment, and courses that deal with the physical aspects of residential housing.

The outlines of experiments at the ends of most of the chapters are intended to emphasize the relation between theory and practice. Buying guides are presented for major equipment and small appliances. These can be used by the consumer when selecting equipment or by the student in laboratory assignments to develop a comprehensive understanding of desirable features of equipment.

An attempt is made to emphasize principles that will be useful in the future as well as the present. Applications of these principles are made to equipment on today's market. Many new illustrations bring the reader an up-to-date view of major and small appliances, kitchen utensils, tools, and kitchen plans.

It is a pleasure to acknowledge the courtesy extended to us by the many home economists and manufacturers who furnished technical information and photographs. Acknowledgment is also made to Helen Le Baron Hilton, who served as editor of the first two editions.

Florence Ehrenkranz
Lydia L. Inman

EQUIPMENT
IN THE HOME

CHAPTER 1
ELECTRICITY
IN THE HOME

INTRODUCTION

This book describes home equipment available to consumers. The utility of the equipment selected to make activities such as dishwashing easier and/or better or to improve livability by good kitchen design, good lighting, air conditioning, and so forth, involves informed choice for the family life style, appropriate facilities in the home, and use of the equipment in the manner and for the purposes for which it was designed.

The first chapter deals with electricity in the home for two reasons: First, adequate wiring is needed in order for electrical equipment to perform properly or as promised in the specifications, and second, nameplate data on electric appliances indicate electric characteristics that have significance for an informed consumer.

Familiarity with some terms and with a few elementary principles will help in understanding many factors associated with the use of electricity in the home. The material in this chapter covers some of the basic concepts needed for understanding why different recommendations are made for particular circuits in the home, for interpreting information given on nameplates of electric appliances, for estimating costs of operating different types of appliances, and so on. In addition, characteristics of the actual electrical parts used in home wiring and recommendations for convenient, safe wiring are considered. The third part of the chapter deals with electrical symbols and recommendations for wiring different areas of residences.

Electricity used in the home is not a stored quantity. Rather, whenever a path for the flow of electric charge is completed, such as by closing a switch, small electric charges known as electrons move along the completed path or circuit, provided the circuit includes a source of electric energy. The source of electric energy used in the home is the electric generator in the city or the rural power plant that serves the residence.

As electric charges move in a wire, certain effects occur that are utilized in household appliances. One effect is generation of heat. Heating appliances, such as hand irons and surface units on electric ranges, are designed to use the heat associated with the motion of electric charges.

Another effect is the creation of a magnetic field around the wire caused by the movement of electric charges in the wire. By means of this magnetic field, the wire can be made to rotate under appropriate conditions. In motor-driven appliances, such as mixers, washers, and motors of electric refrigerators, part of the wire in which the electric charges move is coiled on a shaft which in turn is mounted in bearings. A magnetic field, provided by another coil in the motor, exerts force on the wire coiled on the shaft and thus causes the shaft to rotate in its bearings. This rotation causes beaters of mixers to revolve, agitators of washers to oscillate, pistons or cams of electric refrigerator compressors to operate, and so on. Rotisseries and electric dryers utilize both the magnetic field and the heating effect associated with moving electric charges.

Controls for equipment may depend on phenomena associated with electric current. Of special interest is a solid-state control sometimes used in electric mixers. As the load increases—for example, batter becomes stiffer—the electric current in the mixer circuit increases and the electrical resistance of the solid-state component of the control decreases. The change in resistance in the control causes the beaters to increase in speed again, *provided* the circuit in the mixer is designed to do this.

PHYSICAL PRINCIPLES AND APPLICATIONS

Some terms or words used in a discussion of electricity have a quantitative aspect; so many amperes are required by an appliance, or an appliance is rated for a given number of volts. Other terms do not have a quantitative aspect but nevertheless need to be defined; examples are electric circuit, open circuit, and short circuit.

An *electric circuit* is a complete or closed path for the flow of electric charge. An *open circuit* is interrupted or broken at one or more points; the electrons in the circuit wire on one side of the break do not cross the break and continue along the circuit. In a *short circuit* at least part of the moving electric charge traverses a smaller portion of the circuit than it should. This may happen when two parts (two wires or a wire and a switch, for example) that should not make electrical contact do make contact. A short circuit or a "short" is usually accidental; sometimes, however, an electrician shorts parts of a circuit purposely in testing.

CURRENT, RESISTANCE, AND VOLTAGE

Electric current is a flow or motion of electric charges in a definite direction. The electric current in a wire that is part of an electric circuit is actually a motion in a definite direction of the loosely bound or "free" electrons of the atoms of which the wire is made. When we turn on an electric appliance, we do not wait for electrons from the generator at the power plant to reach the appliance; rather the free electrons in the wire in the appliance start moving almost instantly.

The practical unit of electric current is the *ampere*. The number of amperes flowing in a circuit depends on the electrical resistance of the circuit and the voltage applied to the circuit.

Electrical resistance is a physical property of materials. It is, as the name implies, the opposition or resistance that materials offer to a flow of electric charge. The practical unit of electrical resistance is the *ohm*.

For most household applications, electrical resistance depends on the material, its length, and its cross-sectional area. Materials are classified as conductors, insulators, or semiconductors. Good conductors of electric current such as copper and aluminum have low electrical resistances. Good insulators (poor conductors) such as porcelain and Bakelite have high electrical resistances. Semiconductors are materials intermediate between conductors and insulators; the electrical properties, including the resistance of different semiconductors, can be changed markedly by

different parameters such as heat, pressure, or sound. At this time, these materials are utilized in household appliances only for special applications such as solid-state controls.

Electrical resistance of the commonly used conductors is directly proportional to length and inversely proportional to the cross section available for flow of electric charge. For house circuits copper wire is usual, partly because it has a lower resistance than many conductors. If house circuit A is twice as long as house circuit B and each uses wire of the same material and diameter, the electrical resistance of circuit A is twice the electrical resistance of circuit B. If the diameter of the wire used in house circuit C is twice that used in house circuit D and the lengths of the two circuits are the same, the resistance of circuit C is one-fourth that of circuit D.

The voltage difference or potential difference between two points in an electric circuit is the work done when one unit of electric charge moves from one of the points to the other. The practical unit of potential difference is the *volt*.

Potential difference often is compared to water pressure. Indeed, it is sometimes described as the "electrical pressure" required for the flow of electric charge. In some cases this analogy is fruitful. For example, it suggests correctly that electric charges will always move when a potential difference exists between two points connected by a conductor. Also, electric charges will not move between two points of a conductor when there is no potential difference between the two points. In an electric circuit, of course, the potential difference is due to a source of electric energy such as a battery or a 115-volt supply.

Ground or a good connection to ground is arbitrarily assumed to be at 0 volts. The potential difference between ground and a circuit wire is either plus or minus—for example, plus 115 volts or minus 115 volts. Good connections to ground in the home include metal water supply and drain lines. If plastics are used to make joins on water or drain pipes then the joins are not good ground connections.

OHM'S LAW

Ohm's law states that the current that flows through an electrical resistance is equal to the potential difference divided by the resistance.

$$amperes = \frac{volts}{ohms}$$

$$ohms = \frac{volts}{amperes}$$

$$volts = amperes \times ohms$$

The application of Ohm's law to household appliances is direct. Assume that a hand iron with a resistance of 13 ohms is connected to a 115-volt supply. Assume also that the resistance of the wires in the house circuit to the hand iron is negligible. (This last assumption is not completely valid, as is shown in the third example below, but it is a reasonable one for the present illustration.) According to Ohm's law, the amperes that will flow through the iron equal volts divided by ohms, and 115 divided by 13 is approximately 8.8 amperes.

As another example of Ohm's law, consider that the nameplate of a portable electric broiler indicates 10 amperes, 115 volts. The resistance of the coil in the broiler is volts divided by amperes, and 115 divided by 10 is 11.5 ohms.

As a third example, Ohm's law can be used to compute the *voltage drop* in a house circuit. Assume that an outlet in which we are interested is 100 feet from the electric service entrance in the basement where electricity is "delivered" to the house at 115 volts. Assume that 9 am-

peres actually flow through a small appliance plugged into this outlet. Then 9 amperes also flow through the two wires between the service entrance and the outlet.

To determine the voltage drop between the electric service entrance and the outlet, we must know the total resistance of the wires between the service entrance and the outlet. Wire used for small-appliance circuits in the home has an approximate resistance of 0.16 ohms per 100 feet. Applying Ohm's law, the voltage drop in each of the two wires between the service entrance and the outlet is amperes times ohms, or 9 × 0.16, which is 1.44. The total voltage drop in the two wires is 2 × 1.44 or 2.88 volts. Therefore the voltage at the outlet into which the appliance is plugged is not 115 volts; rather, it is 115 minus 2.88 or 112.12 volts.

POWER, ENERGY, AND COST OF ELECTRIC ENERGY

Electric power in a circuit or any part of a circuit is the rate at which work is done or energy is given up. The practical *unit of electric power* is the *watt*. Another unit often used is the *kilowatt*, 1 kilowatt being equal to 1,000 watts.

In circuits that contain resistances only, such as those of heating appliances, watts equal volts times amperes. For example, one model of a nonautomatic coffee maker uses 4.4 amperes at 110 volts. Since the coffee maker is a heating-type appliance, the electric power it uses is 110 × 4.4 or 484.0 watts (0.484 kilowatts). As another example, we return to our two 100-foot lengths of wire and the outlet for a small appliance. If it is a heating appliance plugged into the outlet, the power used by it is 112.12 × 9 or 1,009.08 watts. The power used in *each* of the 100-foot lengths of wire is 1.44 × 9 or 12.96 watts. The power used in the entire circuit is 115 × 9 or 1,035 watts.

Electric energy is the work done or the energy expended in a circuit or part of a circuit in a given time. Electric energy is power times time, and this is true for any kind of electric circuit. One unit is the *watt-hour*. Most simply, watt-hours are watts times hours. For example, when a 100-watt lamp is turned on for one hour, the lamp uses 100 watt-hours and in two hours it uses 200 watt-hours. A 1,000-watt automatic toaster that is on for three minutes (3/60 of an hour) uses 1,000 × 3/60 or 50 watt-hours. A 300-watt refrigerator that is on for 15 minutes uses 300 × 1/4 or 75 watt-hours.

A *kilowatt-hour* is 1,000 watt-hours. Thus the 100-watt lamp uses 100 watt-hours or 0.1 kilowatt-hours per hour. The same lamp uses 0.2 kilowatt-hours in two hours. The 1,000-watt automatic toaster uses 50 watt-hours or 0.05 kilowatt-hours in three minutes.

To distinguish between electric power and electric energy note the hyphenated name for energy: kilowatt-hours, meaning kilowatts times hours. A 2,000-watt or 2-kilowatt unit operating continuously for one hour uses the same amount of energy as a 1-kilowatt unit operating continuously for two hours and less energy than the 1-kilowatt unit operating for three hours.

The cost of operating appliances and our monthly electric bills are calculated by multiplying kilowatt-hours by cost per kilowatt-hour. At 3 cents per kilowatt-hour the cost of operating a 100-watt lamp for one hour is: 0.1 kw × 1 hr × 3¢ = 0.3¢ or $0.003. For two hours the cost would be: 0.1 kw × 2 hrs × 3¢ = 0.6¢ or $0.006.

A 1,000-watt toaster operated for three minutes would cost: 1 kw × 3/60 hrs × 3¢ = 0.15¢.

A 2-kilowatt unit operating continuously for one hour would cost: 2 kw × 1 hr

Table 1-1. Electric Rates—Residential Minneapolis

first 50 kilowatt-hours	5.0¢ per kilowatt-hour
next 150 kilowatt-hours	2.7¢ per kilowatt-hour
next 500 kilowatt-hours	2.0¢ per kilowatt-hour
over 500 kilowatt-hours	1.5¢ per kilowatt-hour

× 3¢ = 6¢ and a 1-kilowatt unit operating continuously for two hours would be the same (1 kw × 2 hrs × 3¢ = 6¢).

Cost per kilowatt-hour varies in different communities. Also, cost is likely to be on a "sliding scale," that is, cost per kilowatt-hour decreases as more kilowatt-hours are used. For example, residential cost might be 5 cents per kilowatt-hour for the first 100 kilowatt-hours and 2 cents per kilowatt-hour for all over 100. Thus if a family uses 120 kilowatt-hours per month, the cost that month is 100 × 5¢ + 20 × 2¢ or $5.40.

At the time this text is being prepared, a local utility company quotes rates as shown in Table 1-1.

The number of kilowatt-hours used per month is obtained by "reading the meter," actually a kilowatt-hour meter, at the beginning and end of the month and subtracting the initial reading from the final one.

ALTERNATING CURRENT, INDUCTANCE, AND POWER FACTOR

Direct current—that is, current that flows in one direction only—is supplied by dry cells and batteries. Some cordless, portable appliances are battery operated. Most household equipment, however, uses an alternating-current supply. Alternating current varies in magnitude and reverses in direction. The alternating current now commonly supplied in the United States is 60-cycle current, although 50-cycle current may still be supplied in some areas. In each cycle, the current increases from a value of zero to a maximum value; it

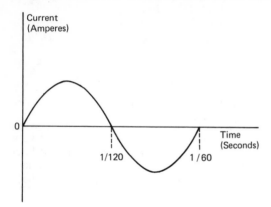

Figure 1-1. Variation of 60-cycle alternating current during one cycle.

then decreases to zero, increases again to a maximum value, but this time in the opposite direction, and then decreases once again to zero. The cycle repeats 60 times per second for 60-cycle current or 50 times per second for 50-cycle current (Fig. 1-1).

During one-half of each cycle, electrons are moving in one direction along a wire; during the other half of the cycle they move in the opposite direction. At the instant the motion reverses, the current is zero.

Inductance, Inductance-Type Motors, and Universal Motors

The *inductance* of a circuit or part of a circuit is the opposition offered to a change in magnitude of electric current. The mechanical analog of inductance is inertia—opposition to a change in motion such as starting or stopping.

Inductance is relatively unimportant in heating appliances, but it is of considerable importance in some motor-driven appliances. Coils of wire are used in motors of motor-driven appliances; they usually are wound on cores of iron or other magnetic material. Such a coil may offer a significant opposition to a change in magnitude of current. And alternating current definitely changes in magnitude.

Motors used in most motor-driven household appliances either are *inductance-type* or *universal* motors. Inductance-type motors are used in washers, dishwashers, food waste disposers, elec-

tric refrigerators, freezers, electric dehumidifiers, dryers, room air conditioners, and in some other appliances. Universal motors are used in vacuum cleaners, some mixers, and in a few other appliances. Inductance-type motors are intended to be used at the *frequency* specified on the nameplate, for example, 60 cycles per second or, for some, 50–60 cycles per second. Universal motors may be rated to operate at any frequency from 0 cycles per second (direct current) up to 60 per second; Hertz (Hz) may be used instead of cycles.

The power developed in inductance-type motors is *not* equal to volts times amperes. Rather, it is equal to volts times amperes times a number which is different for different designs of inductance motors, but is always less than one. This number is the power factor of the motor.

Power Factor

An explanation of the physical basis for a power factor is beyond the scope of this text. It is given in texts on electricity. *Power factor* may be defined, however, as the number obtained when actual (observed) watts are divided by the product of volts and amperes.

The power factor of motor-driven household appliances is not stated on the nameplate (for room air conditioners it is sometimes included on the specification sheet). Measurements made by students in classes of one of the authors have indicated power factors from 0.2 to 0.8, approximately, for different inductance-type motors used in household appliances and from 0.9 to practically 1.0 for universal motors used in household appliances.

The actual or approximate value of the power factor is needed for estimating power and energy used by household appliances with inductance-type motors. For example, the nameplate on one model of an automatic washer specifies: 115 volts,

60 cycles, 1/2 horsepower, 8.5 amperes. Class experiments showed that with a load of clothes the washer used approximately 8.0 amperes and 524 watts at 115 volts during the spin part of the regular cycle. Watts calculated from nameplate data on the incorrect assumption that the power factor were one, would be 115 × 8.5 or 978 watts.

The actual power factor as calculated from the experimental data is: actual (observed) watts/volts times amperes or 524/115 × 8 = 0.57. The electric energy and hence operating cost likewise depends on *actual* watts used.

TRANSMISSION OF ELECTRICITY TO AND WITHIN HOUSES

Electricity is transmitted by high-voltage wires from the power plant where it is generated; the actual value of the high voltage depends on the area in which the electricity is transmitted. A three-wire transmission system is usual, and the three wires are connected to *step-down transformers* at many locations. In cities one

or two transformers on poles often serve for one city block. In some farm areas one transformer per farm is usual. The transformer decreases or steps down the high voltage used for transmission to the voltage used in homes.

From the low-voltage side of a transformer, three wires, or for older houses two wires, connect with the residence kilowatt-hour meter (Fig. 1-2). If a house has a two-wire supply, one wire is at a nominal 115 or 120 volts and the other wire is at 0 volts; that is, the other wire is connected to ground or is grounded (Fig. 1-3). If a house has a three-wire supply, one wire is at nominal plus 115 or 120 volts, one wire is grounded, and the third is at nominal minus 115 or 120 volts. Three wires show that 115 *and* 230 volts are delivered to the house.

Some rural cooperatives, utility companies and cities supply 120 volts; others continue to supply 115 volts. The expression "nominal 115 or 120 volts" is used to indicate that the voltage is not exactly 115 or 120; also the voltage does not have a constant value. The expressions 230 volts and 240 volts likewise are

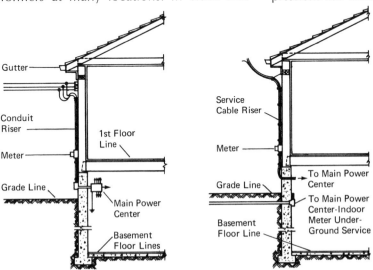

Figure 1-2. Three types of electrical service installations for homes. (Live Better Electrically)

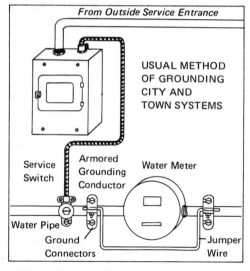

USUAL METHOD
OF GROUNDING
CITY AND
TOWN SYSTEMS

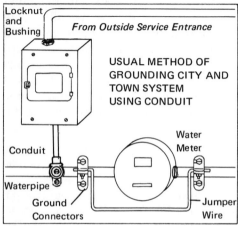

USUAL METHOD OF
GROUNDING CITY AND
TOWN SYSTEM
USING CONDUIT

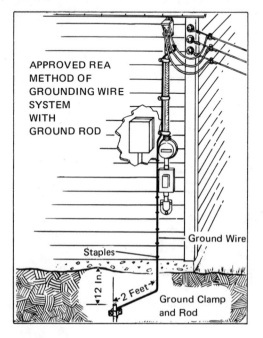

APPROVED REA
METHOD OF
GROUNDING WIRE
SYSTEM
WITH
GROUND ROD

nominal. For simplicity either 115 and 230 *or* 120 and 240 will be used in this text. The actual voltage delivered to a house depends on several factors, one of which is the variable demand for electricity from other houses served by the same transmission system.

From the kilowatt-hour meter the wires go to a main power center or service entrance panel or box. The separate electric circuits in the house start in and have their fuses or circuit breakers in the service entrance panel. For example, one circuit which starts at the service entrance may serve lighting and convenience outlets for a bedroom and part of a living room, another circuit may serve outlets for a laundry area, and so on.

House circuits have two or three *current-carrying wires*. The circuits in the home that have two current-carrying wires are 115-volt circuits, except those for electric water heaters and some power tools, which are 230-volt circuits. The circuits that have three current-carrying wires are always 230-volt circuits. These are used for electric ranges, electric dryers, and some air conditioning equipment.

A 115- or 230-volt circuit with two current-carrying wires is fused with one fuse or one circuit breaker. A 230-volt circuit with three current-carrying wires is fused with two fuses or one double circuit breaker.

The National Electrical Code (NEC) requires that, in addition to the two current-carrying wires, 15- or 20-ampere (115-volt) circuits or circuit raceways such as aluminum-sheathed cable shall provide a grounding connector to which the grounding contacts of three-slot outlets shall be connected. (This requirement is not usually met in houses built before 1962 unless they have been rewired.)

Figure 1-3. Grounding the electric service. (Sears, Roebuck and Company, *Simplified Electric Wiring Handbook*, F5428, Rev. 11-69, p. 14)

HEATING EFFECT AND OVERLOAD PROTECTION

When an electric current I flows through a resistance R for t seconds, the heat developed is proportional to the square of the current times the resistance times the time; that is, the heat developed is proportional to I^2Rt. Expressed in units,

calories = 0.24 (amperes)^2
$\times$ ohms $\times$ seconds

or

calories = 0.24 volts $\times$ amperes
$\times$ seconds

or

calories = (volts)^2 $\times$ seconds/ohms

or

calories = 0.24 watts $\times$ seconds

(The quantity 0.24, or more precisely 0.239, is a conversion factor for changing heat expressed in electrical units to heat expressed in calories.) As stated at the beginning of the chapter, this heat is utilized in heating appliances.

Also, since heat is developed wherever electric current flows through a resistance, it is developed in the wires leading to the outlets into which appliances are plugged. The wires that lead to the electrical outlets in a house are installed where we do not see them. They are likely to be attached to the framing of the house and to pass through walls. Clearly, if enough current should flow through them, the wires *could* become sufficiently hot to be a fire hazard. This is the reason for overload protection, in the form of fuses or circuit breakers, in the separate circuits in a house.

Fuses and circuit breakers are rated in amperes. The ampere rating of the fuse or circuit breaker to be used in a circuit depends on the size of the wire used in the circuit.

The metallic alloy in fuses melts when current above the rated value of the fuse, 15 amperes, 20 amperes, 30 amperes, and so on, flows through it. When the alloy melts, or the fuse "blows," the circuit is open and current no longer flows through it. Thus when a fuse of proper rating is used in a circuit, excessive heat will not develop in the hidden wires, and the house is protected from a fire due to overheating of these wires. The protection is lost when an unwise or uninformed person replaces a "blown" fuse with one whose ampere rating is too high for the size wire in the circuit.

A circuit breaker is similar to a fuse in that it "trips" and opens a circuit when excess current flows. A circuit breaker can be reset manually. A blown fuse must be replaced. Special characteristics of fuses and circuit breakers are discussed on pages 13 and 14.

EFFECTS OF LOW VOLTAGE

Earlier in this chapter, Ohm's law was used to calculate the voltage drop due to the resistance of 200 feet of wire between an outlet and the electric service entrance in a house. For the example given, the voltage at the outlet was approximately 3 volts less than at the service equipment. With poor house wiring (inadequate for electric current demand), the voltage drop could be much greater. Actual voltage at an outlet may also be decreased due to the fact, indicated earlier, that the voltage *supplied* at the service equipment is not constant.

Low voltage has undesirable effects on the performance of both heating and motor-driven appliances. A heating appliance rated to deliver a certain number of calories in a given time at 115 volts will clearly deliver less than the rated amount when the appliance is used on 105 or 110 volts. Suppose an iron rated at 115 volts is used at 105 volts. The heat developed will be proportional to 105 squared instead of 115 squared. The iron will require a longer time, for instance, to reach a linen setting than it was designed to take, and it may not deliver the heat output appropriate to the linen setting during ironing. Thus, for heating appliances, low voltage causes slow heating, and in some cases failure to deliver the designed heat output.

The undesirable effect of low voltage on motor-driven appliances comes when the motor starts or tries to start. If the voltage is low, an inductance-type motor may "labor" on start and its moving part may not reach operating speed. If the labored start is prolonged, excess heat develops and a thermal overload protective device, when one is provided in the motor circuit, opens the circuit. If a protective device is not included, the starting circuit or the motor may burn out. Thus, the effect of low voltage can be more harmful for motors than for heating appliances.

Design voltage for electric appliances is given on the nameplate of the appliance. Design voltage for an incandescent lamp is marked on the lamp.

Care of Appliances in a Brownout

A brownout is a planned reduction in electrical voltage to a given area for a specified time—usually three or four hours. (A blackout is a total loss of power.)

When notified of a brownout (usually by radio or television), avoid using electrical appliances.[1] Do not bake cookies, vacuum, or launder clothes. Unless the house normally has low voltage due to its location at the end of a power line or to poor house wiring, appliances with "continuously" running motors such as electric refrigerators and freezers should be left connected. If the house has excessively

[1] Adapted from National Rural Electric Cooperative Association, "A Word from Jeannie," *Home Service Letter*, April 1971.

low voltage, those appliances and others with continuously running motors—ice maker, air conditioner—should be turned off. Five minutes or so after the brownout ends, the appliances should be turned on one at a time so as not to throw a heavy load on the wiring at once. The five-minute wait is to avoid damage resulting from possible wide fluctuations as correct voltage is returned.

Under reduced voltage, the picture of a correctly operating television set will suddenly shrink. You may notice a black line 1/4 inch to 1 inch wide around the screen. Under extreme conditions, the picture disappears.

NAMEPLATES

The National Electrical Code states that each electric appliance shall be provided with a nameplate giving the identifying name and the rating in volts and amperes, or in volts and watts. Also, when the appliance is to be used on a specific frequency or frequencies, it shall be so marked. (An example would be 60 cycles per second.) The code also specifies that the marking shall be located so as to be visible or easily accessible after installation. (Students and others who look for nameplate data sometimes question the interpretation of "easily accessible.")

"Replaceable heating elements rated more than one ampere which are part of an appliance shall be legibly marked with the rating in volts and amperes, or in volts and watts." Certain heating appliances use different wattages at different settings. A nonthermostatically controlled surface unit of an electric range, for example, may use different wattages at different settings. The wattage given on the heating element for such a unit is the wattage used at the highest setting. For consumer information purposes (not to meet code requirements), horsepower is given on nameplates of some motor-driven appliances. One horsepower equals 746 watts. By convention, the horsepower usually is nominal output whereas wattages on nameplates are input wattages.

The horsepower rating of some vacuum cleaners is an input rating. This is an exception to the general rule that horsepower ratings refer to output.

A nameplate marked 1/3 horsepower states that the appliance will deliver approximately 250 watts. The wattage input will always be higher since the efficiency (ratio of output to input) is always less than one.

Fuses, outlets, switches, and other wiring devices usually are marked for amperes and volts.

SAFETY CONSIDERATIONS

Electrical safety in the home involves the safe use of electric appliances, safe appliances, and safe wiring. Reference has already been made to the fire hazard caused by overfusing. Where extension cords are used, they should be out of the way of floor traffic to prevent walking on wires and possible shorts. Frayed appliance and extension cords should be repaired or replaced.

SAFE USE OF ELECTRIC APPLIANCES

Probably the three most important precautions to follow in using electrical appliances are these: (1) Never handle at the same time an electrical appliance or wiring device such as a switch and a good connection to ground such as a water faucet. (2) Handle only one electrical appliance or wiring device at a time. (3) Do not handle electrical appliances or wiring devices with a wet or even a damp part of the body.

While it is probably true that more people are killed annually by 115 volts than by any other voltage, it is fortunately also true that fatal accidents usually require an unusual set of circumstances. For example, a person can violate the first precaution and handle an electrical appliance and a good connection to ground safely, *provided* the appliance is in good condition.

But if the insulation in the appliance between the "live" wire and the housing is defective and a person touches the housing near the defective insulation and a good ground at the same time, he has placed himself across 115 volts. In the same way, one can handle two electrical appliances safely *if* both appliances are in good condition, but it is not predictable when insulation inside an appliance will become defective.

In case a person has accidentally contacted a live wire and is unable to pull away, a rescuer should, if possible, disconnect the source of power before trying to aid the victim. If this is not practical, the rescuer should use an insulating material such as a dry wood stick or dry clothing to separate the victim and the wire. If he fails to do this, he himself is likely to make contact with the live wire through the victim. If necessary, the rescuer should call a physician and start artificial resuscitation while waiting.

SAFE APPLIANCES

In the United States the consumer depends in part on Underwriters' Laboratories, Inc. (UL), for tests of appliances for safety. Underwriters' Laboratories, Inc., maintains and operates laboratories for the

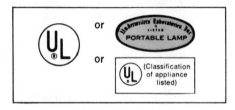

Figure 1-4. Underwriters' Laboratories, Inc., labels. (Underwriters' Laboratories, Inc.)

examination and testing of devices, systems, and materials and construction in their relation to life, fire and casualty hazards, and crime prevention. It is chartered as a nonprofit organization. It contracts with manufacturers and others for the examination and classification of submitted materials and reports and circulates the results to insurance organizations, interested parties, and the public by publication of lists, by provision for attachment of labels, or in some other manner.[2]

Manufacturers submit products voluntarily. The UL label indicates that the product on which it appears is reasonably free from fire, electric shock, and related accident hazards. The UL label usually appears on the body of the product or on the carton (Fig. 1-4).

Since the UL seal indicates that the article has been tested for safety, it is wise to look for it when purchasing equipment and electrical parts. Electrical toys for children should most certainly carry the UL label. An approval seal on a removable electric cord does not indicate approval of the toy or appliance.

In Canada, the Canadian Standards Association (CSA) offers certification services for various products covered by CSA standards. The Canadian Standards Association is an autonomous, nonprofit, nongovernmental organization. Certification services have been established for such widely diversified classifications as electrical products, fuel-burning equipment, plumbing products, factory-built homes, and many more.[3]

[2] Underwriters' Laboratories, Inc., *Testing for Public Safety*, 1963, p. 7.

[3] The principal offices and laboratories of the Canadian Standards Association are located at 178 Rexdale Blvd., Metropolitan Toronto, Rexdale, Ontario.

CSA certification service is a national product-testing and inspection service, designed to provide industry with a voluntary and self-regulated system of checking its products to ensure that they meet the standards of the industry. Certification of certain product classifications (for example, all electrical products) is mandatory by provincial legislations. Further, most such mandatory legislation is national in its scope. Products certified by CSA are eligible to bear the certification monogram:

SAFE HOME WIRING

Safety in U.S. wiring is obtained by using wiring materials and devices that carry the label of the Underwriters' Laboratories and by installing them in accordance with requirements of the National Electrical Code and local codes. When no local code is imposed, as may be the case in some rural areas, the power supplier usually is willing to aid in interpreting applicable provisions of the National Electrical Code.

The National Electrical Code is sponsored by the National Fire Protection Association under the auspices of the United States of America Standards Institute. The aim of the code is "the practical safeguarding of persons and of buildings and their contents from hazards arising from the use of electricity for light, heat, power, radio signaling, and for other purposes."[4] The stated purpose also says that "compliance therewith and proper maintenance will result in an installation essentially free from hazard, but not necessarily efficient, convenient, or adequate for good service or future expansion of electrical use."

In the previous section on transmission

[4] 1968, *National Electrical Code*, National Fire Protection Association, Article 90-1.

of electricity, the National Electrical Code requirement on grounding of 115-volt circuits or circuit raceways was noted. Other examples of the application of the code are given in later sections on outlets and switches.

RESIDENTIAL WIRING DESIGN

ELECTRICAL PARTS

The electrical parts used in home wiring include cables, wires, fuses, circuit breakers, convenience outlets, switches, and other devices.

Cables, Wires, Fuses, and Circuit Breakers

As an electrical part, a *cable* is an assembly of two or more insulated conductors enclosed in a sheath of metal or nonmetallic material such as plastic, braided cotton, or treated paper. When metal is used for the sheath, the cable is described as *armored* cable. Both armored and non-metallic cables may be constructed to be flexible.

A moisture- and corrosion-resistant type of nonmetallic cable or conduit commonly is used between a kilowatt-hour meter located outside the house and the main power center or service entrance panel in the house. (See Fig. 1-2.) The wiring between the main power center and the various outlets in the house may be one of several types, such as nonmetallic cable, armored cable, knob-and-tube wiring, and flexible conduit. At present nonmetallic cable is accepted as a good practical choice from the point of view of cost and time required for installation. Flexible conduit may be more convenient for adding additional circuits.

Knob-and-tube wiring was at one time almost always used for homes in some areas. In this type of branch-circuit wiring, the conductors are supported by knobs and tubes made of an insulating material, usually porcelain, mounted at intervals of about 41/2 feet in hollow spaces of walls and ceilings.

The diameter of the wire used for the conductors in the cable is measured by a device known as an American Standard Wire Gauge and is expressed as an AWG number. The diameters for the AWG numbers listed below were taken from a wire gauge manufactured by the Starrett Company.

AWG (number)	Diameter (inches)
0	0.325
1	0.289
6	0.162
8	0.128
10	0.102
12	0.081
14	0.064
16	0.051
18	0.040

Note that diameter decreases as gauge number increases. *Plug fuses* are standard or tamper resistant. The standard type of fuse is made so that fuses rated at 15, 20, 25, or 30 amperes may be used interchangeably in a single fuse socket. This type permits an uninformed person to use a fuse with a higher rating than the circuit should have. Tamper-resistant fuses have adapters which restrict the possibility of overfusing.

Both standard and tamper-resistant fuses that are made with a time-delay feature provide a short time allowance for overloads due, for example, to starting of motors. With continued overload or with excessive instantaneous load, the fuse blows.

Circuit breakers are thermal, electro-magnetic, or thermal-electromagnetic. A thermal breaker includes a bimetallic element that expands and flexes when heated by the current in the circuit. With enough heat, the flexing of this element opens the circuit contacts. The thermal type thus depends on current and time, since the heating effect of electric current depends on current and time.

An electromagnetic breaker is essentially a switch operated by an electromagnet. When the current through it reaches a predetermined value, the breaker opens the circuit. The electromagnetic type thus depends on current only.

A thermal-electromagnetic breaker utilizes both the thermal and electromagnetic actions. The thermal part depends on current and time; the electromagnetic part depends on current alone. The thermal part provides the time-delay feature needed for the instantaneous, somewhat high, starting currents of some electric appliances. (A circuit breaker that tripped every time a refrigerator started, for example, would be a nuisance.) The electromagnetic part provides for instantaneous tripping of the breaker due to the very high currents characteristic of short circuits.

Fuses or circuit breakers of proper ratings provide overload protection. The choice of which to use depends partly on cost versus convenience. It is easier to reset a circuit breaker than to replace a fuse, but circuit breakers are more expensive to install initially than fuses. Some manufacturers suggest that circuit breakers occasionally be turned on and off several times.

Wiring Devices

Wiring devices include outlets and switches.

A *lighting outlet* is a means by which house wiring is connected to fixtures, portable lamps, light sources in valances, and so on. A *convenience outlet* is a plug-in receptacle used for portable electric appliances, radios, and other electric housewares. A *split-wired*, duplex convenience outlet, as the name implies, is connected

 Standard Duplex—"U" slot devices recommended to permit grounding. Split-wired receptacles permit switching of one outlet.

 Weatherproof Outlet—Single or duplex devices have screw-on caps to keep out moisture.

 Dryer Outlet—30A-250V receptacle may be flush mounted in standard box. Also suitable for work bench power outlet.

 Cover-Mounted Outlets—Several types of receptacles can be obtained mounted on covers for use with exposed boxes in basement and garage.

  Range Outlet—3 wire—50-amp, 250-V polarized receptacles available for flush or surface mounting

Figure 1-5. Wiring devices. (Live Better Electrically)

to two circuits. Such a receptacle is connected to a three-wire 115/230 volt circuit with contacts in one outlet connected to the neutral wire and the plus 115-volt side of the 230 volts, and the contacts in the other outlet connected to the same neutral wire and the minus 115-volt side of the 230 volts.

A *special-purpose* outlet is a plug-in receptacle or a point of permanent connection to the wiring system of a particular appliance, and normally the outlet is reserved for the exclusive use of the appliance. Special-purpose outlets used for dishwashers, ranges, and dryers usually are installed in the walls or floor. Special-purpose outlets for electric water heaters, fuel-fired furnaces, and built-in space heaters may be part of the equipment; in this case the equipment is connected permanently to the house wiring system.

Available outlets include standard duplex and triplex convenience outlets, floor outlets, weatherproof outlets with protective caps, hanger outlets for clocks and fans, outlets that provide a grounding connection, lock-type outlets, raceways with convenience outlets spaced relatively close together, and special-purpose outlets. Some of these outlets are illustrated in Figures 1-5 through 1-12.

A standard duplex outlet is illustrated in Figure 1-5. A triplex or even a quadruplex outlet is advantageous in a bedroom. A triplex outlet is illustrated in Figure 1-6. A weatherproof outlet is useful outside the house for decorative lighting, outdoor cooking, and electrical garden tools (Fig. 1-5). A threaded cover may be included to provide a weatherproof seal. A recessed hanger outlet for an electric clock or fan eliminates long wires (Fig. 1-7). A grounding type outlet is illustrated in Figure 1-8.

Lock-type outlets require a partial turn of the plug to complete the circuit. Such outlets are especially desirable for freezers since they prevent accidental disconnec-

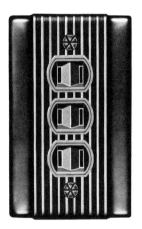

Figure 1-6. Triplex convenience outlet. (The Bryant Electric Company)

Figure 1-7. Recessed hanger outlet. (The Bryant Electric Company)

Figure 1-8. Grounding convenience outlet. The outlet has two current-carrying contacts and one grounding contact. (The Bryant Electric Company)

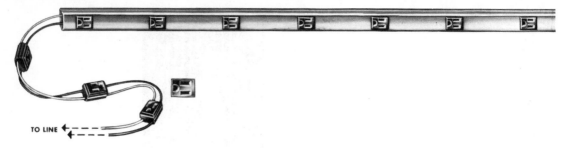

TO LINE

Figure 1-9. Raceway with built-in outlets at regular intervals. (The Wiremold Company)

tion of the freezer. The plug on the appliance cord must be appropriate for this outlet.

Raceways with outlets at frequent intervals permit flexibility in placement of lamps, radio, television set, and so forth, in the living room when the raceway is in-

Figure 1-10. Range outlet. (The Bryant Electric Company)

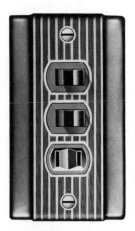

Figure 1-11. Interchangeable device consisting of two switches and a receptacle. (The Bryant Electric Company)

stalled in the baseboard (Fig. 1-9). They are useful in workshops when installed above the counter. Raceways are available with different types of outlets, such as standard duplex or grounding outlets.

The special-purpose outlets for electric ranges up to 21-kilowatt rating are three-wire, 50-ampere, 115/230-volt, polarized receptacles (Figs. 1-5 and 1-10); for built-in ovens and built-in surface units the outlets are the same—three-wire, 30-ampere, 115/230-volt type. The polarized special-purpose outlets for 230-volt electric dryers also are usually rated at 30 amperes.

The straight toggle switch is probably the most common switch in the home. However, various special switches are available. *Interchangeable devices*, for example, are combinations of one, two, or three switches, pilots, and convenience outlets that fit into one box. The interchangeable device shown in Figure 1-11 consists of two switches and an outlet. An interchangeable device for a night light may consist of a box with a cover plate, the night light, and two switches—one to control the night light and the other to control hall or room lighting. Such a device is useful in a bathroom to prevent groping for a switch in the dark.

A multicontrol light switch controls one or more lights from more than one point. A three-way switch controls one or more lights from two separate points. One use of a three-way switch is to control a garage light from both house and garage. A four-way switch provides control from three separate points. They are called three-way and four-way switches because the electrician must connect them with multiple wires.

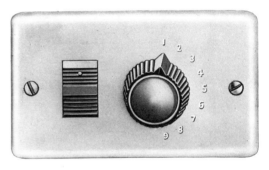

Figure 1-12. Master selector switch. (The Bryant Electric Company)

A *multicontrol master switch* is somewhat complex in wiring, unless a *low-voltage control* is used. Such a switch in the master bedroom, for example, may be wired to control outdoor and indoor lights in selected areas of the house, as well as equipment such as a space heater (Fig. 1-12).

BRANCH CIRCUITS, SERVICE ENTRANCE, AND SPECIAL WIRING

Branch Circuits

Three types of branch circuits are considered in requirements and recommendations for safe and convenient electric wiring in the home.[5] *General-purpose branch circuits* supply lighting outlets throughout the home. The National Electrical Code requirement is 3 watts per square foot and this is interpreted as one 20-ampere circuit for not more than each 500 square feet, or one 15-ampere circuit for a 375-square-foot area. Minimum wire size (gauge) for 15-ampere circuits generally is #14 and for 20-ampere circuits, #12.

Small-appliance branch circuits are used for electric appliances. NEC requires a total of two or more 20-ampere circuits for the convenience outlets in the kitchen, dining room, dining area of other rooms, such as a breakfast nook, and family room. Wire size is #12 gauge. The minimum requirement is met by one three-wire, 115/230-volt circuit equipped with split wiring or two separate 115-volt circuits. Note that having two circuits or a split-wired circuit serve several rooms increases flexibility. When the kitchen has two appliance cir-

[5] Requirements considered here are from the current edition of the *National Electrical Code*. Recommendations in this section and in the section on Outlets and Switches generally follow the *Residential Wiring Design Guide* published by the Edison Electric Institute, New York. EEI Pub. No. 69-53, December 1969.

cuits the homemaker can safely use several small electric appliances in the kitchen at once, provided appliances are not also in use at the same time on these circuits in another room.

NEC in addition requires one 115-volt, 20-ampere branch circuit for the laundry receptacle(s). This may serve for a wringer or automatic washer or other use.

Individual-equipment branch circuits serve for one piece of equipment and are required for the following: electric range or built-in oven plus built-in surface units, 230-volt electric clothes dryer, dishwasher and food waste disposer (if necessary plumbing is installed), gas- or oil-fired fixed heating equipment (if installed), electric water heater if installed.

Spare circuit equipment shall be provided for at least two future 20-ampere, 115-volt circuits in addition to those initially installed.

The Residential Wiring Design Guide suggests that consideration be given to providing circuits for the following equipment: food freezer; room air-conditioners, central air-conditioning unit, or attic fan; water pump (where used); bathroom heater; work bench. The authors think an individual-equipment circuit should be considered for a large refrigerator-freezer in a kitchen. Finally, the guide suggests consideration of extra circuits for patios or "outdoor" living rooms *and* for exterior decorative or flood lighting.

Service Entrance

The *service equipment* is the main power center (switches and fuses or circuit break-

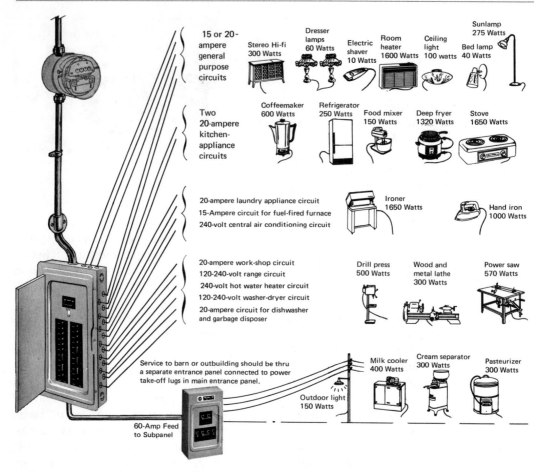

Figure 1-13. Individual circuits connected to main panel and one subpanel. (Slightly modified from Sears, Roebuck and Company, *Simplified Electric Wiring Handbook,* F5428, Rev. 11-69, p. 6)

ers and their housing and other accessories) located near the point of entrance of the electrical supply to the house. The *service entrance conductors* are the portion of the supply wires between the service equipment (main power center) and the meter for overhead supply, or between the service equipment and the point outside the home where the supply wires are joined to the street mains for underground supply.

The size of the service entrance conductors and the rating of the service equipment should be adequate for the total electric load—that is, for lighting, for portable electric appliances, and for the fixed electrical equipment that will be used in the house.

A three-wire, 230-volt supply to the service equipment is specified even when no 230-volt appliance is installed. The three-wire service permits connecting individual 115-volt circuits across each half of the supply and split-wiring of convenience outlets. This lessens the voltage drop in the residence.

Common wiring practice usually has all the separate circuits in a house controlled at the service equipment. However, in some houses one or more subcontrol centers served by relatively heavy feeder lines are used between the service equipment and one or more different areas of the house or yard (Fig. 1-13).

The Residential Wiring Design Guide gives the capacity of the service entrance conductors and rating of the service equipment (see Table 1-2).

Table 1-2. Service Entrance Recommendations (Residential Wiring Design Guide)

Square-Foot Floor Area	Minimum Service Capacity
up to 1,000 sq ft	125 amperes
1,001–2,000 sq ft	150 amperes
2,001–3,000 sq ft	200 amperes

These capacities are sufficient for lighting, portable appliances, the equipment listed above to be served by individual appliance circuits, and for air conditioning or electric space heating of the individual-room type or both. A larger service may be required for larger houses.

Disconnecting means at service entrance. The National Electrical Code prescribes that with certain exceptions buildings shall be supplied through only one set of conductors. The code also requires that the service entrance conductors be provided with a *readily accessible means of disconnecting all conductors* from the source of supply. The broad, general requirement is that no more than six hand operations shall be necessary to disconnect all conductors. This requirement can be met, of course, by installation of one master switch.

Special Wiring

Special wiring may be planned for low voltage electric front and rear door entrance signals operated through a transformer, for intercommunication devices between rooms or between someone in the house and a caller at a front or rear door, and for decorative yard lighting. (The intercommunication service should be operated from a power unit recommended by the equipment manufacturer.) Additional special wiring might include lead-in connections for an outdoor television antenna and/or FM radio reception, or for telephone outlets.

OUTLETS AND SWITCHES FOR DIFFERENT AREAS

Some general recommendations apply for many areas in the home. Perhaps the one best mentioned first is: The wiring installation should be fitted to the structure and if compliance with any recommendation on outlets or switches would require architectural alteration the homeowner and electrical contractor should plan for an alternative.

Generally, convenience outlets should be located near the ends of wall space to reduce the likelihood of being concealed behind large pieces of furniture. Also, unless special considerations dictate a different location, they should be 12 inches above the floor line.

Wall switches for lights normally are located at the latch side of doors within the room or area to be lighted, and at a height of about 48 inches above the floor line. An exception might be control of lights from an access space nearby.

All spaces which have more than one entrance shall be equipped with multiple-switch control at each principal entrance in order that a person need not walk in darkness to a switch. This requirement is waived for principal entrances so situated that two multiple-switch controls would be located within 10 feet of each other. (The authors would not waive that requirement in a home with a handicapped or elderly person.)

Recommendations for particular areas are summarized below.

Exterior entrances. One or more wall-switch-controlled lighting outlets should be at each entrance.

A weatherproof convenience outlet, preferably near the front entrance, should be located 18 inches or more above grade and controlled by a wall switch inside the house. Any additional outlets along the exterior of the house for decorative lighting or electric garden tools should also be controlled by a wall switch. Special purpose convenience outlets may be desirable for certain equipment such as heating cable to melt snow in the driveway or on roofs.

Living room. General illumination is desirable. It may be supplied by ceiling or wall fixtures, by lighting in coves, valances, or cornices, or by portable lamps. Whichever method or methods are used, wall switches should be located near entrances. Outlets are also suggested for use of an electric clock, television set, radio, decorative lighting, and so on, near built-in shelves, and at other appropriate locations.

Convenience outlets should be placed so that no point along the floor line in any usable wall space is more than 6 feet from an outlet. Split-wired convenience outlets are desirable. Where windows extend to the floor, floor outlets may be used.

Dining area. At least one wall-switch-controlled lighting outlet is necessary in the dining area.

Convenience outlets should be placed so that no point along the floor line in any wall space is more than 6 feet from an outlet. If the table is to be placed against a wall, one of the outlets should be installed at the table location just above table height. Also, if a counter is to be built in, an outlet above counter height is convenient for portable appliances.

Kitchen. Wall-switch-controlled lighting outlets are necessary for general illumination and for lighting at the sink. Local switch controls should be within easy reach for lighting fixtures under wall cabinets.

There should be one convenience outlet for an electric refrigerator and one for each 4 linear feet of work-surface frontage, with at least one outlet for each work surface, placed approximately 44 inches above

the floor line. If a planning desk is to be installed, an outlet should be near it. Also, a convenience outlet is recommended in a wall space if an iron or electric roaster might be used near that wall.

One special-purpose outlet each is suggested for an electric range and a ventilating fan, if both are used. There should also be an outlet for a dishwasher and/or food waste disposer, if appropriate plumbing facilities also are installed, one for an electric clock in the kitchen, and a special-purpose outlet for a freezer either in the kitchen or in some other location.

Laundry area. At least one wall-switch-controlled lighting outlet is essential. (Additional laundry lighting outlets should also be wall-switch controlled.) For complete laundry areas, one or more light outlets should be provided for laundry tray(s), sorting table, sink (if there is one), and ironing, washing, and drying areas.

According to the National Electrical Code, convenience outlets should be connected to a 20-ampere branch circuit which serves no other area. A special-purpose outlet might be provided for an iron or ironer.

A convenience outlet serves for a hand iron, or ironer. The convenience outlet becomes a special-purpose outlet when installed at a height and/or in a space that normally would be reserved for use of an iron or ironer.

If a 230-volt electric dryer is used, a special-purpose outlet will be needed. In the case of a gas dryer, a convenience outlet serves for the motor, interior light, and electric control. A special-purpose outlet will be needed for an electric water heater if one is to be installed in this area.

Bedroom. There should be wall-switch-controlled outlets for general illumination purposes. As noted earlier, a multicontrol master switch in the master bedroom is suggested for selected interior and exterior lights and equipment.

A convenience outlet should be placed on each side and within 6 feet of the center line of each probable individual bed location. Additional outlets should be installed so that no point along the floor line in any other usable wall space is more than 6 feet from an outlet in that space. It is recommended also that a convenience outlet be provided at one of the switch locations for a vacuum cleaner or other portable electric appliance.

A special-purpose outlet is required for a room air conditioner if one is installed.

Bathroom. All lighting outlets should be wall-switch controlled. There should be outlets to provide illumination on both sides of the face at mirrors, and a ceiling outlet or outlets for general room lighting and for combination shower and tub. A switch-controlled night light is desirable. (See p. 16.)

If an enclosed shower stall is planned, the outlet for the vaporproof luminaire in the stall should be controlled by a wall switch *outside* the stall.

There should be a convenience outlet near the basin mirror, 3 to 5 feet above the floor, and one in any wall on which a mirror might be placed for use when shaving with an electric razor. Also recommended are wall-switch-controlled special-purpose outlets for a built-in type space heater and built-in ventilating fan.

Hall. Wall-switch-controlled lighting outlets should provide for appropriate illumination of the entire area.

At least one convenience outlet is needed for each hall with a floor area greater than 25 square feet. For long halls, the recommendation is one outlet for each 15 linear feet, measured along a central line. Also, it is recommended that a convenience outlet be provided at one of the switch outlets for a vacuum cleaner, floor polisher, and so on.

Closet. One outlet should be positioned in the closet or in an adjoining space so that the interior of the closet is illuminated. Use of wall switches near the closet door or door-type switches is recommended.

Stairway with finished rooms at both ends. Lighting outlets with multiple-switch controls should be located at the head and foot of the stairway. *No* switch should be located so that a fall might result from a misstep while reaching it.

A convenience outlet is suggested at intermediate landings of a long stairway.

Recreation, television, or family room. Wall-switch-controlled lighting outlets are recommended for general illumination purposes.

Convenience outlets should be installed so that no point along the floor line of any usable wall space is more than 6 feet from an outlet in that space. These should be of the split-wired type.

Utility space and basement. Lighting outlets are needed for the furnace area and work bench, if one is planned. In unfinished basements, the light at the foot of the stairs should be wall-switch controlled near the head of the stairs. Other lights may be pull-chain controlled.

At least two convenience outlets should be provided, one near the furnace and another near the work bench. Additional convenience outlets might be provided for equipment such as an electric dehumidifier, electric hobby items, and other tools. A special-purpose outlet for an electric control or for other equipment used with a furnace is needed. A special-purpose outlet for a freezer might also be provided in the basement.

Accessible attic. One wall-switch-controlled outlet at the foot of the attic stairs should be provided for general illumination. (A pilot light in conjunction with the switch controlling the attic light is recommended.)

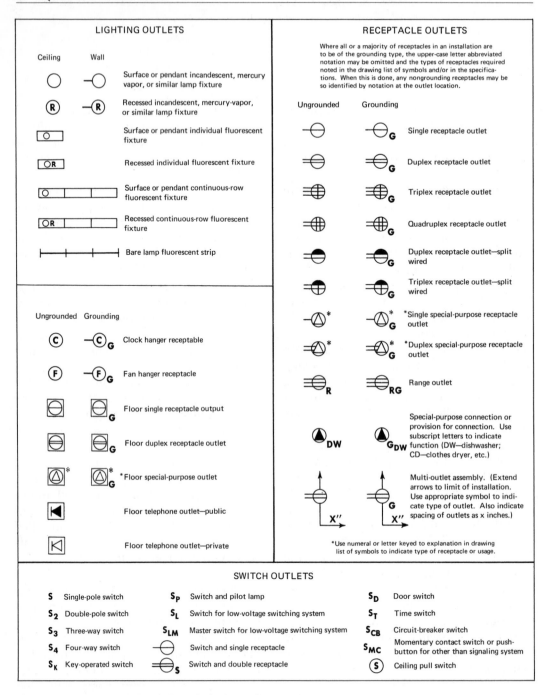

Figure 1-14. Graphical electrical symbols for architectural and electrical layout drawings.

One convenience outlet is suggested for general use, such as for cleaning equipment. Also, if a stairway leads to future attic rooms, a junction box connected to the main service panel is desirable. If stairs are not installed, a pull-chain-controlled lighting outlet over the access door may be used.

Porches and breezeways. A wall-switch-controlled lighting outlet is necessary if cover (roof) is provided over more than 75 square feet of floor area. Multiple-switch control is recommended when the area is used as a passageway between the house and garage.

At least one convenience outlet, weatherproof if exposed, is desirable for each 15 feet of wall bordering porch or breezeway if the area is planned for informal eating. (It is recommended that one or more of these outlets be controlled by an interior wall switch.)

Terraces and patios. Recommendations on lighting, switches, and outlets are similar to those for porches and breezeways planned for eating. The convenience outlet(s) should be 18 inches above grade line.

Garage or carport. At least one wall-switch-controlled ceiling lighting outlet is needed for a one- or two-car storage area. If the passageway between house and garage is not covered, an exterior garage light is required for one- or two-car storage. Special purpose outlets might be provided for special equipment such as electric door opener, car heater, and so on.

ELECTRICAL SYMBOLS, ROOM LAYOUTS, AND HOUSE WIRING CHART

The symbols shown in Figure 1-14 include those from the American National Stan-

dard Y32.9–1962. Familiarity with symbols is helpful in checking a wiring plan for a room or for a complete house.[6]

The kitchen plan in Figure 1-15 has a hood over the range with two incandescent bulbs controlled by a single switch and the blower or fan in the hood controlled by another single switch. These two switches might be supplied with the hood. A 230-volt special-purpose range outlet and a 115-volt special-purpose hood outlet are provided in the house.

A single grounded convenience outlet is shown at the sink location for the disposer and dishwasher. (Both are on the same circuit.) A grounded duplex outlet is

[6] The 1967 revision of the standard (ANSI Y32.2–1967) shows graphic symbols that cover a broad spectrum—analog computers, mechanical functions, readout devices, as well as fundamental symbols for contacts, mechanical connections, and so forth. The graphic symbol represents the *function* of a part in the circuit. For example, when a resistor is employed as a fuse, the fuse symbol is used. An oil, high-voltage primary fuse cutout has a different symbol than a dry component that serves the same function and a general fuse has still a different symbol.

The earlier symbols have been retained in this edition because they seem more directly appropriate to house wiring.

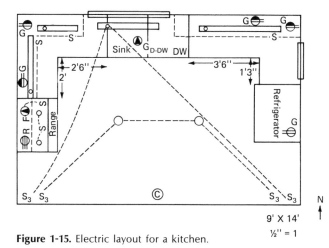

Figure 1-15. Electric layout for a kitchen.

9' X 14'

½'' = 1

N

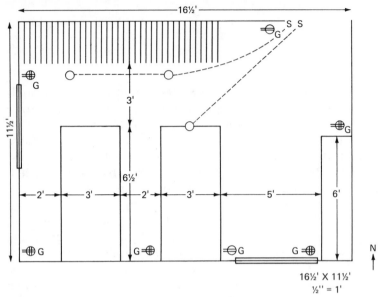

Figure 1-16. Electric layout for a master bedroom.

shown for the refrigerator. A clock hanger receptacle is on the south wall.

The two ceiling incandescent fixtures are connected to each other and three-way switches at the two entrances to the kitchen turn both ceiling lights on or off. Two other three-way switches turn the fluorescent light over the sink on or off. The undercabinet fluorescent lighting on the two walls is operated by individual switches.

Four grounded duplex convenience outlets—split wired—are provided at convenient locations above counter height not too close to the sink. (Locations somewhat removed from the sink are selected because most portable appliances used in United States kitchens have two-prong plugs.)

The 16 1/2 by 11 1/2 foot master bedroom layout in Figure 1-16 shows lighting outlets (two for the closet and one in the ceiling for general illumination) and convenience outlets that will be appropriate for the space use generally provided by architects, that is, for beds, night tables, clothes storage, and necessary clearances. The drawing, however, does include a 6-foot long dresser as well as space for night tables near the beds, and wall space for benches. Where it seems reasonable to do

so the grounded quadruplex convenience outlets are located near the corners, as well as at possible night table locations on the south wall. Although a specific beds-dresser arrangement is shown, occupants may choose to rearrange furniture; convenience outlets near the corners may permit doing this without having the outlets behind the furniture. The two closet ceiling lights are connected together and operated by a switch near the door; the central ceiling light is operated by another switch near the door.

If other activities are planned for the bedroom such as reading in a chair or applying makeup at a table, several additional or different kinds of lighting might be planned.

The Residential Wiring Design Guide referred to previously uses slightly different electrical symbols. The guide shows wiring diagrams for various rooms and areas in the home, as well as for two complete houses.

The guide also gives a sample specification form for small and medium-size dwellings (Chart 1-1). All outlets, the locations of wall switches, and the outlet(s) controlled by each switch should be shown on floor plans which are an essential part of the specifications.

Chart 1-1. Specifications for Electric Wiring in the Dwelling to Be Erected at _____ for _____. (Taken from the *Residential Wiring Design Guide*, loc. cit. p. 32.)

1. *General*—The installation of electric wiring and equipment shall conform with local regulations, the National Electrical Code, and the requirements of the local electric power supplier. All materials shall be new and shall be listed by Underwriters' Laboratories, Inc., as conforming to its standards in every case where such a standard has been established for the type of material in question.

2. *Guarantee*—The contractor shall leave his work in proper order and, without additional charge, replace any work or material which develops defects, except from ordinary wear and tear, within _____ months from the date of the final certificate of approval.

3. *Wiring Instructions*—Outlets, switches, and control shall be installed as shown on the plans. Contractor shall furnish switch and outlet bodies and plates and lampholders where indicated. Unless specifically contracted for, the hanging of lighting fixtures is not included.

4. *Wiring Methods*—Interior wiring shall be _____.

(Fill in wiring method)

No exposed wiring shall be installed, except in unfinished portions of basement, utility room, garage, attic, and other spaces that may be unfinished.

5. *Service Entrance* conductors shall be three No. _____ wires.

(Fill in wiring method)

6. *Service Equipment* shall consist of _____

7. *Distribution Panel(s)*—Branch-circuit equipment shall be _____

(Fill in type of equipment)

Distribution panel shall provide for termination of all initial branch circuits, plus two future branch circuits.

8. *General-Purpose Branch Circuits*—At least _____ general-purpose <div align="center">(Number)</div> branch circuits of _____ ampere capacity shall be installed to supply all lighting outlets and all convenience outlets not otherwise provided for below. Outlets shall be divided as equally as possible between these circuits. In living room and bedrooms, outlets shall be divided between two or more circuits.

9. *Appliance Branch Circuits*—At least _____ appliance branch circuits <div align="center">(Number)</div> of 20-ampere capacity, three wires, shall be installed to supply convenience outlets exclusively in the kitchen, family room, and dining spaces. At least one appliance branch circuit of 20-ampere capacity, two wires, 115 volts, shall be installed to supply convenience outlets exclusively in the laundry area.

10. *Special-Purpose Outlets and Circuits* shall be installed as follows:

Circuits for	No. of Wires and Ampere Rating	Termination
_____	_____	_____
_____	_____	_____
_____	_____	_____

11. *Signal and Communication Wiring*—Complete signaling system, including bell-ringing transtormer, push buttons, and audible signal equipment, shall be installed as shown on plans.

NAMEPLATE INSPECTION EXPERIMENT AND WIRING PLAN EXERCISES

Nameplate Inspection

Locate nameplates or rating information on heating appliances, motor-driven appliances, combination heating and motor-driven appliances, wiring devices, and fuses.

For all appliances, wiring devices, and fuses record the following:

1. Manufacturer's name and address.
2. Model number.
3. Underwriters' Laboratories seal. (Is it *on* the appliance or device? Where?)
4. Electrical quantities stated: volts, watts, amperes, cycles or Hertz.

   For motor-driven appliances, such as electric refrigerators, note whether horsepower rating is given.

Wiring Plan Exercises

1. Prepare a wiring layout for a living room or family room. Draw the room to scale and show such architectural features as doors, windows, fireplace.
2. Get a house wiring plan from a contractor, homeowner, utility, electric cooperative, or other source. Check the plan. (If a plan is obtained from a homeowner, you may be able to get information on how convenient it actually is.)

CHAPTER 2
HOME LIGHTING

LIGHT AND LAMPS

Light is radiant energy visible to the eye. The spectrum includes other forms of radiant energy not visible to the eye (Fig. 2-1). The light section of the spectrum includes all colors of the rainbow—violet, blue, green, yellow, orange, red—and each color blends into the colors adjacent to it. When these colors of light are mixed in proper proportions the result is white light.

Light is essential for seeing. Poor lighting can cause eyestrain and good lighting can enable many individuals with poor sight to see more clearly. Good lighting is an aid in making working conditions safer, decreasing fatigue and nervous tension, improving sitting posture, and making the home a more pleasant and interesting place.

The objectives in home lighting are (1) good lighting for visual activities and (2) coordination of the lighting sources with the decorative scheme. The first is achieved by having enough light and light of desirable quality. The second is achieved by selection of appropriate fixtures and portable lamps, wise use of structural lighting, and judicious placement of the lighting sources. Lighting should not be obvious in itself but it should make the room more attractive and facilitate seeing tasks.

QUANTITY OF LIGHT

The quantity of light necessary for seeing is influenced by the size of the object or material, the amount of contrast between the object and its background, and the time allowed for the eyes to focus on the object.

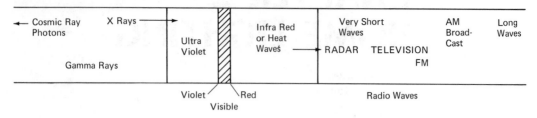

Figure 2-1. The electromagnetic spectrum.

If the print on a page of a book is very small, more light and more time will be needed to read it than if the print is large (Fig. 2-2). If the print is small the tendency is to bring the page closer to the eyes. It is common knowledge that one of the essentials in advertising is to make a sign large enough and with enough contrast so that it will be easily read.

If black letters are printed on a stark white page, the print will stand out and be seen quite easily. Less light will be needed than if the print were grayed and the page an off-white color. In other words, the contrast between the object viewed and its immediate background influences how much light is needed.

If an object must be seen in quite a short time, more light is needed than if viewing time is longer. If there is more time for the eyes to focus and it is possible to look at the object long enough, the eyes often can see detail that at first escapes the view. It is harder to see at dusk than in broad daylight—and this is one of the reasons more care should be taken while driving a car during the twilight hours of the day.

QUALITY OF LIGHT

The quality of light depends primarily on how much light surfaces emit or reflect and the ratio of the amount of light associated with the surfaces to the amount of light in the surrounding area or space. From a practical point of view, a good quality of lighting is achieved by minimizing glare.

SIZE   SIZE   SIZE   SIZE   SIZE   SIZE

Figure 2-2. Size is an important factor in seeing.

Glare

Glare has been defined as "light out of place." When the amount of light is such that it is uncomfortable for the eyes, it interferes with the process of seeing, or is in any way annoying, the light is said to be glaring. Glare can be caused in a number of ways. Sources of light of very high intensity are a cause of glare. Even though the eyes may not be focused directly on the source, the light may be so bright that it causes glare when it falls on other surfaces. Light from a gooseneck study lamp falling on a book often causes glare this way. This is accentuated if the pages of the book are glossy. This type of glare often is referred to as reflected glare.

Glare occurs when the source of light is exposed. This is true when a lamp is not shaded, when the shade is too small, or the lamp is placed incorrectly in relation to its use. The lamp may be so placed that the user cannot help but see the bright lamp or diffusing bowl inside the shade.

Glare may be caused by contrast. If a white paper is laid on a dark desk, the difference in values of the two surfaces will cause glare. Bright red, bright blue, and other intense colors seem to be popular for use as desk blotters in some students' rooms at college. From the standpoint of seeing, it would be much better to introduce color into the room in some other fashion and choose for a blotter a grayed color of less intensity. The dark wood of many desk tops may be a source of glare when papers are on it. When this wood, dark or light, is highly polished or covered with glass, another source of glare has been introduced.

Glare occurs while the eyes are adapting to a change in seeing conditions. When an

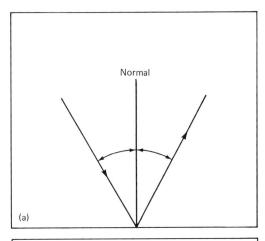

intense light is turned on in a dark room, the eyes experience glare until they adapt to the surroundings. This same adaptation is necessary if reading is done with a restricted source of light in a darkened room. Every time the eyes move away from the printed page, this adaptation must take place. It is impossible for the eyes not to look away even though a person sometimes does not realize that this happens. Some bed lamps and desk lamps that shed light only on one surface are not good for the eyes, if this is the only source of light in the room. Authorities say there should be one-tenth to one-fifth as much general light in the room as is provided for any specific seeing task. The desirable luminance in the area immediately surrounding the specific task is higher than for the general light further away. A desirable ratio of luminance in this area is one-third to one-fifth that in the specific task area.

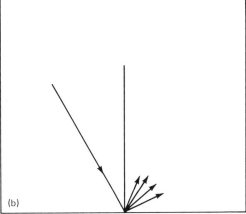

REFLECTED LIGHT

Light is reflected by either direct, spread, or diffuse reflection (Fig. 2-3). In direct reflection, the angle of incidence equals the angle of reflection. In spread reflection, the source of light is only partially mirrored. Spread reflection takes place when light falls on enameled surfaces in kitchens. It is sometimes possible to see the reflections of objects on these surfaces, though not as clearly as if one were seeing them in a mirror. Light that falls on mat surfaces is spread equally in all directions; this is diffuse reflection. Since this type of reflection is easy on the eyes, wall surfaces should preferably be of a mat finish. Enameled wall finishes, such as are often found in kitchens and bathrooms, have higher reflectance and more glare.

Reflectance ratios are determined by measuring the amount of light falling on a surface and the light reflected from it. The

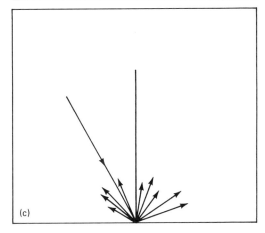

Figure 2-3. Types of reflection: (a) direct (angle of incidence = angle of reflection); (b) spread (light is scattered); (c) diffuse (light is scattered in all directions).

incident light is measured in foot-candles and the reflected light is measured in foot-lamberts. A foot-lambert is a measure of reflected lumens per square foot. For practical purposes a foot-candle meter can be used to measure reflected light. The greater the diffusion of the surface the more nearly the foot-candle reading will approximate foot-lamberts. Reflectance is determined by dividing the amount of reflected light by the amount of incident light. A high reflectance value indicates uncomfortable seeing conditions.

MEASUREMENT OF LIGHT

Light at the source (luminous intensity of the source) is measured in *candlepower.* For many years the composition of a standard or international candle that would have a total light output of 1 candlepower was specified by international agreement. This standard has been replaced, but the term candlepower has remained as a unit for light output.

Light sources and the light they emit also are described in *lumens,* which are units of light energy. If one imagines a point source of light placed inside a sphere, one can visualize that the inside of the sphere would be lighted by the source. The amount of light from a source of 1 candlepower on each square foot of the inside surface of this sphere is called a lumen. The surface area of a sphere is found by the formula $4\pi r^2$, and so the area of a sphere with a 1-foot radius would be 12.57 square feet. A 1-candlepower source, then, emits 12.57 lumens. A 100-candlepower source has a light output of approximately 1,250 lumens.

Inverse Square Law

The amount of light per unit of area (*illuminance*) falling on a surface may be measured in *foot-candles* or lux. If a piece of paper is placed I foot from a point source of 1 candlepower, the paper has an illuminance of 1 foot-candle. If the paper is placed 1 foot from a 40-candlepower source, the illuminance of the paper is 40 foot-candles. Candlepower divided by the square of the distance of the source of light from a surface equals the number of foot-candles of illuminance on that surface. For a given point source of light, the illuminance of a surface decreases as the square of its distance from that source.

The *lux* is a quantitative unit for measuring illumination when the metric system of measurement is used. The illumination on a surface 1 square meter on which there is a uniformly distributed flux of 1 lumen is a lux. Ten lux are equal to 1 dekalux.[1]

Units used for the light energy emitted or reflected per unit of area by a surface are foot-lamberts, candles per square centimeter, and candles per square inch. A *foot-lambert* is the amount of light energy per square foot reflected by a surface that diffuses light equally in all directions when the surface is illuminated with 1 foot-candle. One candle per square inch equals 452 foot-lamberts. White gives the luminance of a fluorescent lamp as 2 c/cm^2, a frosted mazda lamp as 5c/cm^2, and the sun as 50,000 c/cm^2.[2]

Light sources and illuminated surfaces are described subjectively in terms of *brightness,* which is actually a subjective measure of the light energy coming from a surface. Brightness of light is produced in three ways: it may come from the source of light itself, it may come from an object that transmits light, or it may be a result of reflection of light. Objectionable brightness may result from any one of these methods. An unshaded light bulb is an example of the first, light shining through a very translucent diffusing plate would be an example of the second, and light from

[1] Residence Lighting Committee, *Design Criteria for Lighting Interior Living Spaces.* New York: Illuminating Engineering Society, 1969, pp 24, 25.
[2] Harvey E. White, *Modern College Physics.* New York: Van Nostrand Reinhold, 1948, p. 378.

a shiny page of a book would be an example of the third.

NATURAL AND ARTIFICIAL LIGHT

Natural light or daylight may be very pleasant and comfortable, or it may be just the opposite. It is difficult to control because the day may be cloudy, sunny, or fluctuating from one to the other with great rapidity. The amount of light next to a window may be quite different from the amount a few feet away. The amount of light is influenced by the cleanliness of the windows, and by the use of screens, window shades, and draperies.

Natural light adds beauty to the home and makes a room a more cheerful and pleasant place in which to live. Good light in the home is a result of wise use of natural light and discriminating, selective use of artificial light.

Artificial sources of light used in the home are incandescent and flourescent lamps.

Incandescent Lamps

The incandescent lamp bulb has a tungsten filament. In most lamp bulbs the tungsten filament is a coiled wire that has been recoiled. The coil is so tight and the wire so tiny that to the naked eye it may look like a straight wire. Tungsten filaments operate at temperatures between 4,000 and 5,400°F. Lamps are either vacuum or gas-filled. Those of 40-watt size and larger are filled with an argon-nitrogen gas mixture. The glass bulbs usually are made of standard lime glass, but lead and borosilicate glass is used where heat resistance is important. The latter permits higher wattages and smaller bulbs and is good for outdoor use. The inside of a lamp bulb is frosted by flushing out the bulb with an acid solution that etches the inside surface. Some lamp bulbs have a silica coating on the inside to improve the diffusion of the light.

After incandescent lamps are used for a time the tungsten filament becomes thinner, the bulb darkens, and less light is produced. If operated at higher than rated voltage, lamps wear out sooner but give more light. The average life of an incandescent lamp is approximately 750 to 1,000 hours, depending upon the individual lamp and how it is used.

A 100-watt lamp has an output of approximately 1,750 lumens or 17.5 lumens per watt. The efficiency is slightly lower for smaller-wattage bulbs and higher for larger bulbs. Most lamps have a lumen output between 11 and 22 lumens per watt. A great number of kinds and types of lamp bulbs are manufactured for home use.

Silvered bowl lamps have had silver, copper, and aluminum applied to the bowl of the lamp, making that part opaque. This also serves to reflect light from the filament. The silver is applied to the glass on the outside of the lamp. It is sealed on with an electrolytic coating of copper, and aluminum is applied over the copper for further protection. These lamps are used in some indirect lighting arrangements, to protect the eye from glare in certain locations, as in a closet, and for other special uses. They are a little more expensive than the ordinary inside-frost lamp.

Some lamps are colored. These are designed for decorative or specific uses such as the yellow lamp that does not attract insects as readily as white or blue light. Color may be fused to the lamp in the form of a ceramic enamel. It is applied before the bulb is made into a lamp and the color on these is not easily scratched. Sprayed lacquer is applied to the finished lamp and is somewhat less resistant to scratching. The enamel finish is recommended for outdoor uses.

Three-way lamps have three connec-

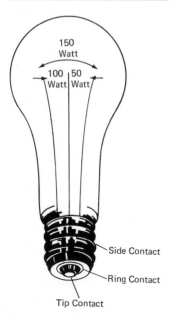

Figure 2-4. Three-way incandescent lamp. (Adapted from General Electric *Home Lighting Bulb Guide,* TP-120)

tions in the base and two filaments (Fig. 2-4). The low-wattage filament is switched on first, then the high wattage, and the last setting uses both filaments. The socket in which this is used must be a three-way socket if the lamp is to function as intended. Wattages vary for three-way lamps; some are 50-100-150, others 30-70-100. The 100-200-300-watt lamps use a mogul base. Other combinations of wattages are also available.

Lumiline lamps have tungsten filaments and the tubular shape of the fluorescent lamp. The filament extends the full length of the lamp and is connected to disc-like bases at each end. Lumiline lamps are used in cove lighting and for lighting of mirrors and pictures where the overall length is short. They give off more heat than fluorescent lamps of equal lumen capacity.

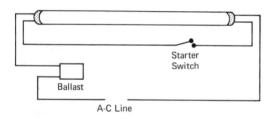

Figure 2-5. Preheat-start fluorescent lamp circuit.

Fluorescent Lamps

The fluorescent lamp consists of a sealed glass tube that contains small amounts of mercury and argon gas and has an electrode at each end. The interior of the glass tube is coated with fluorescent chemicals known as phosphors. When an electric current is sent through the tube, the mercury vapor emits an ultraviolet light that is converted to visible white light by the phosphors on the inside surface of the tube.

The older type of fluorescent lamp is known as the *preheat-starting* type. A few seconds after the light switch is turned on, the light comes on. The circuit for this type of lamp includes a ballast and a starting switch as shown in Figure 2-5.

The starter switch may be either manually operated or automatically controlled. An example of a manually operated starter switch is the one on many desk-type fluorescent lamps. The starter switch on fluorescent lamps used in ceiling luminaires usually is automatically controlled. There are several types of automatically controlled starter switches.

When the starter switch is closed, electricity flows through the terminals and heating coils at one end of the lamp, through the starter switch to the heating coils and terminals at the other end of the tube, and finally, through the ballast. When the heating coils are hot enough, the mercury within the tube is vaporized and the ends of the lamp light. The operator then releases the starter switch. The magnetic field of the ballast collapses when the starter switch is released, and this collapse gives a momentary high voltage which starts the lamp arc. The life of the lamp is affected by the starter because it determines the amount of time the starting current is applied to the cathode.

The ballast serves to limit the size of the current within the lamp and hence to protect the life of the lamp. It consists of a coil of insulated copper wire around a laminated iron core. The layers of thin iron stampings have a tendency to hum. This hum can be reduced in varying amounts by impregnating the core and coils with a special compound. Ballasts are rated according to the amount of hum they make. In industrial uses where the noise level may already be high, ballast hum is unimportant. In a home which is relatively quiet, a slight hum may be noticeable. Not all ballasts have a hum. The way in which the ballast is installed within the fixture also influences the amount of hum. Ballasts are made for the specific lamp size and type as shown on the label of the ballast. Several hundred different types of ballasts are available. If the incorrect ballast is used, then the life of the lamp can be shortened.

In determining the cost of operation of a fluorescent lamp, the wattage of the lamp and the ballast should be added together. This is also true when computing the efficiency of lamps—lumens per watt.

Fluorescent lamps may be operated without separate starter switches. The *instant-start* type of lamp has the advantage of having no momentary delay while waiting for the light to come on. A wiring circuit for this type of lamp is shown in Figure 2-6. Instant-start lamps are not preheated. Ballasts are designed specifically for use in this type of circuit, and the lamps are made differently from preheat-type lamps. No cathode preheating is required, and so only one terminal is needed at each end. This is the characteristic feature of slimline lamps. Instant-start lamps have a specially designed socket. Both ends of the lamp must be in place before voltage can be applied to the lamp terminals. This is a safety precaution necessary because of the high voltage required to start these lamps.[3]

[3] C. E. Weitz, *General Electric Lamps*, General Electric Company, 1956, p. 69.

Rapid-start fluorescent lamps require no starter and start almost instantly—in less than a second. The cathode is designed for continuous heating from special windings in the ballast. Higher ballast voltage is used to start the lamps when preheat starters are not used.

With a few exceptions, most fluorescent lamps of 15 watts and above have a lamp life greater than 7,500 hours. Some ratings are made on lamps operated for three hours per start; others on more extended use per start. Some lamps with ballasts, operated continuously, have had a lamp life of 20,000 hours or more. If lamps are to have maximum life it is important that ballasts provide the proper electrical values. When lamps are operated for less than three hours per start shorter life can be expected.

The *realized* efficiency of higher-wattage lamps is 70 to 80 lumens per watt. Low-wattage lamps have an efficiency of 25 lumens per watt. There are a number of reasons why the *realized* efficiency is lower than the theoretical efficiency. It seems that fluorescent lamps are near their limit of efficiency unless some new principle, at present unknown, should be discovered.[4] Fluorescent lamps operate best at an ambient temperature of approximately 70° F. Luminaires should be designed to prevent overheating of lamps as this decreases their efficiency.

[4] Ibid., p. 62.

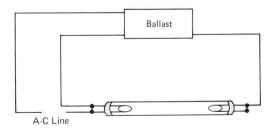

Figure 2-6. Instant-start fluorescent lamp circuit. (Adapted from *General Electric Lamp Bulletin*, LD-1)

The material with which fluorescent tubes are lined is not harmful. Except for some lamps manufactured before 1949, the usual precautions in disposing of broken glass are sufficient. Presumably few if any of these lamps would still be in use.

Fluorescent lamps get darker as they are used. Because this darkening is usually quite even over the entire lamp, it is seldom noticeable until a new lamp is compared with one that has been used for quite a long time. This darkening is, however, the main cause for depreciation of light output during the life of the lamp.

As a general rule, when a lamp begins to flicker, either it or the starter should be replaced. Other reasons for flickering are low-circuit voltage, low-ballast rating, low temperature, and cold drafts. Working under a flickering light is very irritating to many people.

Fluorescent lamps sometimes cause radio interference, which can often be eliminated by moving the radio and lamp farther apart. Radio interference filters can also be installed to minimize this interference. Filters may be placed in various locations, depending upon the specific cause of the interference. A filter may be placed at the radio or at each panel box feeding fluorescent lamp circuits.

Fluorescent lamps, like incandescent lamps, should be shaded. The louvered type of reflector has proved quite satisfactory. Light from a lamp without proper shading or diffusing may be uncomfortable for the eyes, particularly when doing close work, such as reading or sewing.

Dimming of fluorescent lamps requires that special dimming ballasts be used with the dimmer control. Forty-watt lamps are usually used but systems can be designed for 30-watt lamps. However, the two wattages cannot be used on the same system.

LIGHT SOURCES

Good lighting contributes to home safety. Glare may inhibit seeing or darkness may camouflage toys or articles accidentally left in the path of travel through the house. Many activities are made not only safer but easier if one can really see what is being done.

The decision on the type of lighting to be used in a room depends on the type of activities carried on there, which of course varies from one family to another. A few general guides may serve to make home lighting more useful as well as pleasing.

General lighting from any light source in a room should be shaded and well diffused. The quantity of light is often less in living room areas than in the kitchen, laundry, or utility-room areas. General lighting should be adequate for the potential activities in a room, except for special tasks to be carried on where local light sources are available.

The light source should not shine into the eyes. This can happen from a poorly designed or incorrectly placed portable lamp, from a luminaire under a cabinet, over a sink or range, or on the ceiling, or from poorly designed structural lighting. Light should fall on the subject that is to be seen and not on the eyes of the person doing the seeing. Activities such as reading or ironing, done while facing a window, may often cause undue eyestrain because of the glare from the strong natural light. Light over a dining table should highlight the table settings and enable the people at the table to see each other easily. A candlelit table is often uncomfortably lighted because the flame height of the candles is directly in the line of vision.

Shadows make a room more attractive and pleasant but the shadow should not interfere with the seeing process. The overall well-lighted classroom or laboratory may lack some of the warmth and charm of the home because it has more light and fewer shadows. But wherever close work for an extended period of time is done the light should have very few shadows.

The color of the lamps chosen should be considered in relation to the effect that is desired. The regular fluorescent lamp brings out the green and blue colors of the spectrum and generally dulls reds and yellows. The deluxe warm or standard warm lamp enhances all colors, but especially the warm colors. Complexions appear more attractive under this warm lamp light than under light from the standard fluorescent lamp.

The amount of light given off by any luminaire or structural lighting source is cut down if the lamp, diffuser, or shade are dirty. It is most important to choose a luminaire in the kitchen, bathroom, laundry, or utility room that can be cleaned easily. Lamps over a range get dirty very quickly because a film of grease soon settles on them and dust clings to the oily base.

In general, ceilings should be light in color but not glossy. A glossy ceiling tends to make the reflections of light glaring. Dark ceilings create a contrast between the light and the ceiling and also absorb the light. Quite generally, dark colors and dark furnishings absorb light. Artificial light that has been quite adequate in a room with pastel furnishings may be totally inadequate if the room is redecorated in darker colors.

Good general light makes it much easier to clean a room because it is easier to see dust and dirt. It is easier to clean drawers in well-lighted kitchens and bedrooms.

There is also a psychological value associated with working in a room that is adequately and pleasantly lighted.

There should be at least one light in a room that can be controlled by a switch at the entrance. This should give sufficient illumination to enable a person to walk into the room safely. If there is more than one entrance and the entrances are more than ten feet apart, the light should be controlled at each entrance. It should not be necessary to retrace one's steps in the dark when leaving a room.

There are a number of places where it is especially important to remember that light should fall on the seeing task. Lamps placed above and at each side of a bathroom mirror facilitate shaving and skillful makeup artistry (Fig. 2-7); light at a dressing table should shine on the face; light shining on the ironing board enables one to see the wrinkles while ironing; light at the house entrance should light the pathway and the house numbers, and help the home occupants identify visitors. Luminaires need not always be located in the center of the room. Often more strategic locations can be determined by careful consideration of the objectives to be achieved by the lighting in the room.

STRUCTURAL LIGHTING

Structural lighting is described as light sources, designed and constructed to fit a particular situation, which are built into the home as part of its finished structure. It can be used for general lighting instead of a luminaire and can be designed to blend with color schemes or decorative plans. The wall surface itself, rather than a relatively small portable lamp or luminaire, becomes the major source of light. With good wall lighting, the walls tend to recede and the room appears larger.

There are various ways of achieving well-

Figure 2-7. Recessed lights give general illumination; twin luminaires cast light on the face; ceiling lamps provide radiant heat. (NuTone Division of Scovill Manufacturing Company)

designed structural lighting. It is somewhat easier to install at the time a house is being constructed, but some forms can be installed in an existing house.

Ceiling Lighting

Cove lighting is used to light the ceiling and provide indirect light for the room. All the light is directed upward in a cove installation, and because of this it is not a very efficient form of lighting (Fig. 2-8). If much light is to be reflected back into the room the ceilings should be white or very nearly white. When a luminous panel forms the cove, the light in the room is not totally indirect light.

If this light is adequate, there is greater flexibility in the use of the room and less need for moving portable lamps. The upward light from portable lamps also serves to increase the general lighting in a room.

Care must be taken in the installation of cove lighting. The lamp should be placed far enough from the ceiling so that the light can spread, and the trough must be wide enough so that the wall does not get too warm when the lights are on for long periods of time.

Cove lighting provides good background light for television viewing, because the light is not reflected from the television screen and at the same time enough light is provided so that the eyes are not strained by the wide difference between room lighting and screen lighting. It is an extremely bad practice to watch television in a totally darkened room.

Luminous ceiling panels are used in various sizes; some cover the entire ceiling; others cover only a section of the ceiling area. Usually the plastic diffusers are dropped 10 to 12 inches from the original ceiling, and so a room must have enough height to avoid creating other problems. Luminous ceilings give a feeling of spaciousness. This same idea is used for lighting wall spaces. Some designs of plastic are particularly attractive when used in this way. Where it is important to obtain a fairly high level of light, polished aluminum reflectors are used over the lamps and a louver is used in the bottom of the panel arrangement. Luminous ceilings provide general light. It is still necessary to use localized light for certain seeing tasks, such as work at the kitchen counter.

Spot and flood lighting are used to emphasize certain objects or areas within a room. The equipment may be recessed, semirecessed, or surfacemounted. Recessed or semirecessed lighting is more easily installed during house construction. Some styles cannot be installed in two-story homes where the distance between floors is not very great. It is also possible to

have this type of light from both luminaires and portable lamps. This lighting is distributed downward. It should be balanced with other lighting from structural installations or portable lamps.

Soffit Lighting

Soffit lighting makes use of furred-down areas over kitchen sinks, bathroom mirrors, and space in niches and between beams. The light is directed either downward or outward. Soffits use two rows of lamps. Often polished aluminum reflectors are used to increase the useful light. The glass or plastic enclosing the lights should be selected carefully so that the brightness is not objectionable, particularly if the light is in the line of vision. Fluorescent channels should, as in other concealed lighting designs, be mounted on fireproof materials or with a surrounding air space (Fig. 2-8).

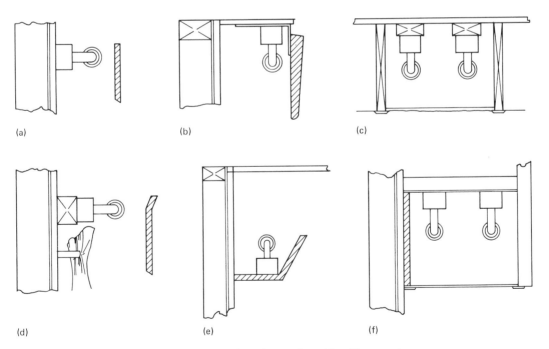

(a)   (b)   (c)

(d)   (e)   (f)

Figure 2-8. Structural lighting forms: (a) bracket; (b) cornice; (c) ceiling panel; (d) valance; (e) cove; (f) soffit. (Adapted from General Electric *Residential Structural Lighting*, TP-107)

Wall Lighting

Valances, cornices, and wall brackets are all used to light large wall areas (Fig. 2-8). *Valances* provide both upward and downward light and are used with windows. Valances can be purchased in prefabricated units or be custommade. If the distance above a window is less than 10 inches a cornice usually is used.

Light from a *cornice* is always downward. The cornices are mounted at the junction of the ceiling and wall and may or may not be used over a window. Cornices are recommended for rooms with low ceilings. When plastic is used to form the cornice, resulting in a luminous cornice, there is very little difference between cornice and cove lighting. The cove is not attached to the ceiling and the cornice is attached to both ceiling and wall (Fig. 2-9).

The *wall bracket* is similar to the valance, but it is used on wall surfaces that do not have windows. Because they do not have to extend beyond the draperies, they are sometimes slightly closer to the wall. Specific instructions for design of any of the lighting structures should be obtained before construction. The illustrations are explanatory only.

Luminous wall panels can be used in relatively small spaces. The space between studs in the house, even those constructed of two-by-fours, can be utilized, although high brightness is sometimes a problem with these installations. Usually a depth of 10 to 14 inches gives better results because the lamps can be placed far enough away from the plastic diffuser to give more even distribution of light and lower brightness.

There are many interesting effects that can be created with lighting in order to obtain light that is not only attractive, but also comfortable and designed adequately for the various tasks or activities conducted in a room. Consult with lighting specialists and such organizations as Live Better Electrically and the Illuminating Engineering Society before making decisions about the lighting in your home.

LUMINAIRES

A complete lighting unit that is installed permanently is technically a *luminaire*. The more familiar term is a lighting fixture. The luminaire should harmonize with the decorating scheme and at the same time give

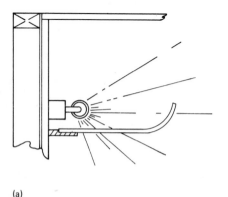

(a)

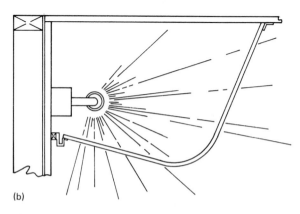

(b)

Figure 2-9. Luminous cove and cornice lighting forms: (a) cove; (b) cornice. (Adapted from Live Better Electrically *Fixture Lighting Guide*, 1.5.598-4, second edition)

good-quality light. A ceiling luminaire often is used for general overall lighting, but it may be used for localized light, as over a dining table or workbench. Wall luminaires often are used for special seeing purposes. Luminaires should be kept clean; most can be cleaned by washing with a detergent, rinsing well, and drying with a towel.

PORTABLE LAMPS

Portable lamps, as the name implies, may be moved from one place to another. A portable lamp may be used to give general overall light, localized light, or both. Every home ought to have at least one good portable floor, table, or wall lamp to be used for reading, sewing, studying, and similar seeing tasks.

Good portable lamps are hard to find. A reading lamp, whether it be a floor or table model, should be tall enough to give a good spread of light, have a fairly wide shade, diffuser, and large wattage bulb. Short and squatty or tall and narrow lamps are not satisfactory to use for seeing purposes. Most pole and bullet lamps are modified forms of the gooseneck lamp. They do not give a good diffused light nor a very wide spread of light. These lamps make an effective means of emphasis for certain decorative objects, pictures, or plants in a room, but should not be substituted for a good reading or sewing lamp.

A portable lamp should be placed so that one does not view the light source or the relatively bright diffuser as he uses the light from the lamp. Often a table lamp by a sofa or chair is just the right height to be annoying to someone sitting on the sofa who is trying to speak with a person standing. Even worse, these lamps often have nothing but a bare bulb under the shade. The bottom of the shade for a floor lamp placed behind a reader should be about 47 inches from the floor; if placed at the reader's side, about 42 inches is considered desirable.

For reading, it does not matter whether the lamp is placed to the left or right of the reader, provided he has reasonably good sitting posture. For writing, the lamp should be placed at the right of a left-handed writer and at the left of a right-handed writer.

Lamp Components

The lamp base should not only be high enough to give the lamp a good spread of light, but it should also be heavy enough so that it is not easily knocked over. It is especially important that floor lamps be sturdily constructed if small children are in the home and may use the lamp as support when learning to walk or when playing games. The cord on the lamp should be long enough so that the lamp can be moved without undue restriction. Both the lamp base and the cord should carry the UL seal of approval.

Sockets on lamps are quite standardized. The medium base socket is used on most lamps under 300 watts; the mogul base is used for lamps of 300 watts or larger. These often are used for floor lamps. The candelabra base is used for some appliance lamps, night lights, and some decorative lamps. Lighting equipment imported from other countries usually is designed for medium-base sockets. However, that purchased in another country by an American tourist is not necessarily designed for sockets commonly used in this country.

The harp on a lamp is the metal part attached to the socket that supports the shade. Harps are different shapes and sizes. If the R-40 lamp is used, a special harp designed for it must also be used. The metal extension that steadies the shade on the harp is called a finial. Clip-on shades have metal supports that fit directly

Figure 2-10. A good shade for a reading lamp. (General Electric Company)

over the lamp bulb. These do not give as firm a mounting for the shades as either harps or the support given by diffusing bowls.

Shades for lamps should be deep enough to cover the lamp socket and also to allow a diffuser to be used on the lamp. Shades should have a white or off-white lining. Colored linings reflect the same color light as the lining. Not only do they make seeing more difficult, but they may also make the furnishing in a room look rather odd. The shade should have a wide opening at both bottom and top, so that light can go upward and give overall illumination, as well as lighting the surface of the seeing task (Fig. 2-10). The shade should be translucent but not transparant. A shade that transmits too much light is uncomfortable for the eyes. Translucent shades should be light in color—white, off-white, eggshell, or cream.

Opaque shades are not as desirable as translucent ones, but are far better than using no shade at all or a shade that is too translucent. An opaque shade should be used if the shade "must" be a dark color. Light shining through a dark translucent shade tends to cast a dark-colored glow over all surfaces on which it falls.

Shallow shades require the use of a plastic diffuser at the bottom, and sometimes at the top, if the lamp is placed so that its top is in the line of vision when a person is standing. The minimum depth for shallow shades is 6 inches. The modified empire shade and the drum shade are large enough so that diffusing bowls can be

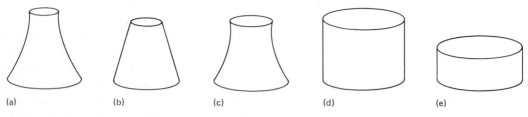

(a)   (b)   (c)   (d)   (e)

Figure 2-11. Some satisfactory shades: (a) bell; (b) cone; (c) modified empire; (d) drum; (e) shallow. (Adapted from Live Better Electrically *Portable Lamp Lighting Guide*, 404-7199)

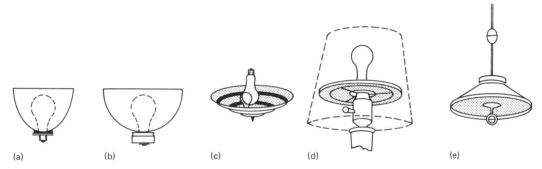

Figure 2-12. Diffusing bowls and diffusers.

used effectively. The cone-shaped shade is less desirable for use on a reading or study lamp but can be used for dresser lamps and lamps that provide accent lighting (Fig. 2-11).

Diffusing bowls and reflectors help to spread light and eliminate glare (Fig. 2-12). The material of a diffuser should allow light to pass through easily without the light source showing through. Glass and plastic are used for diffusers. The material of the diffuser should not be a source of glare. Diffusers used on pull-down lamps should not extend below the bottom of the shade.

Plastic ceiling reflectors or diffusers of a satisfactory type can be purchased relatively inexpensively. They usually clip on to the lamp bulb and hence can be used where a permanent installation is not feasible. Care should be taken to choose a pleasing color, as the light that shines through takes on that color. Size of lamp bulb usually is limited because of the heat produced.

Study Center Lighting

The study center in the home or college room should be well planned. Poor posture, eyestrain, nervous fatigue, as well as poor grades, may result from a poorly planned, poorly lighted study center. The top surface of the desk should have a nonglossy finish of a light or medium color. The desk should not face a window; it is better to place it at right angles to the window or facing a wall. The wall should have a mat finish in a light color. There should be no reflected glare, and the wall should be light enough in color so that useful light is not absorbed by a dark surface. If the wall is finished with a highly patterned paper, it is wise to cover this with a mat-finished wallboard approximately 36 inches high and as long as the desk. Books and papers should be propped up at an angle, as this makes seeing easier. The chair should be comfortable and so designed that the eyes are 14 to 15 inches above the desk top. The bottom of the shade should be slightly below the eye level of the person seated at the desk.

The type of lamp, its placement, and its height are most important. Figure 2-13 shows one good arrangement for a study lamp. There should be no dark shadows on the desk.

Figure 2-13. Good lighting for the study desk. (General Electric Company)

IMPROVEMENT OF
EXISTING LIGHTING CONDITIONS

Existing lighting situations can often be improved immensely with a little ingenuity. The source of light should not be directly visible to the eyes; shades and diffusers should cover bare bulbs. Light should be diffused instead of falling directly on a surface. Many lamps, some very expensive, have only a bulb under the striking shade. If a diffuser is added to the lamp, it will be improved for seeing purposes, provided the shade is of a reasonably good design.

The use of diffusing bowls is not so important in dressing table lamps. The source of light should not be too close to the seeing task. Many desk lamps are too short, and glare is caused because the source of light is too bright for its distance from the seeing task. Study lamps can be made taller by placing them on books, glass bricks, or wood platforms. End table lamps can be raised in the same manner or put on taller tables. A clever person can probably think of other ways to raise the height of a lamp.

Many times the placement of the lamp is poor. Often a worker casts his own shadow over the work he is trying to do. Sometimes moving a portable lamp a few inches will give considerably more light on the seeing task. Measurements for placement of portable lamps are good guides but these should be tempered with judgment that is based on discerning observations.

Consider the background for the lamp. Dark walls, dark draperies, and dark rugs absorb light. Light colors reflect light. Mat surfaces diffuse light; shiny surfaces reflect light in a more glaring fashion. Choose the furnishings of a room and accessories for a desk with these facts in mind.

When lamps are dirty much of the light they emit may be lost. If the shades and the diffusers are covered with dust, the user pays for electricity that is being wasted. Shades can be cleaned with the dusting attachment on the vacuum cleaner, and diffusers and lamp bulbs can be washed. Care should be taken not to subject the diffusers to extremes of temperature; sockets of lamp bulbs should never be placed in water.

As incandescent bulbs get old, they darken on one end, depending upon which end has been in the higher position. Because heat rises, it is the upper part of the lamp bulb that will turn dark. If it is the large end that darkens and not the socket end, the amount of light may be cut down considerably as the bulb ages. It is good practice to put these old bulbs in less critical seeing positions, such as hallways, storage rooms, and basement rooms, and to place new bulbs in lamps used for reading, sewing, and other detailed work.

EXPERIMENTS AND DEMONSTRATIONS

Experiments 1, 2, and 4 can be done without any special equipment, but Experiment 3 requires special equipment.

Experiment 1. The Effect of Different Color Blotters on the Ease of Seeing

Pull the shades, if possible. Use a recommended study lamp. Place it on a blotter, 19 by 24 inches, of neutral color. On this place a sheet of ordinary notebook paper. Turn all room lights off and turn the study lamp on. Look at the notebook paper for ten seconds. Then substitute other blotters such as bright red, bright blue, and yellow.

1. Which blotter showed the least glare by contrast?
2. What would you expect the effect to be over a longer period of viewing?

Experiment 2. The Effect of Different Heights of Lamps and of Large and Small Shades on the Light Falling upon the Study Area of a Desk

Place a light mat-finish blotter, about 19 by 24 inches, on a table. On this place a lamp in the recommended position for a study lamp. Use a lamp shorter than that recommended for a study lamp. It should have a small decorative shade with a diffusing bowl. Use a 100-watt lamp bulb. Pull the room shades and turn off all room lights. Turn on lamp and measure foot-candles of light in at least five positions—near the corners and at the center of the blotter. Record readings.

Raise the height of the lamp so that it meets the recommendations for a study lamp. Again take the readings and record.

Raise the height of the lamp so that it exceeds the recommendations by 3 or 4 inches. Again take the readings and record.

Substitute for the small shade a shade recommended for study lamps. Repeat readings as above and record.

1. Why does the recommended height give more effective light?
2. What effect does the size of shade have on the amount of light on the blotter?
3. What effect does the size of shade have on the spread of light?

Experiment 3. Effect of Voltage on Light Output of a Lamp

Connect a study lamp with a three-way lamp bulb to a source of electricity in which the voltage can be controlled.

Turn off all artificial lights in the room. Close the blinds or pull the shades. Using the 50- or 100-watt setting, take foot-candle readings at the center of the work area when the voltage is at 125, 120, 115, 110, 105, and 100 volts. Repeat at the 150-watt setting.

Connect a fluorescent study lamp and repeat the experiment at the various voltages.

Draw a graph depicting the effect of voltage on light output.

Experiment 4. Colors of Light

Use two fluorescent portable desk lamps and one well-designed incandescent desk lamp with 100-150 watt lamp bulb. In one fluorescent use a daylight lamp; in the other a deluxe warm lamp. Have three replicas of each color sample—minimum size 8 by 10 inches. Construction paper, wall paper, or drapery fabric are quite satisfactory. Separate lamps by cardboard dividers or place 20–24 inches apart. Study each color sample under each of the lamps so that comparisons can be made.

The demonstration can be enlarged by using other colors of fluorescent lamps and more samples of fabric or paper.

BUYING GUIDE

Buying a good lamp for comfortable seeing is not an easy task. It not only requires that the consumer know quality, but considerable stamina and perseverance are also necessary. In the long run, manufacturers supply what consumers want but this is not accomplished without each consumer making her voice heard. Refusal to buy poor-quality lamps and a willingness to pay a reasonable price for a good lamp can be persuasive.

1. Does the lamp carry the UL seal? Both the lamp and the cord should have this seal of approval.
2. Is the lamp base heavy, well-balanced, and resistant to tipping?
3. Is the bottom of the lamp smooth so that it will not scratch table tops or floors?
4. Is the cord at least 7 feet long or long enough for easy placement of the lamp in the home?
5. Is the shade at the right height? Approximately 15 inches from desk top, 42 or 47 inches from floor, depending on placement, are considered desirable. What height is the table on which you plan to use a table lamp? How high will the lamp then be?
6. Is the shade wide enough to give a good spread of light and deep enough to cover lamp bulb, socket, and diffuser?
7. Does the shade have a wide opening both at the top and the bottom?
8. Is the shade white or near white, if translucent, or does it have a white lining, if opaque?
9. Does it have an effective diffuser? The material of the diffuser should not limit the lamp bulb to an inadequate size.
10. Does it have a three-way switch? This makes a lamp more useful.
11. Is the base attractive? Is the lamp well proportioned? The lamp should be attractive when not in use. In many homes lamps serve two purposes—vision and decoration. Do not buy it unless it provides comfortable seeing conditions.

CHAPTER 3
MATERIALS

BASE MATERIALS

Materials considered here are those from which the frame, the article itself, or various parts of the appliance are constructed. Some of these materials are used for many articles of household equipment and others have more limited uses. The material from which an appliance is constructed greatly influences the consumer's eventual satisfaction with it. Much greater satisfaction would result from many purchases if the consumer knew what he was buying and at the same time realized the limitations of the materials used and the care required. As in most other areas there is no one perfect material, and wise selection involves choosing the one best suited to an individual's needs. When labels are provided they should be studied for the information that they contain.

ALUMINUM

Aluminum in both cast and sheet metal form is used for household equipment. Cast utensils are made by pouring molten metal into a mold; articles from sheet metal are pressed or drawn into shape. Sheet metal that has been cold rolled is very hard and durable. Annealing makes the sheet metal more ductile and hence more easily shaped.

Pure aluminum is very soft and does not stain. Aluminum as used in household articles has had some alloy added to it. Aluminum foil is made from an aluminum alloy.

Pressure saucepans, skillets, and griddles usually are made of cast aluminum or a very heavy gauge of sheet aluminum. Saucepans are made of several gauges of sheet metal and of cast aluminum. Refrigerator shelves and ice cube trays, baking pans, gelatin molds, and measuring cups are a few of the many articles made from this sheet metal. About 2 percent of the aluminum now produced is used for pots and pans.

Aluminum is light in weight. Sheet aluminum does not break upon impact, but can be dented. Cast aluminum may be broken by a severe impact. Aluminum does not rust but it is affected by food acids and alkalies. The alkalies leave the pan dark. This darkening can be removed by cooking acid foods, such as tomatoes, in the pan, or a solution of vinegar and water or cream of tartar and water can be boiled in the pan. A solution of 2 teaspoons of cream of tartar to 1 quart of water boiled in the pan for ten minutes is often effective. If the darkening has been excessive the process will need to be repeated. The darkening does not harm foods, and tomatoes or other foods that remove the darkening are not harmed when cooked in darkened pans.

Because aluminum is a very good conductor of heat, it is sometimes used on the exterior bottom of stainless steel pans to improve their heat conductivity. When these pans are new it is difficult—sometimes almost impossible—to tell where the aluminum plating ends, but after some use the line of demarcation shows plainly.

Cast metal is more porous than sheet metal, and for this reason pans will sometimes pit if hard water or foods stand in them for periods of time. Lack of thorough drying may also cause pitting. Pitting is minute or very tiny pinprick holes which usually are observed in the bottom of the pan. Theoretically this could become a problem when cleaning pans. Actually it is not a serious problem, since pitting of utensils does not seem to become excessive even after several years of use. Sheet metal is much more resistant to pitting.

Thickness of sheet metal is measured by its gauge. Too thin a gauge is not desirable for cooking utensils; 10-, 12-, or 14-gauge metals are fairly heavy and are often used. Thinner 18- and 20-gauge metals also are used for cooking utensils. Aluminum is a good conductor of heat, and in the thinner gauges this heat transfer takes place so quickly that sticking may become a problem. When using pans made from the thinner gauges, one should start the cooking process at a relatively low heat unless doing nothing much more complicated than boiling water. Articles made from thinner gauges will dent more easily than those made from a heavier gauge. However, the amount of cold rolling in the manufacturing process partially determines an article's resistance to denting.

Aluminum is sometimes anodized. This electrolytic process produces a comparatively heavy oxide coating on the aluminum. The aluminum oxide is formed by making the aluminum surface the anode in a suitable electrolyte. A 15 percent sulfuric acid solution often is used. About 10 to 15 minutes are required for the process, which takes place at temperatures approximating room temperature. The oxide film is porous and will absorb color readily. The porous surface can be sealed by treatment

Figure 3-1. Saucepan with an anodic coating known as Gem Coat. (The West Bend Company)

in boiling water.[1] Very high temperatures such as may be reached in a mechanical dishwasher or an oven may remove this color.

Uncolored anodized articles may be used in the oven. Pans that have been anodized bake about 30 percent faster than untreated pans. Anodizing makes aluminum stain and corrosion resistant and gives it a harder surface. Anodized aluminum articles do not rub off on light clothing, sinks, and enamel-finished surfaces. This treatment is used for both protective and decorative purposes.[2]

Another finish used on aluminum makes the surface very hard. It is said to be comparable to that of sapphire which is second to the hardness of diamonds. The finish does not crack, peel, chip, or flake. It is resistant to abrasion and corrosion that is common in everyday cooking processes. The electrochemical process takes place at a temperature below freezing and results in a darkened surface—a very dark gray. In current design of pans this gray has been relieved by lines of indentation which reveal a shiny aluminum color (Fig. 3-1). A thick gauge of metal—8—is used for these pans. The material withstands temperatures as high as 800°F.

GLASS

Glass for use in the kitchen is of four main types: (1) glass that will withstand oven temperatures without breaking, (2) glass that can be used for surface cookery, (3) glass that is not heat resistant and is used for accessory equipment such as mixing bowls, and (4) glass used in cabinet construction. The appearance of all glass

is much the same, and care should be taken to select and use the glass for the purpose for which it was manufactured.

The many possibilities for using glass in cabinet construction and kitchens are limited by the weight of the glass, the difficulty of obtaining it, and its cost. Glass for cabinet shelves has the advantage of making the contents of the cabinets easily visible. Sometimes this might be considered a disadvantage. Drawer dividers and sliding panel fronts for cabinets are other uses. The edges of the glass are ground and polished to dull them. Glass is used for mixing bowls, refrigerator dishes, measuring cups, saucepans, cake pans, tea kettles, and other articles (Fig. 3-2). Articles to be used over direct heat must be made from a specially processed glass which has a low coefficient of expansion. Borosilicate glass has a small percent of boric oxide added to the glass "recipe" which helps to

[1] *Encyclopedia Americana,* 1962, Vol. 10, p. 196.
[2] Personal communication, A. G. Vraney, Aluminum Goods Manufacturing Company, December 1956.

Figure 3-2. Glass utensils used for food preparation and surface cookery. (Corning Glass Works)

Figure 3-3. Skillets, in the process of becoming Pyroceram, shown leaving the annealing oven. (Corning Glass Works)

make glass that is less subject to cracking due to changes in temperature. Some mixing bowls, refrigerator dishes, and measuring cups can be put into the oven; others will not stand oven heat without breaking. Be sure to read the label at the time of purchase.

Glass is nonporous and holds heat well, but it is not a good conductor of heat. It will break if subjected to sudden extremes of temperature or to sharp impact. It is not affected by food acids or alkalies, does not rust, and is relatively easy to care for. Glass needs no finish coating of other material. It usually is considered an inexpensive material for small kitchen utensils.

Glass can be scratched by scouring with metal dishcloths and the like. If food sticks to glass, the best procedure is to try to soak it off. If an abrasive is needed, only a very fine one should be used. Cleaning can become progressively more difficult once the glass has become scratched. Some people believe that foods like tomatoes, lemon pie fillings, and other acid foods taste better when prepared in glass utensils. It often is stated that coffee has a better flavor when made in a glass pot.

Glass utensils are particularly well suited for use in demonstrations, as they allow the audience to see how the food is being prepared. However, food cooked in glass saucepans may require quite a lot of watching during the cooking process.

There is always a possibility that liquid in any pan with a cover that does not fit tightly enough will completely evaporate. Covers on glass utensils do not fit tightly and therefore a bit more liquid should be used when cooking foods in glass than when cooking in metal pans with tight-fitting covers. It is usually recommended that a heat protector or metal grid be placed on the surface unit when glass is used over high heat on an electric range. Utensils designed to be used in the oven should not be used on surface units or burners, but utensils for top-of-the-range cookery may be used in the oven.

Pyroceram (pronounced pīe-rō-sīr-răm) is the registered trademark for a certain family of materials that have been changed from a glassy state into crystalline ceramic by the use of heat and nucleating agents.[3]

[3] *Pyroceram*, Codes 9606–9608, PY-3 Progress Report, no. 3, Corning Glass Works.

The glass-ceramics are made from materials such as glass sand, limestone, soda ash, and borax. Glass with a nucleating agent is melted and formed into a transparent glass article, using conventional glass-making techniques (Fig. 3-3). The nucleating agents are precipitated by cooling. Then the article is heated and the nucleated crystals undergo growth. The type of crystallization and the properties of the material are determined by the composition of the glass and the degree of heat treatments.

The Pyroceram article is essentially the same shape and size as the original glass article but it is an opaque ceramic. Altered density and expansion coefficient of the crystalline material cause minute volume changes.

This treatment of glass results in a material that is harder, stronger, and more abrasion resistant than the original glass and one that has improved electrical properties. There can be numerous types of glass-ceramic materials. A large number have been experimentally melted. The Pyroceram material used for Corning Ware[4] articles (Fig. 3-4) is an opaque, white to cream-colored, fine-grained crystalline material. It has a low coefficient of thermal expansion and is nonporous. Because of its dense crystalline structure it has an indentation hardness equal to that of case-hardened steels. It can be heated to 1300°F and plunged into ice water with no effect.

This material has a thermal conductivity of 0.0047 cal per (sec) (cm) (degree C) at room temperature. Comparable figures for aluminum, copper, and iron are 0.49, 0.91, and 0.16, respectively. Window glass is rated at 0.0012 to 0.0024 and asbestos sheets at 0.0004.[5]

Corning Ware dishes can be used for surface, oven, and broiler cookery as well as for cooking on outdoor grills. For surface cookery it is better to use medium to low heat. If a dish boils dry it will be unharmed. It can be taken directly from

[4] *Questions and Answers about Corning Ware,* unpublished leaflet, Corning Glass Works.

[5] Henry Semat, *Fundamentals of Physics,* 3rd ed. New York: Holt, Rinehart and Winston, 1958, p. 344.

Figure 3-4. Pyroceram utensils for food preparation, serving, and storing. (Corning Glass Works)

freezer to range; it is easy to wash, not easily broken, and has no effect on the taste of foods stored in it.

Some metal spoons, egg beaters, or other metal tools may leave grayish marks on Corning Ware dishes. Metal scouring pads will do the same. The gray marks can be removed with a plastic cleaning pad or cleaning powder on a damp cloth. Use of these utensils for frying and pan-broiling can be questioned because of the low heat conductivity of the material. However, when a liquid is used as the heat-conducting medium the transfer of heat by the container is not so important. This is not a good material to use with a thermostatically controlled unit or burner.

Formulas for Pyroceram tableware are not identical to that used for Corning Ware dishes. Use should be guided by information on the label.

IRON

Iron used in household equipment is made either by casting or from sheets of metal. Dutch ovens, skillets, griddles, sinks, bathtubs, radiators, and grates for gas range surface burners are some of the articles made from cast iron. Cast iron is heavy, hard, and somewhat brittle. Skillets may break if dropped on the floor. Iron conducts heat quite evenly and holds heat well because of its high thermal mass. For easy, even browning of foods and for long, slow, carefree cooking it is hard to beat. Cast iron is a porous metal. It is essential to season iron cooking utensils so that these microscopic pores are filled with a covering of fat. Foods can then be cooked without sticking. Though most cast-iron cooking utensils come preseasoned from the factory, reseasoning is sometimes necessary. To reseason, scour the utensil to be treated, wash, and dry. Cover the interior with an unsalted fat. Heat in a slow oven,

250° to 300° F for several hours. Remove, cool, wash, and dry. The utensil is now ready for use.

Sheet metal is used for skillets, pie and cake pans, knife blades, and panels for refrigerators, ranges, and other large equipment. It is less porous than cast metal and has greater impact resistance. It does not hold heat as well as cast iron does nor transfer heat as evenly. Extreme sudden changes of temperature may cause the metal to warp, and for this reason, as well as the fact that it is thinner and hence transfers heat faster, sheet iron is not as desirable as cast iron for skillets. Newly purchased skillets made of sheet metal usually have a coating of lacquer which must be removed before the skillet is used.

Both cast and sheet iron may rust if not treated or finished to prevent rusting. Various processes are used to give a rust-resistant finish to metals. Utensils should be dried carefully before storing. Sheet iron used in large equipment may be Bonderized before the finish is applied. Bonderizing is a chemical treatment which makes the metal rust resistant by forming a base for the paint that prevents lifting or peeling. Bonderizing creates a nonmetallic phosphate coating on the metal which essentially becomes an integral part of the base metal and serves as an anchor for the finish. This phosphate coating inhibits the spread of corrosion if the finish is scratched or chipped and the base metal is exposed.[6] The term Bonderite is used with a number of treatments which may be used on steel, aluminum, zinc, and cadmium.[7]

For some uses iron is finished with a coating of tin or zinc, neither of which will rust.

[6] Personal communication, R. Dhue, Jr., Parker Rust Proof Division, Hooker Chemical Corporation, January 1964.

[7] Bonderite—Registered U.S. Patent Office, Parker Rust Proof Co.

Cast-iron utensils are somewhat difficult to keep looking attractive. If cared for after each use, a skillet does not need to acquire the black crust of burned grease that is often evident on the exterior and sometimes on the interior. The outside of the skillet may be scoured. It is better not to scour the inside of the skillet because scouring removes the seasoning, but the skillet can be washed thoroughly.

MAGNALITE

Magnalite is an alloy of metals of which aluminum is one. Magnalite cooking utensils are made by casting. Magnalite is a good conductor of heat, and because of its high thermal mass, it holds heat well. It is affected by alkalies, but this discoloration can be removed by use of steel wool. The amount of work involved is in direct proportion to the amount of darkening that has taken place. It is not discolored by heat, is light in weight, and is attractive in appearance.

PLASTICS

Plastics in use in the home are included within two main groups—thermosetting and thermoplastic. Several types of plastics are included in each of these main divisions. Some plastics have been misused not only by the homemaker but also by some manufacturers. It is, therefore, essential that care be given not only to the use of plastics but to wise selection.

Information about plastics has been adapted from many sources, but a few sources have provided the greater part of the background material.[8]

"A plastic is any one of a large and varied group of materials which consists of, or contains as an essential ingredient, a substance of large molecular weight which, while solid in the finished state, at some stage in its manufacture has been or can be formed into various shapes by flow, usually through application of heat or pressure, singly or together."[9]

There are a number of methods of processing plastics. Various types of molding, laminating, calendering, casting, coating, and fabricating are some of the processes used. Plastics can be made into a variety of forms—definite shapes, flexible film, sheeting, rods, tubes, filaments, and coatings. They are used also as binders, adhesives, and in lacquers and paints.

There are some standards for plastic products which are called Voluntary Commercial Standards. However, in general these do not include products designed for home use. An exception are standards for melamine dinnerware.

Thermoplastics

Thermoplastic is a type of plastic that softens when heated to temperatures normally used in forming it, without much if any chemical change, and quickly becomes more rigid upon cooling. This process of heating to soften and cooling to harden can be repeated almost indefinitely.

There are ten groupings or types of thermoplastics with which the homemaker might be most concerned and almost as many groupings that are not used for

[8] Ruth Hutcheson, *How to Select and Treat Plastics,* Cooperative Extension Work in Agriculture and Home Economics, Purdue University, USDA Cooperating, No. 298-2; *Plastics in Your Home*; The Society of the Plastics Industry, Inc., *Plastics Today*; *The Story of*

Styron in the Home, The Dow Chemical Company, 1952; *Get Acquainted with Plastics,* U.S. Testing League, U.S. Testing Company, Inc., 1949; *When You Buy Plastics,* National Consumer-Retailer Council, Inc., No. 104, 1950; *Plastics: The Story of an Industry,* The Society of the Plastics Industry, Inc., 10th rev. ed., 1962, 12th rev. ed., 1970 (entitled *The Story of the Plastics Industry*).

[9] *Plastics Today: The Story of Styron in the Home,* The Dow Chemical Company, 1952, p. 4.

household products. *Acrylics* are used in articles like clock cases, salad bowls, lamp bases, combs, and backs for hair brushes. They are tasteless, odorless, nontoxic, have good resistance to cracking or breaking, and can be made colorless or in many colors. They can be made transparent, translucent, or opaque. In crystal-clear form acrylics can pipe light—even around curves. Acrylics can be scratched by abrasive powders, but are unaffected by water that is not too hot for the hand. They are unaffected too by salt, vinegar, animal and mineral oils, and foods commonly found in the home. Cleaning fluid and nail polish removers should not be used on them.

The *cellulosics* group has five subgroups, the characteristics of which are all similar but not quite the same. Celluloid was the very first plastic in this group—created in 1868 in the United States. Cellulosics are used for toys, lamp shades, vacuum cleaner parts, pens and pencils, optical frames, telephone housings, tool handles, and flashlight cases. They are electrical insulators, light in weight, unlimited in color. They will take fairly hard knocks without breaking. They are harmed by nail polish and nail polish remover. Certain ones also are harmed by cleaning fluids. Cellulosics may be washed with warm water and mild soap, but abrasives should not be used. They are not affected by freezing temperatures but should be kept away from high heat.

Fluorocarbons have great stability over a wide temperature range, a zero water absorption rate, a high resistance to chemicals, and one of the lowest coefficients of friction known for solids. Fluorocarbons are used widely in industry but one is most familiar to homemakers. This is a fluorocarbon resin known as Teflon.[10] Teflon was discovered in 1938, but it was not introduced into America's kitchens until 1961. The United States Food and Drug Administration approved Teflon as safe for conventional kitchen use. Teflon II[11] is also nonstick but more scratch resistant than Teflon. It is applied by a four-step process which consists of roughening the surface of the pan, fusing on a ceramic or metal frit coating, spraying on a primer coat of Teflon and baking at about 450°F, then spraying on the top coat and baking at 750–800°F.

Fluorocarbon-coated utensils do not prevent scorching of foods. Foods will scorch in a nonstick pan, just as they do in any ordinary pan if too much heat is used. However, it is easier to remove the burned food than it would be from another utensil. At temperatures above 450°F a nonstick finish may discolor or lose its nonstick property. Therefore when using utensils with this finish, preheat the oven before inserting the pan and use medium or low heat for surface cookery.

There are commercial stain removers available for cleaning of nonstick finishes. It is best to select one recommended by the manufacturer of the nonstick finish. A home remedy which will lighten or remove stain is 2 tablespoons baking soda mixed with 1/2 cup of liquid household bleach and one cup of water. Simmer this solution for five minutes in the stained utensil. The utensil should then be washed, rinsed thoroughly, dried, and wiped with salad oil. This treatment may cause the dark colors of a nonstick finish to become lighter, and if the solution boils over there is a chance it will stain the outside surface of the pan.

Ionomers, introduced in 1964, are a family of thermoplastic resins which contain inorganic as well as organic materials.

[10] Reg. U.S. Pat. Off. for DuPont's nonstick finishes.

[11] DuPont's Certification Mark for scratch-resistant "Teflon"-coated cookware which meets DuPont standards.

Molded houseware items, tool handles, toys, and trays are some of the articles produced from this plastic. Ionomers are transparent, tough, unaffected by oils and grease, tasteless, odorless, and more resilient than polyethylene.

Nylon is the generic name for a family of polyamide resins with related but not identical compositions. It is used for tumblers, kitchen funnels, brush bristles, washers, and gaskets. It can be sterilized. It comes in a range of soft colors; is odorless, tasteless, and nontoxic. It can be scratched by abrasives. It is resilient so it has high impact resistance and flexural strength. It is stained fairly easily by coffee and tea.

Polycarbonates are used to provide a combination of structural rigidity and outer housing in one piece. Housings for electric can openers, coffee pots, electric knife handles, and panels for diffusing light are among the household uses made of polycarbonates. Parts made from this resin have exceptionally high impact strength—they can take hammer blows without shattering—and are stain and oil resistant.

Polyethylene is used in film, semirigid and rigid forms, or as a coating. Flexible mixing bowls, ice cube trays, and raincoats are some of the items made of polyethylene (Fig. 3-5). It is odorless, tasteless, and nontoxic. Some polyethylene formulations will withstand boiling water and even sterilization. Others should not be used in boiling water. It should not be used in an oven or over an open flame. It will retain its flexibility even when cold. Abrasives should not be used on it. It is moisture proof but allows the passage of oxygen. It is not affected by food acids and household solvents. It comes in a range of colors in transparent, translucent, or opaque material.

Polypropylene plastics were introduced into the United States in 1957. This plastic is used for tubs for washing machines, re-

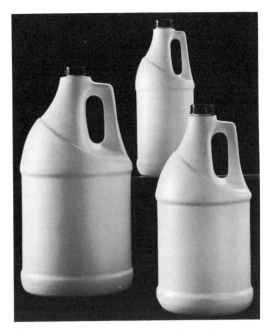

Figure 3-5. High-density polyethylene blownware. (The Dow Chemical Company)

frigerator parts, and a variety of houseware items (Fig. 3-6). It has good heat resistance and unusual chemical resistance. It will resist acid, alkaline, and saline solutions even at elevated temperatures.

Polystyrene is used for refrigerator dishes, utility trays, canister sets, wall tile,

Figure 3-6. Molded polypropylene is used for the percolator; the base is constructed of phenolic. (The West Bend Company)

an assortment of other kitchen items, and disposable dishes. It comes in many colors, is tasteless, odorless, and nontoxic. It is not affected by citrus fruit juices, but citrus fruit rinds may be harmful to it. Cleaning fluids and nail polish remover should not be used on it. It is hard and rigid and will stand up under ordinary household use. It does not withstand severe impact or bending. Although special types may have considerable resistance to heat, most should not be subjected to boiling water temperatures. If exposed to direct flame it will burn slowly. It repels moisture and is light in weight.

There are seven major types of *vinyl* and it is available in film, sheeting, semirigid and rigid forms, as a coating, and as foam. Familiar uses of the vinyl group are for such articles as floor coverings, raincoats, upholstery, draperies, garment bags, electric plugs, and garden hose. Vinyl products may have a slight odor when boxed, and some types will stick to lacquered surfaces. They give satisfactory performance at food-freezing temperatures. Most types are unaffected by water, oil, food, common chemicals, and cleaning fluids. They are affected by chlorinated solvents, nail polish and remover, and moth repellents. They are generally recommended for indoor use. Vinyls should be washed in lukewarm water. They are very strong and tough, stain resistant, and not easily scratched. However, abrasives should not

Figure 3-7. Sheet aluminum saucepan with phenolic handle. (Wear-Ever Aluminum, Inc.)

be used on them. Vinyls can be made in a wide range of colors.

Thermosetting Plastics

Thermosetting plastics are permanently set into shape when formed. Heat is used in this process. The thermosetting plastics do not soften or melt by heat after they have been "set." Included in this group are resins of phenol called phenolics, resins of melamine called melamine, resins of urea called urea, and casein.

The *phenolics*, developed in 1909, are electrical insulators, heat resistant, and are not adversely affected by water, alcohol, oils, mild acids, or common solvents. Cast phenolic is available in all colors. Other phenolic products are usually of dark colors. Phenolics will take hard use and do not break or scratch easily. However, abrasives should not be used on them. Among the uses for phenolics are washing machine agitators, radio cabinets, toaster bases, appliance handles, and handles for cooking utensils (Fig. 3-7).

Melamine and *urea*—amino plastics—are colorfast, breakresistant, odorless, tasteless, and nontoxic. They are used in the manufacture of certain types of tableware, cutlery handles, counter coverings, range knobs and handles, and appliance housings. Products made of urea may be washed in water of temperatures up to 170° F and those of melamine up to 210° F. They give satisfactory performance in temperatures as low as −70° F. Charring or discoloration may occur if either material is used in an oven or over a flame. Abrasives should not be used on them even though they are scratch resistant. They are not affected by detergents, carbon tetrachloride, nail polish and remover, alcohol, oil, or grease.

Casein, developed in 1919, is the product of skim milk protein and formaldehyde. It is used for such articles as beads, buttons, and knitting needles. Casein articles will withstand dry cleaning but not washing. Immersion in water can cause casein articles to absorb water, swell, soften, and break. As the temperature of the water is increased the time necessary for absorption of water decreases. A wide range of colors is available.

STEEL

Steel is made by refining molten pig iron. Hundreds of different recipes are used in making steels. All steel requires some manganese.[12] Manganese is important in hardening steel and in giving it a refined grain structure. Silicon, chromium, and nickel are also important alloying elements. Sheet steel is widely used in household equipment. The gauge will vary. Walls of refrigerators, ranges, freezers, cabinets, and cutlery blades are some of the items for which it is used.

Steel is extremely strong and durable. It will rust if not thoroughly dried or properly treated in the manufacturing process. Bonderizing often is used on steel that is to be finished with synthetic enamel. Steel also is finished with porcelain enamel, tin, and zinc. Steel is not as good a heat conductor as iron.

Steel is manufactured as *high-carbon* or *low-carbon steel*. High-carbon steel has in it 1 percent or more of carbon. Low-carbon steel has 1/10 percent or less of carbon. High-carbon steel is important in the manufacture of knives, as it will take and hold a very sharp edge.

Stainless Steel

Stainless steel is made from recipes combining molten steel, chromium, and nickel. Manganese and silicon also are added. Steel must contain at least 11.5 percent chromium if it is to be stainless steel.

[12] *Steelways*, American Iron and Steel Institute, January 1963, vol. 19, no. 15.

Stainless steel is widely used in household equipment for wash tubs, sinks, counter tops, liners in freezers, cutlery blades, saucepans, flatware, and trim on counters and large equipment. It can be highly polished during manufacture and therefore is very attractive both for trim and for complete utensils for the kitchen. As a counter top it may have a high polish, a satin finish, or perhaps a corrugated pattern.

Stainless steel is resistant to staining; some stainless steels are almost stainproof. It needs to be thoroughly dried or it may water spot. It does not rust. If overheated, it turns dark. Sometimes it is possible to remove minor darkening by scouring, but if the overheating has been very extensive, nothing can be done about the darkening but to decide the steel is more attractive dark! It is extremely durable, not easily dented, and practically impossible to break. It can be scratched if a counter surface is used as a cutting board and this is equally hard on a knife so used. Stainless steel is not resilient and seems noisy when used as a counter top.

Because it is not a good conductor of heat other metals may be used with it in cooking utensils to improve the heat conductivity of the pan. Stainless steel utensils that are not covered on the bottom with a good heat conductor such as copper or aluminum will often develop hot spots which in turn result in scorched food and pans that are difficult to clean. This can be avoided by careful use of heat. If a pan is heated slowly and not overheated these hot spots do not develop.

Although stainless steel is usually lower in carbon content than the steel from which it is made, it is possible to buy *high-carbon stainless steel*. This not only has the characteristics of ease of upkeep and resistance to staining but it also has the ability to take and keep a sharp edge.

Knives made of this high-carbon stainless steel are somewhat more expensive than comparable knives of chromium-plated, high-carbon iron or steel.

Aluminum and stainless steel are used in the production of stainless-clad aluminum sheet. Stainless steel 1/100 inch thick is bonded to an aluminum sheet which can vary from 3/100 to 1/4 inch thick. The thickness of the aluminum sheet depends on the specifications of the utensil manufacturer. Utensils made from this aluminum-clad stainless sheet have an interior of stainless steel and an exterior of aluminum. According to the manufacturer, the sheets can be shaped without deforming or separating the aluminum and the steel. Manufacturers of Durantel[13] do not recommend washing these utensils in an electric dishwasher.

MATERIALS USED FOR FINISHES

The type of finish used on a base material is determined by the end product desired and the specifications that must be met. Finishes may be used to prevent rust and corrosion, to make an article more attractive, to make it a better heat conductor, or to make it easier to maintain.

Finishes may be divided into two groups —mechanical and applied. Either may be used to make an article more attractive or easier to maintain. Mechanical finishes are commonly used on materials that do not rust or corrode and do not affect or are not affected by the foods or other materials with which they may come in contact. Applied finishes are used on materials that rust, corrode, or are affected by foods; they are used also to make utensils better conductors of heat.

MECHANICAL FINISHES

Mechanical finishing consists in polishing or buffing the material until it has reached the desired luster. The two more common mechanical finishes are known as high polish and satin finish. The former is very bright and shiny and the latter has somewhat more the patina often associated with sterling silver. A third mechanical finish is known as hammered or pebbled. In this the metal is slightly dented to give it an appearance that is different from the totally smooth surface.

APPLIED FINISHES

An applied finish is one in which a second material is added over the base material, usually by the use of heat or by an electroplating process.

Chromium

Chromium-finished base metal is a very popular trim for cabinets, refrigerators, and ranges. Chromium also is used as a covering for small appliances such as coffee makers, waffle bakers, and irons. Chromium is a shiny, hard metal which requires little care except washing with soap and water and thorough drying. Fat that has splattered and burned on a chromium finish can be removed with whiting—calcium carbonate, $CaCO_3$. Abrasives should not be used on a chromium finish.

A good chromium finish will have a layer of copper and one of nickel underneath it. A layer of copper is plated on to the steel base, then a layer of nickel is added, and lastly the chromium finish.

[13] Trademark of Aluminum Company of America.

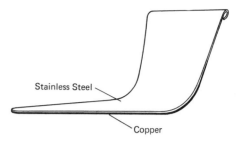

Figure 3-8. Cross section of a copper-clad stainless steel utensil. (Revere Copper and Brass, Inc.)

Copper

Copper is used as a finish on stainless steel pans to improve their heat conductivity. It usually extends beyond the angle of bend so that heat is spread up the sides of the pan more evenly (Fig. 3-8). Copper is very attractive when new but it oxidizes and darkens with exposure to air and during use. It is not difficult to polish, and there are a number of cleaners that can be used for this purpose. Vinegar also works well. The dark copper absorbs heat more efficiently than bright shiny metal which reflects heat. The difference between dark and shiny copper either in time or expense of fuel to the average homemaker is not appreciable. The fact, however, that there is a difference serves as a good excuse for not polishing the pans! Some copper or copper-finished utensils are coated with a lacquer which prevents darkening. This is not permanent, however, and the utensils can look most unattractive when the lacquer is in the process of wearing off. Overheating is one reason that the lacquer finish deteriorates.

Nickel

Nickel has been used as a finish more in the past than it is at the present. It now is used widely as an alloy to add strength and heat resistance. Nichrome wire (a nickel-chromium alloy) is used as the heating element in many appliances. Nichrome wires withstand cherry-red heat. As stated previously, nickel is an essential ingredient in stainless steel.

Nickel looks very much like chromium when it is new but takes on a slightly yellow cast with age. It requires polishing to keep it looking new. Nickel finishes can be scratched if abrasives are used on them.

Porcelain Enamel

Porcelain enamel is essentially a glass material which has been made white or colored by the addition of coloring and opacifying agents. After this glass is ground very fine and mixed with a liquid, it is sprayed onto a base metal, or the metal may be dipped into the liquid porcelain enamel (Fig. 3-9). The glass literally is fused into the pores of the metal as the glass and metal are heated in ovens to approximately 1,500° F. Temperatures as low as 850° F are used for porcelain enamel on aluminum and as high as 1,850° F for some porcelain on stainless steel. Dipping, spraying, and flow-coating are methods most widely used in the application of the enamel to a metal surface. An electrostatic process is used by a few enamelers but its use is restricted to certain sizes and shapes of metal parts. The basic principle on which the electrostatic system of spraying is built is that electrically charged particles of one polarity are attracted to an object of the opposite polarity. The atomized

Figure 3-9. After spraying porcelain enamel coating on these oven door panels, it is dried by infrared heat to bisque condition before final high-temperature firing. (Porcelain Enamel Institute, Inc.)

spray is charged and the piece to be sprayed is of the opposite polarity. The power for creating the electric field is obtained by using as much as 150,000 volts.[14]

Porcelain enamel can be made resistant to abrasion, acids, alkalies, atmospheric corrosion, heat, impact, or combinations of these agents. There are thousands of formulations for specific service conditions. It is nonporous and is easily kept clean and sanitary. Soap and water are practically the only cleaning agents needed. Should food be allowed to stick on it, a solution of soda water or a weak ammonia solution will aid in its removal. Porcelain enamel should not be scoured, as its glaze may be scratched. It is long wearing and except for chipping from sharp impact will last indefinitely. Poorer qualities are less chip resistant. The resistance to chipping is dependent upon a number of factors—the chemical composition of the enamel, the coefficient of thermal expansion of the metal and of the enamel, proper application, the type of metal and its preparation, the thickness of the metal, and thickness of porcelain enamel. The average thickness of porcelain enamel on household equipment is about 0.006 inch.[15] Porcelain enamel can be crazed by a sudden change of temperature such as could be caused by putting ice water in a hot porcelain-enameled pan.

Some manufacturers use porcelain enamel as the finish on ranges, refrigerators, washers, dryers, and other appliances —both inside and out. When titanium oxide, an opacifying agent, is used in the porcelain mixture, the result is a very acid-resistant porcelain with a high degree of opacity. The titanium coat may be applied directly to special low-carbon steels for some appliances or over a ground coat of cobalt enamel in others such as sinks, bathtubs, refrigerator liners, and cookware. Cobalt is a coloring agent that promotes the fusion of glass and metal by a complex chemical reaction during the enamel firing process which produces a stronger bond between the metal and the glass. Acid-resistant porcelain is highly desirable on range tops (working surfaces), the bottom surface of refrigerator interiors, refrigerator vegetable pans, table tops, and sink and drainboard surfaces.

Porcelain enamel also is used for saucepans, baking pans, tea kettles, casseroles, and other small equipment. Foods prepared in such utensils have no color or flavor change due to the material of the pan. Porcelain enamel is not a good heat conductor, and since the base metal is often steel, which is also not a particularly good heat conductor, some care must be used to see that foods do not stick. Porcelain enamel applied to the exterior of aluminum utensils provides a surface that remains attractive even when the utensil is washed in a mechanical dishwasher.

Porcelain enamel is not resilient, thus, it is noisier when used as a table top than a more resilient covering would be.

Porcelain enamel commonly is fused onto sheet metal, but also is used on cast iron, cast aluminum, stainless steel, and copper. It adds color and variety to kitchen utensils.

Porcelain enamel comes in a variety of colors as well as white. Each major manufacturer of household equipment has his own selection of colors. Some care needs to be taken to choose colors that harmonize or results are apt to be most disillusioning.

Synthetic Enamel

Synthetic enamel is a plastic resin-base paint that is sprayed onto a base metal and baked at temperatures between 150° and

[14] *Spraying,* Bulletin P-301, Porcelain Enamel Institute, Inc., p. 21.

[15] Personal communication, Edison T. Blair, Porcelain Enamel Institute, Inc., 1963.

400° F. The electrostatic application process also is used with synthetic enamel. It is not fused to the metal in the same way that porcelain enamel is.

Various types of plastics are used in these enamels; sometimes they are known by the name of the plastic and other times are referred to as synthetic or baked enamel. Alkyd enamel, which may be used on large appliances, is baked on at approximately 350° F and is chip resistant but not scratch resistant. Acrylic enamel is bonded to aluminum or stainless steel at approximately 450° F. It too may be used on large appliances and also on the exterior of cookware. Polyimides came into general use in 1962. They have no measurable melting point, but share the linear structure of the thermoplastics. Polyimides are used on the exterior of cookware.

Synthetic enamel is lighter in weight than porcelain enamel and so an appliance finished with synthetic enamel weighs less than if it were finished with porcelain. Synthetic enamel does not chip but it can be scratched. Only an expert could consistently tell by appearance when new whether an appliance was finished with synthetic or porcelain enamel. Because of the complexity of finishes that can be used on cookware as well as on large appliances it becomes more and more important to buy from a manufacturer on whom one can depend to have made good choices for the materials used in the products placed on the market.

Tin

Tin is used as plating and not as a base metal; hence tinned pans is a more accurate term than tin pans. Tin is a very soft metal. The quality of a tin-covered article depends upon the quality of the base metal, the quality of the tin, and the manner in which the article is covered. A thin layer of tin or a layer which has tiny bubbles in the surface is not of the best quality.

Tin is bright and shiny when new, and looks very much like aluminum. It darkens as it ages, and eventually, with use and age, it may become quite dark. Tin should not be scoured as it is fairly easy to scrape off some of the coating, thus allowing the base metal to rust.

Zinc

Zinc is a hard metal. Articles finished with zinc are known as galvanized. Technically, only articles coated with zinc by electroplating should be so called. This finish protects the base metal from rusting. Zinc is applied after the article is made. It can be scoured or worn off. Zinc finishes are used chiefly for water pails, garbage cans, laundry trays, and certain other articles. Zinc is not used for cooking utensils, but may be used for counter tops. Such a surface is neither resilient nor stain or acid resistant, but it is greaseproof and waterproof. It will tarnish and darken with use. It is not as attractive as many other counter-top finishes.

COUNTER AND FLOOR COVERINGS

A number of materials are now in use as counter and floor coverings, most of which are quite satisfactory, but there is no one completely ideal covering.

ASPHALT TILE

Asphalt tile is made of asbestos, asphaltic or resinous binders, color pigments, and plasticizers. Only small amounts of asphalt are used in most asphalt tile. Asphalt tile is available in many colors, both plain and patterned.

Asphalt tile is neither oil nor grease resistant. There is, however, a tile available which has been specially processed to make it more resistant to oil and grease than regular asphalt tile. Asphalt tile does not stand up under continued high temperatures above 85-90° F, which might be reached if the sun shone through glass onto the floor, or if carpeting were laid over asphalt tile in a home having a floor heating installation. Asphalt tile is a good conductor of heat. It is not considered to have much indentation resistance.

Asphalt tile should be cleaned with a mild detergent and water, rinsed, and waxed. Only a water-base wax should be used.

CERAMIC TILE

Ceramic tile is made from clay which has been especially treated and fired in kilns at a temperature of approximately 2,000° F. Both glazed and unglazed tiles are available. Glazed tiles are easier to clean and are more resistant to staining.

A great variety of sizes, colors, designs, and shapes are available. Small tiles usually are mounted on paper sheets, which makes installation quicker.

The installation of ceramic tiles requires the services of an experienced ceramic tile worker. It is one of the more expensive finishes to install, but some authorities consider it one of the easiest to maintain when properly installed.[16] The grouting between tiles can become difficult to clean, particularly if the tile is not installed correctly.

Ceramic surfaces are waterproof if the

[16] Ben J. Small, *Flooring Materials*, Circular Series F4.6, Small Homes Council, University of Illinois, March 1955.

tile has been properly installed. Ceramic tile is very hard and not easily scratched. It can, however, be shattered by severe impact. It is resistant to water, acids, oil, grease, and heat. Because it is hard, care is required when placing dishes and glassware on it to avoid chipping or cracking the dishes and to cut down on noise.

CORK TILE

Cork tile is made from cork shavings, either with or without resin binder, compressed into molds, and baked. It comes in many sizes and several thicknesses.

Cork tile is not a counter covering and is not recommended for floors where the traffic is heavy. Cork may become porous and brittle if washed too often or if strong cleaners are used. It is not oil and grease resistant. It is very quiet and resilient when walked on.

LAMINATED PLASTICS

Laminated plastics, belonging to the thermosetting group, are sold under a variety of trade names. They consist of several layers of paper or cloth which have been impregnated with a resin and dried, a design layer impregnated with melamine resin, and a final top layer of a protective overlay sheet which has been saturated with melamine. These layers are compressed together under pressure and heat. A pressure of approximately 1,200 pounds per square inch at 280° F for 60 minutes is followed by gradual cooling and lastly a release of the pressure. After this process has been completed the resulting product cannot be soaked or split apart. When lower pressures are used, the resulting product is known as low-pressure, thermoset, laminated plastic.

Thicknesses of 1/32 inch are made specifically for vertical surfaces. They are not designed for counter use. Preferably, counters should be installed by skilled workmen or the process can be done at the manufacturing plant. Laminate-bonded plywood can be sawed, cut, and shaped to fit counter requirements in each home. The edges may be finished with an edging strip. One-piece counter tops with back-splash can be made of these laminates.

Weaver[17] found that the laminated plastics were resistant to stain, heat, moisture absorption, impact, and abrasion, and had good color retention. A good grade of rigid, thermosetting plastic is more durable than the low-pressure flexible type.

LINOLEUM

Linoleum is one of the "old-timers" in the field of counter-top finishes and floor coverings. It is made from powdered cork or wood flour, linseed oil, color pigments, and resins that are bonded to a backing of felt or burlap.

Linoleum can be had in a number of different weights or gauges and patterns. Heavy gauge is approximately 1/8 inch in overall thickness, standard gauge 3/32 inch, and light gauge 1/16 inch thick. Linoleum is resilient and quite resistant to indentation.

Battleship linoleum is more durable than other linoleums and is made especially for industrial uses. Powdered cork is added to the regular linoleum formula and the backing is usually finely woven burlap. It comes in both heavy and standard gauges in plain dark colors.

Jaspé, marbleized, and *inlaid* linoleums are made in all gauges and in all of them the pattern extends through the covering on the surface to the backing. Jaspé is a variegated pattern usually of two colors

which has a striated effect. Marbleized linoleum resembles marble and is a multi-colored material. In both of these, the pattern extends at random to the backing. In inlaid linoleum it extends through to the backing in a pattern. This pattern may be made by pieces cut from various colors, fitted together, and then pressed onto the backing; or the linoleum mix may be sifted through stencils onto the backing and then pressed in. *Embossed* linoleum has a raised design and is usually found in standard and light gauges. It is somewhat difficult to keep clean because of the raised design.[18]

Linoleum should be washed with a mild detergent and rinsed. The surface can be kept cleaner and brighter if it is waxed periodically, depending upon the extent of wear upon it. Either a water-base or naphtha-base wax may be used. Old wax should be removed before new wax is put on.

A special linoleum tile without backing is sold in blocks of various sizes. It is a long-wearing covering which has been specially processed.

RUBBER TILE

Rubber tile is made by vulcanizing rubber and pigments under pressure. It is available in several degrees of hardness and in thicknesses of 5/64 inch, 3/32 inch, and 1/8 inch. It comes in plain colors and marbleized designs.

Rubber tile is affected adversely by oil, grease, and solvents, and so it is not used for kitchen counter tops. Lacquers, shellacs, varnishes, or alkaline cleaners used on it may cause it to dry out, crack, or curl. It should be installed where there will be traffic as it keeps its resiliency better if

[17] Elaine K. Weaver and Velma V. Everhart, *Work Counter Surface Finishes,* Research Bulletin 764, Ohio Agricultural Experiment Station, 1955, p. 44.

[18] *The Cornell Kitchen,* New York State College of Home Economics, W. F. Humphrey Press, Inc., 1952, p. 40.

used. It is long wearing, very resilient, fire resistant, and has a high indentation resistance.

Rubber tile needs to be buffed with very fine steel wool to remove dirt and stains that do not come off with ordinary washing. An untreated dust mop should be used for daily cleaning. Only water-base waxes should be used on rubber tile.

VINYL

Vinyl is made from vinyl resins, plasticizers, and color pigments. It may have a backing of felt, cork, or degraded vinyl cemented to it, or it may have no backing. It belongs to the thermoplastic group of plastics. It comes in a wide range of bright, clear colors and in many designs and patterns.

Vinyl is quite resistant to indentation; it is more resistant to heat than linoleum but not as much as laminated plastics. It is resistant to oil and grease. The care of vinyl is much like the care of linoleum. Either a water-base or naphtha-base wax may be used on it, and it should be waxed to protect the glossy finish.[19] Vinyl does not harden with age.

Vinyl asbestos tile has had asbestos added to the vinyl formula. It has no backing. The tile size is a 9-inch square. Vinyl tile is not as resistant to indentation as is the vinyl. It is resistant to oil and grease and to moisture from concrete floors. More noise is made upon impact with the tile than with the vinyl.

WOOD

Wood used for counter tops without a covering such as vinyl, linoleum, or laminated plastic should be very resistant to wear, stains, water absorption, slivering, and warping. Beech, birch, and hard maple—all hard woods—are considered most desirable. Hard maple is the least grease absorbent. The end or edge grain of wood is best for chopping blocks because tool marks are less apparent.

Before using a new hardwood work top, it should be carefully dusted with a dry cloth and then rubbed with a soft cloth saturated with a little mineral oil until it shines. This oil should be left on overnight and the treatment repeated. After four to six hours, the excess oil should be wiped off with a soft, dry, clean cloth. Daily care should consist of wiping with a damp cloth. Ordinary soil can be removed with a damp sudsy cloth.

INSULATION MATERIALS

Insulation is used for many purposes: refrigerators and freezers are insulated to keep the interiors cold; ranges are insulated to keep the heat from escaping to the room; rooms are insulated to inhibit the transmission of heat and decrease the transmission of sound; wires carrying electricity are insulated so that the electricity will not travel on unwanted pathways.

Heat always moves from a warmer to a cooler body. A perfect insulator would completely stop this transfer of heat. The ideal insulation would not burn; it would be light in weight, and unaffected by mold, bacteria, or insects; it would not rot and would have the ability to remain permanently in place.

[19] Small, op. cit., p. 5.

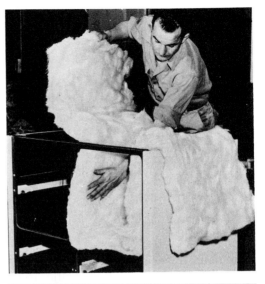

FIBERGLAS

Fiberglas is the registered trademark for a variety of products made of or with glass fibers.[20] The thermal conductivity of Fiberglas is very low. It is a good sound absorber, is moisture resistant, will not burn, and will not rot. Fiberglas has no tendency to warp, expand, or contract. Vibration does not cause the material to settle. It is clean, odorless, noncorrosive, resilient, and very flexible.

Fiberglas wool is fabricated in batts of numerous sizes (Fig. 3-10). Exact shapes and sizes can be made for special insulation projects. It is used in home appliances for both thermal and sound insulation. Fiberglas cloth is used as a backing for mica. Fiberglas also is used as a base for material of laminated products which can be formed into shapes as desired. Shatterproof windows are a glass-fiber reinforced plastic.

MICA

Mica is an essential part of igneous rock and is widely used as an electrical insulator. It will not burn and, depending upon the grade, may be quite transparent. Small sheets of mica may be cemented with shellac or other insulating cement onto cloth or paper. Nichrome wire wound around mica, in the sheet form, forms the heating units of some small appliances.

MINERAL WOOL

Mineral wool is made from melted rock, slag, or glass blown into fibers by steam or air blast. These fibers form a fluffy mass that can be made into different forms. In this loose fluffy form it is used to fill hard-to-reach corners and as a fill-type insulation for walls and floors and similar

[20] Owens-Corning Fiberglas Corporation.

Figure 3-10. Fiberglas in two forms as used in refrigerators and ranges. (Owens-Corning Fiberglas Corporation)

structures. In a granulated form it is commonly installed pneumatically, through a hose, to otherwise inaccessible spaces. In a felt form it has had a binder added and is a flexible, semirigid sheet or roll. In the board or block form it is a rigid insulation which is easily sawed or cut to fit into place.

Mineral wool has a very low thermal conductivity and is extremely light in weight. The fibers are incombustible. They even tend to smother fire. Mineral wool is resistant to changes in structure when exposed to moisture and freezing. It is also very resistant to the growth of mold or bacteria. Mineral wool insulation is practically indestructible.

PLASTIC FOAMS

Polystyrene foam is a man-made insulation that repels moisture, does not settle, and is very lightweight. Common items made of this material are the insulated containers for carrying foods to picnics and lightweight ice buckets.

Urethane foam is also a synthetic insulation. It resists water, rot, and vermin, and can be made flame resistant. Both foams are used in appliance construction. Urethane foam also is used for articles such as mattresses, insulated clothing, rug underlays, packing material, and sandwich panels for house construction. The space required for urethane foam insulation is less than for some other types which were used for many years. Therefore, usable space in refrigerators can be increased without increasing exterior size.

The K factor of sheet asbestos is 0.097; that of urethane foam is 0.12, which is about twice as efficient as other insulation materials in common use in household appliances. Use of urethane foam necessitates additional expense for special machinery to apply it. The foam may expand 30 to 40 times after it is applied to a cabinet liner.

CHAPTER 4
KITCHEN UTENSILS

Chiefly nonelectric kitchen utensils, such as pans, and tools such as measuring cups, beaters, and knives are discussed in this chapter. The small electric appliances used in kitchens and eating areas are discussed in Chapter 5.

UTENSILS

TOP-OF-RANGE UTENSILS

Saucepans

A saucepan is a utensil with one handle; a saucepot has two handles on opposite sides of the pan; and a kettle is equipped with a bail. Saucepans are somewhat more commonly used at the present time than the other two. However, most of the points considered in the study of saucepans also apply to saucepots and kettles.

Sizes of saucepans usually range from 5/8-quart to 4-quart capacity. Measurements are in liquid measure to overflow full capacity.[1] Saucepots are available in larger sizes.

Although it is possible to cook satisfactorily with almost any kind of pan in any condition, it is far easier to cook well with a wisely selected pan. It is especially important when pans are used on thermostatically controlled surface units that the pan be well designed and of material that conducts heat evenly and quickly.

The saucepan should be flat on the bottom and have nearly straight sides with a rounded bend where the sides join the bottom. A rounded bend is much easier to clean than a 90-degree angle would be, and it is also far more usable when stirring puddings or similar mixtures.

---

[1]Metal Cookware Manufacturers Association Standard, pp. 6, 10.

The finish for the top edge of the pan should be smooth, not sharp, and so constructed that there are no crevices where dirt and food particles can collect. The top may be cut off smooth and the edge polished slightly. This is probably the most satisfactory because cleaning is so simple. Cast utensils and some sheet metal utensils are finished in this manner. Or the edge can be bent down to the side of the pan. This often is referred to as a "beaded" edge. If it is turned down tightly enough, it does not leave much room for dirt to get up under the edge. This also serves to make the edge of the pan sturdy. If the gauge of the pan is thin, a beaded edge helps the pan to take wear.

The material should be one that conducts heat evenly and rather quickly. If it does not conduct heat well, hot spots may develop and any thickened mixture will nearly always scorch unless great care is taken to regulate the heat from the unit or burner and the food is stirred often and thoroughly. The material should have no effect on food cooked in the pan, even though the food may darken the material. The material should be easily cleaned and durable enough to withstand ordinary household use without denting. Weights of pans should be compared only between pans that have been made of the same gauge of metal.

The saucepan should be well balanced. If the handle is too heavy for the pan, it can easily be tipped over when only a very small amount of food is in the pan. To check the balance of a pan, tap the handle lightly. It will, of course, tip, but if it is well balanced it will right itself. This should be done without the cover, because the cover will make the pan heavy enough to maintain balance. The pan should also be balanced when filled so that carrying it is easy.

The handle of the pan should be se-curely attached. A handle that turns or loosens can cause a bad accident if a pan is filled with hot food. One of the most satisfactory methods for attaching the handle is by electric spot welding. Several tiny spots are evident on both the handle and the inside of the pan where the handle has been welded, but they are almost perfectly smooth, so cleaning is no problem and the joining is quite secure. Rivets also are used and give a secure fastening. It is necessary to clean around the rivets which protrude on both the exterior and interior of the pan. The shape of rivets vary; some are more rounded than others.

Soldering is sometimes used, which means that a flux or third material has been used in the process of joining the handle to the pan. This is more satisfactory on an enameled pan than any other because the enamel covers it all. Solder is composed largely of lead and tin. Since these alloys are of low strength the solder should not be expected to withstand great strain. Soldering generally is thought to be less permanently secure than welding.[2] Brazing is also a process of joining two metals by use of a third nonferrous metal of a lower melting point. It is done at temperatures above 800° F and soldering is done at temperatures below this. Brazed joints are considered to be quite strong. Filler materials commonly used are copper, copper alloys, silver alloys, and aluminum alloys.[3]

In utensils made by casting, the shank of the handle is made as an integral part of the pan. To this the handle proper is attached, usually by a screw. These handles may tend to loosen with use but for a time they can be tightened by turning the screw. The threads may wear eventually and the screw then cannot be tightened.

---

[2] *The Encyclopedia Americana*, 1962, vol. 25, p. 238f.
[3] Ibid., vol. 4, p. 451e.

If a wood or plastic handle is put into a square shank, it cannot turn even if it does loosen. Screws often are countersunk into the handle, and this is better than if they protruded. No matter how a screw is put in, more effort is needed to keep it clean than a rivet which is smooth. Should a handle wear out, break, or be ruined by excessive heat, it is replaced quite easily if it is attached with a screw.

Handles should be made of a nonheat-conducting material. They should be designed so that the user does not have to grasp metal in the form of rivets, screws, or metal supports, since they can be hot enough to burn the hand even though these parts may be small. The material used for handles should be impervious to moisture and resistant to slipping. The design of the handle varies, and some shapes are more comfortable for one person than another.

All-metal handles are found on less expensive utensils. Some of these are hollow with an opening at one end or sometimes at each end. These present a cleaning problem because dishwater gets inside, and it is difficult to wash the interior of the handle. If a metal handle is not hollow, its edges may not be satisfactorily turned under or beaded down, and dirt collects in the crevices. Some are of a single piece of metal with smooth and dull edges.

Some handles are ovenproof; others are not. One should read the directions that come with the pan. Phenolic handles can be used safely in the oven if the temperature does not exceed 425° F. An oven should be preheated before a pan with a phenolic handle is placed in it. Only all metal handles can withstand broiler heat.

The cover of the pan should also have a nonheat-conducting handle that is sufficiently large so that one can grasp it without fear of burning one's fingers. It should be attached so that it will not easily loosen but can be replaced if necessary. The cover should fit securely but still allow slight space for steam to escape. Either the pan or the cover should be fitted with a lip which will keep condensed steam from dripping on the outside of the pan.

Skillets

Skillets or frypans should be made of a material that is a good heat conductor, that holds heat relatively well, and that is fairly easy to keep looking attractive. A fairly heavy gauge of material is preferred for skillets. Cast-iron skillets should be seasoned before use, either at the factory or at home.

Skillets should be flat on the bottom and finished with a smooth edge. If the edge is beaded, the beading should be such that grease and dirt do not collect under it. If either the handle or its shank has been made in one piece with the skillet, cleaning is less of a problem. A skillet that is to be used in the oven should have a handle that cannot be harmed by oven heat. If the handle can be removed when the skillet is to be used in the oven, the method of attaching the handle to the shank should be such that it is not easily worn down by the repeated removing and replacing. Other factors to consider in choosing a good handle for a skillet are the same as for saucepans.

It is possible to warp a skillet by misuse. Skillets should not be subjected to sudden extremes of temperature. A warped skillet will not brown foods evenly.

Skillets should be chosen with the intended use in mind. A size should be chosen that will hold the amount of food usually cooked.

Pressure Pans

Pressure pans have become an accepted utensil in many kitchens (Fig. 4-1). The pressure canner was forerunner of the

Figure 4-1. Pressure saucepan with domed cover, Porcelain coated. (Mirro Aluminum Company)

pressure saucepan. Pressure canners are available in 8- and 16-quart sizes. Pressure saucepans are made of both drawn and cast aluminum.

Cooking food under pressure is fast because it is cooked at high temperatures. Care needs to be taken not to overcook foods because they do cook much quicker than usual. When pressure is above that at sea level, water boils at a higher temperature than 212° F. In the pressure pan 5, 10,

Figure 4-2. Weight: Pressure registered by sound. (Mirro Aluminum Company)

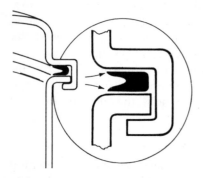

Figure 4-3. Cross section showing gasket construction. (Mirro Aluminum Company)

and 15 pounds of cooking pressure are common. Perhaps the most widely used is the latter. At 15 pounds of pressure the boiling temperature is about 250° F, at 10 pounds of pressure about 240°F, and at 5 pounds of pressure about 228°F. At high altitudes these temperatures are lower.

A pressure pan has a heat-resistant handle, a lid that fits tightly with a rubber gasket or a sealing ring, a safety fuse or automatic air vent, and a weight that indicates the amount of pressure within the pan. In the newer models a special clamping lock eliminates the need for a gasket.

Weights are classified in two groups: those that register pressure (1) by visual means and (2) by sound. It may in some ways be easier for the beginner to use the visual type of weight which is more often found on canners than on saucepans. It is necessary to learn what to listen for when using a weight which jiggles (Fig. 4-2). A certain amount of jiggling per minute indicates that a specified pressure is being maintained. Some weights are washable; others are not. Some can be harmed by dropping; others will not be damaged.

The gasket or sealing ring is most important if a tight seal is to be obtained (Fig. 4-3). The sealing ring and the parts of the pan and cover with which it comes in contact must be clean if a good seal is to be made. After the pressure pan has been used for some time, the gasket may lose some of its resiliency and should be replaced. Gaskets will wear longer if washed, rinsed, and dried carefully after use.

The vent should always be open and free from any particles of food. It is good practice to check the vent after the cover is washed and again as it is placed on the pan of food to be cooked. The vent must be clean if the pressure pan is to operate properly.

The safety fuse or automatic air vent is made of rubber or an alloy of metals. A metal safety plug will melt when the pressure within the pan becomes too high, and a rubber safety fuse will blow out when the pressure becomes excessive.

If the vent is kept clean, the recommended amount of water used in the pan, and the heat controlled, there is practically no possibility that the pressure will ever become dangerously high. A few foods are not recommended for cooking in a pressure pan. The manufacturer's directions should be followed.

Electric pressure pans also come equipped with thermostatic heat controls (Fig. 4-4), which eliminate much of the watching previously necessary. Once the correct cooking pressure has been reached, the heat selector is turned down, and from then on the cooking is automatic. However, when the cooking is complete, pressure must be released as in the non-thermostatic models.

Pressure can be released by placing the pan in cold water, by running cold water over it (but not over the automatic air vent), or by removing the pan from the heat and allowing the pressure to go down gradually. The cover should never be removed before the weight is removed, and the weight should not be removed before the pressure is down. If in doubt as to whether the pressure is down, tilt the weight slightly with a fork. If steam escapes the pressure is not yet down.

The weight should be stored in the pan and the cover placed upside down in the pan. Never store the cover on the pan in a sealed position.

OVEN UTENSILS

Cake Pans

Cake pans come in various sizes and materials. Loaf pans are usually rectangular; layer pans are square or round. The American National Standards Institute, Inc., in cooperation with the American Home Economics Association and other organizations, has set up standard sizes and depths for these pans. No recipe, however wonderful, will produce equally satisfactory products in all sizes of pans. Not only sizes for pans but recommendations as to how these sizes should be measured have been made. Measurements are made on the interior of the pan. Interior and exterior measurements are not the same; bottom and top measurements are often different. Recommendations have been made for loaf and layer pans. Some manufacturers now label or inscribe the exact size of the pan on the bottom of it. A boxed cake mix will yield a more satisfactory cake in the right size pan.

Aluminum baking pans are widely used. In many test kitchens they are considered the standard baking pans, and many recipes are temperature tested with them. The inside and bottom of the pan should have a satin or rather dull finish and the exterior sides a polished finish. Rounded corners rather than square or folded ones are desirable both for ease in removing cake and for cleaning. Some cake pans and cooky sheets have handles that are integral parts of the pan and are also large enough to

Figure 4-4. Heat-controlled pressure saucepan. (National Presto Industries, Inc.)

make it easy to grasp the pan with a bulky potholder. Cakes baked in aluminum pans do not have crusty sides. As a general rule, it is slightly easier to get an even cake in a round pan than a square one because the heat penetration is the same all around the cake. In a square or rectangular pan the heat penetration through the corners is somewhat more intense than on the sides, and the batter may be set before complete action of the leavening agents has taken place. This may result in a domed cake. (Overbeating and/or too much flour may also cause doming.) To avoid this rounded product push a little more batter into the corners than in the center, and a level cake will usually result.

Porcelain enamel cake pans are more popular in the loaf design. Cakes baked in enamel pans are more crusty on the sides.

Glass cake pans also result in crustier cakes, and cakes baked for the same time and at the same temperature as recommended for aluminum pans may be quite dark. It is recommended that oven temperature when baking cakes in glass pans be 25° F less than when baking them in aluminum pans. Glass pans do not stack easily; hence they may require more storage space than aluminum pans.

Tin darkens with use and age, and therefore its baking qualities change. A dark tinned pan will yield a much browner product than a shiny one. A shiny tinned pan produces a cake much like one baked in aluminum. A waffleized tin pan is supposed to give better baking results, but insofar as the authors have been able to determine it is merely a little harder to clean. Tinned pans will warp if subjected to sudden temperature changes, and a warped pan will not bake evenly.

Pie Pans

Pie pans have not changed much over the years. Some are available with an extra trough to catch juices that may run over from berry or fruit pies, thus making it easier to keep the oven clean. Glass and dark tin make quite satisfactory pie pans, as a browned pie crust is highly desirable. Glass is fine for baking a filled crust, since it absorbs heat readily and conducts it slowly. Care should be taken in cutting pies with extra-sharp knives because it is easy to scratch the bottom of the pans.

Cooky Sheets

Cooky sheets are made especially for baking cookies and biscuits. A cooky sheet is a flat sheet of metal that may have slightly raised edges on one, two, or three sides. It should not have all sides raised; and except for ease in removing the pan from the oven, it would be preferable not to have any sides raised. Because cooky sheets are used to bake products in a short time, the lightweight materials that transmit heat readily are preferable.

Muffin Pans

Muffin pans commonly come in sets of 6 and 12 cups, but 8 cups are also available. A metal that does not change color or one in which the finish cannot be removed by scouring will be more satisfactory over a period of time. It is possible to buy muffin pans made from one piece of metal, which thus have no places for dirt to collect and are easy to clean. In some muffin pans the cups are made separately and inserted in a frame. This method produces a somewhat stronger muffin pan, but if care is used in storage, pans made in one piece are quite satisfactory. Muffin pans will not bake evenly if warped or bent out of shape.

Casseroles and Custard Cups

Casseroles are quite satisfactory when made of a material which holds heat well but is not too heavy to manipulate easily. Materials used are glass, heat-resistant china, pottery, porcelain-enameled cast iron, and Pyroceram. They are somewhat more useful if they are attractive enough to be used for serving as well as cooking. Covers that fit increase their versatility, though for nicely browned products the cover is usually removed during the last part of the cooking time. Casseroles should not be scoured when washing as the material may be scratched. If food has stuck, it should be soaked off. The American National Standard Institute says that capacities should be stated in level full liquid measurements.

Custard cups are made of glass, heat-resistant china, or pottery. They are defined by the American National Standard Institute as small, deep, individual, bowl-shaped utensils especially designed for oven use. The size should be given in liquid measurements. Usually custard cups do not have covers. A perfectly smooth cup is easier to clean than one which has lines or other ornamentation.

Roasters

Roasters should be made of fairly heavy-gauge sheet metal. If they are to be used on top of the range as Dutch ovens, very heavy-gauge sheet metal or cast metal is preferred. If they are to be used as an uncovered pan only, the broiler pan of the range may serve as the roaster. A pan about 10 by 15 by 2 inches can be used for baking large loaf cakes and larger quantities of scalloped dishes such as potatoes or corn, as well as for roasts.

Roasting is a method of cooking by dry heat; but since many cuts of meat or fowl are more tender when baked in moist heat, the pan should have a cover that fits. It is not always necessary to use the cover but it is sometimes desirable. A rack serves to keep the meat out of the fat and juices from the meat. Sometimes the rack makes it easier to remove the meat from the roaster.

The size roaster chosen should depend upon the type of food for which it is to be used. It may be desirable to have one roaster for the ordinary needs of the family and a second large one that can be used for the holiday turkey. An extra-large roaster is inconvenient for everyday use and difficult to store conveniently.

TOOLS

MEASURING, CUTTING, AND MIXING UTENSILS

Measuring Cups

Measuring utensils should meet the standards of the American National Standard Institute. Most of the recipes in current books specify standard measures. It is rather pointless to follow a recipe with meticulous care, to follow all recommended procedures for measuring, and then to use cups and spoons that are not standard. Look for the words U.S. Standard Measurement on labels or for the capacity in ounces, since the latter can be compared with standards. Eight fluid ounces equal 1 cup or 237 milliliters.

Measuring cups of transparent material have the advantage of being easier to use for liquids. Metal cups are resistant to breaking and can be used over low heat for melting small amounts of chocolate, fat, etc. Plastic cups should be made of a heat-resistant plastic; otherwise they will often be warped out of shape by hot foods measured in them or by washing in a dishwasher. Pliable plastic is not desirable for measuring cups as it is difficult to get an accurate measure. Plastic measuring cups seldom carry the U.S. Standard seal of ap-

Figure 4-5. Nested measuring cups—U.S. Standard. (Mirro Aluminum Company)

proval. Measuring cups should be strong and rigid enough so that they are not bent, dented, misshapen, or otherwise damaged by ordinary use.

For dry ingredients the full-cup measure should be at the top of the cup so that the food can be leveled with a spatula. For liquids the cup should extend beyond the full-cup measure. The cup for measuring liquids should have a good pouring lip, so placed that it is convenient for pouring to the right or left.

Nested measuring cups are convenient for measuring fractions of a cup of dry ingredients or fat. The contents can be leveled off, which makes it quite easy to get accurate measures. American National Standard Institute standards state that dry measures shall be 1-, 1/2-, 1/3-, and 1/4-cup capacities (Fig. 4-5). Some measuring cups are marked in the metric system as well as in cups and/or in ounces.

All measuring cups should have broad enough bases so that they do not tip easily. The bend at the bottom should be rounded for ease in removing food. The handle should be an integral part of the cup or securely fastened by welding in such a manner that there are no crevices in which spilled food will collect. Glass or wood handles are better than metal ones for use with hot food. Measuring cups are easiest to clean if the top edges are merely cut off and smoothed down, rather than beaded under as some are. Some foods are not easily removed from rectangular measuring cups.

The homemaker who prepares food in large quantities will find that 2- and 4-cup measures are a convenience. The 4-cup measure is convenient for mixing some foods. Waffles, for example, can be mixed in a glass 4-cup measure and then poured directly onto the waffle grid.

Measuring Spoons

Measuring spoons should be made of a material that is sturdy—not easily bent out of shape, broken, or melted. A distorted spoon, like a misshapen cup, is not an accurate measure.

Measuring spoons usually come in sets of 1 tablespoon, 1 teaspoon, 1/2 teaspoon, and 1/4 teaspoon. It is a good idea to separate them and hang them on individual hooks, since it is then not necessary to wash four spoons when you use only one. Because measuring cups and spoons are relatively inexpensive, and often may be used in different working areas in the kitchen, it may be wise to invest in more than one set and store them at strategic points.

Knives

The construction of a knife determines to a great extent satisfaction in use. A poor knife is a constant annoyance, a safety hazard, and an energy user. Careful and wise use add to the life of any knife, especially a good one.

There are three major parts in the construction of a knife: the blade, the handle, and the method of attaching the blade to the handle. A well-constructed knife is likely to be made from good materials, although this is not a hard and fast rule.

Materials commonly used for blades are steel, iron, high-carbon steel, stainless steel, high-carbon stainless steel, and vanadium steel. High-carbon steel is considered to be one of the best materials for knife blades. It can be sharpened to a very fine, thin, sharp edge and will keep this sharp edge better than a low-carbon steel.

Stainless steel is desirable in that it is resistant to stains and is easy to keep looking attractive. It never darkens any foods that may be cut with it. However, it is usually lower in carbon content than high-carbon steel and therefore its ability to take and keep a sharp edge is not as good.

High-carbon stainless steels are available and have the good characteristics of both stainless and high-carbon steels. Cutlery blades of high-carbon stainless steel that have been hardened and tempered at a temperature 100 degrees below zero are called Frozen Heat[4] knives. The resulting blades are tough and resist dulling.

Vanadium steel is a high-carbon steel with other elements added such as vanadium, chromium, and molybdenum. These alloy steels are tough for longer wear.

Some very inexpensive knives may be made from sheet steel or sheet iron. Although they need to be sharpened often, they can be satisfactory for certain uses. However, their use is limited because they are so easily stained and in turn may stain foods.

Much high-quality cutlery is chrome-plated, which makes it very attractive and nearly stainproof. Chrome plate can be scratched, though, and then the exposed steel may rust or tarnish. The cutting edge

[4] Registered Trademark, Robeson Cutlery Company.

of a blade is not chrome-plated and usually appears dark after the knife has been in use. Chrome-plated knives should be washed and dried after use because fruit juices or salt on the blade have a deteriorating effect on the plating.

Forging, beveling, and stamping are processes used to make knife blades. A forged knife blade is hammered into shape by hand or by machine and is more expensive to make than a knife made by beveling or stamping. The blade of a forged knife will taper from the handle to the point of the knife and from the back to the cutting edge. This makes a knife that is strong and firm at the handle and flexible near the point. The tapering from the back to the cutting edge makes it easier to produce a very thin cutting edge.

Beveled knives are made by cutting in half a bar of metal that is thicker at one edge than the other. They taper from the back of the knife to the cutting edge but the blade does not taper from the handle to the point. Some very high-quality beveled knives are made.

Stamped knives are made by stamping the blades from a sheet of metal. The blade is the same thickness at all points until the cutting edge is ground. Stamped knives are not as good as beveled or forged ones, though a stamped knife may sometimes be of a better quality than the method of making would indicate. The quality of steel used and the method of grinding determine to a great extent the quality of this type of blade.

Grinding is one of the more costly steps in manufacturing a knife. It is very difficult for the average person to evaluate the quality of a grind. In general, there are two types of grind—the "flat" grind and "hollow" or concave grind. These types are sometimes broken down still further, but the general shape of grinding falls within these two classifications.

The flat grind, sometimes called "V" grind, extends from the back to the cutting edge in a flat plane. It gives a sturdy blade and can be used for heavier duty cutting than the hollow ground blade. Hollow grinding usually starts below the back of the blade and extends to the cutting edge. A concave area on each side of the blade gradually reduces the blade's thickness until it reaches its thinnest point at the cutting edge. A hollow ground knife blade has a thinner cutting edge which makes it easier to pare, carve, or slice.

Tungsten carbide, an extremely hard metal, is used on one side of the cutting edge of some knife blades. It is applied at a speed said to be ten times the speed of sound and at a temperature of 6,000° F. Because the tungsten carbide edge is so very hard, knives stay sharp a long time if given reasonable care. Some knives carry a guarantee to repair or replace free any knife that becomes defective within 20 years. The blade can be sharpened by using a sharpening steel, stone, or electric sharpener but it must be sharpened on the steel side only. The tungsten carbide edge is so hard it cannot be sharpened at home.

A saw-toothed or scalloped-edge knife may be of several lengths or patterns. At one time scalloped and serrated edges were only found on knives designed for cutting bread, steak, or grapefruit. Now there are knives for many uses that have a scalloped or serrated edge. The scalloped knives may be hollow ground. The scalloped edges are designed to stay sharp longer than a smooth edge because the point of the scallop tends to take the dulling if the blade comes in contact with surfaces other than food. The cutting edge within the scallop is not harmed. Some of these knives carry a guarantee that the knife will not need sharpening for five years. Scalloped edge knives can be sharpened by the use of a sharpening steel or oilstone but they should be sharpened on the nonscalloped side only.

The length of blade is determined by the knife's use and its comfort in manipulation. The handle and blade should be well balanced so that the knife will be comfortable to use.

The handle of a knife is of almost equal importance with the blade. If the handle is not right, the knife will be almost useless. If it is of poor construction, it will shorten the usable life of the knife.

The handle should be of a moisture- and grease-resistant material. Hard woods—hickory, beech, rosewood—often are used and are more desirable than softer woods finished with paint or varnish. Wood handles finished by smoothing and polishing are more durable than handles finished with paint or varnish. Varnish will wear off and paint will often flake or chip.

Plastic, hard rubber, metal, and plastic-impregnated wood handles also are used. The latter is growing in popularity as it is a very attractive handle. It is made by impregnating hardwoods with resins and then subjecting them to intense heat and pressure. These handles are resistant to heat, stain, moisture, food, and chipping.

In plastic handles, the tang of the knife is set in the molten plastic which when cooled forms a solid piece. These handles are moisture resistant but may not be resistant to heat. However, if moisture penetrates at the point of insertion of the tang, the plastic will eventually crack apart. A nylon resin is used for some cutlery handles. This resin comes from the same basic materials as the yarns for nylon hosiery. These handles do not chip, crack, or peel and can be washed in the dishwasher.

The shape of the handle should fit the user's hand, be easily grasped, and held securely without strain. It is especially important that the handle of a paring knife be comfortable for the user.

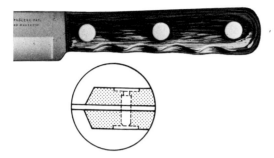

Figure 4-6. Cross section view of compression rivets in knife handle. (W. R. Case & Sons Cutlery Company)

The handle should be securely attached to the blade. Probably the most effective means is by use of a full tang and medium-sized rivets (Fig. 4-6). The tang is the part of the blade that extends into the handle. A full tang extends throughout the length and the width of the handle. In half-tang construction it extends into at least one third of the handle and is held in place by rivets. Full-tang construction is desirable for knives that will have heavy use, such as butcher knives. Half-tang construction is usually quite satisfactory for most household knives.

A less desirable method of attaching the handles is by the collar and pin or push-tang method. The tang, which is often much less than a half tang, is inserted into the handle. A metal collar is slipped over the blade end of the handle, and a very small pin is inserted through the handle to help hold the tang in place. Often the pin is omitted and only the collar used. Neither method is very satisfactory, as the collar soon loosens and moisture and dirt get between it and the handle. Also, the tang often slips out of the handle.

Knives are designed for different purposes and for most efficient use the proper knife should be selected. Paring knives have a short blade, usually 2 1/2 to 3 inches. Blades may be of different shapes. Some women prefer one and some another, or several paring knives of different shapes may be desired for different uses.

The utility knife has a 6- or 7-inch blade and is used for halving oranges and grapefruit, slicing small roasts, tomatoes, cake, and nut bread. In a kitchen where only a few knives are available, this knife has numerous uses.

The butcher knife is a heavy-duty knife used for cutting meats, fowl, squash, melon, and large heads of cabbage. It has a broad, sturdy blade.

The French cook or chef's knife is used for chopping food on a board. The blade is usually 8 or 9 inches long and fairly wide at the handle end. It is set into the handle in such a way that the handle can be grasped and food chopped without the knuckles touching the board. It is highly recommended for chopping large quantities of food.

Slicers have a fairly narrow blade which is quite long—8 to 12 inches. The end is shaped in various ways, depending upon whether it is a ham slicer, roast slicer, narrow slicer, or carver. Some authorities list only the fairly straight blade knives as slicers; those that are somewhat curved and have a pointed end are called carvers.

The grapefruit knife is curved to fit the inside of half a grapefruit and may or may not have a serrated blade. It is used to loosen the grapefruit segments from the rind, to remove seed pods from green peppers, and to core tomatoes for canning.

The bread knife has a blade 8 to 10 inches long with a serrated edge. It will cut bread without crushing it and is especially desirable for slicing fresh bread.

The cleaver is a heavy-duty tool which has been described as a cross between a knife and a hatchet.[5] It is useful for cutting joints in meat and poultry and for cracking soup bones.

The kinds of knives available to the homemaker are numerous (Fig. 4-7). A few well-chosen knives will be far more useful than a galaxy of poor knives not suited to the use to which they are put. Knives

[5] Elizabeth Beveridge, "Good Knives for the Kitchen," *Kelvinator Kitchen Reporter*, November 1951.

Figure 4-7. Household knives: Ham slicer, utility knife, and paring knives. (W. R. Case & Sons Co.)

are used often in food preparation, even in this day of mixes and partially prepared foods. The Associated Cutlery Industries of America recommend the "Basic Six"

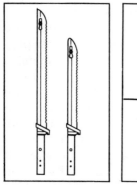

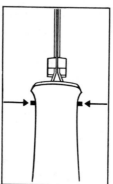

Two sets of blades: one for slicing, one for paring.

Safety blade release.

as an absolute minimum—a paring knife, utility knife, 8-inch narrow slicer, 8-inch cook's knife, 7- or 8-inch long-handled pot fork, and a sharpening steel.[6]

Good knives deserve good care. (1) They should be stored individually so that the blades do not come in contact with other blades and hard surfaces. Knife holders are of numerous types, but a small block of wood fitted with slots for blades is quite satisfactory. (2) A cutting board should be used in chopping and slicing operations. (3) Good knives should not be used to cut string or wire or as screw drivers or levers. (4) They should never be left standing in water or allowed to get hot.

It is not only exasperating to use a dull knife but it is also unsafe. A good knife can be kept sharp for a long time by good care and intelligent use, but eventually all knives become dulled.

Electric knives have two counter-reciprocating scalloped-edged blades that cut with a shearing action as the scallops of one blade meet the scallops of the other (Fig. 4-8). The blades are driven by a series wound motor with a wattage rating of less than 100 watts. Some women may find the handle is not comfortable for any extended use because of its bulk and weight, although most manufacturers have improved design of handles since the electric knife was first introduced. Cutting is extremely quick and it is a wise precaution to place the food on a cutting board. When

[6] Lewis D. Bement, *The Cutlery Story*, The Associated Cutlery Industries of America, 1950, pp. 26, 27.

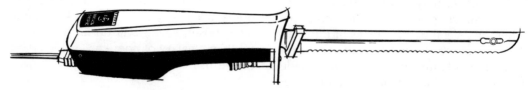

Figure 4-8. Electric slicing knife with safety blade release; two sets of blades. (General Electric Company)

the knife is not in use it should be disconnected. Because this appliance could be
very dangerous in the hands of an unreliable operator it should be kept out of the
reach of children at all times. Make a practice of always putting it at the back of
the .counter. Be sure to follow directions
explicitly in removing and inserting the
blades. Blades should be slid apart, not
pulled apart.

The knife cuts an assortment of foods—
angel cake, cheese, cold meats, roasts, and
fruit cake—with great speed and consistently good results.

Steels for sharpening knives are in common use today; at least they often are sold
with carving sets and sets of knives. Steels
are used to perfect the cutting edge of a
knife.

The conventional method is to hold the steel
horizontally slightly slanted away from you in
the left hand, hold the knife by the handle in
the right hand, rest the edge at the heel lightly
near the point of the steel at an angle of about
20 degrees, and draw the blade toward you
against the edge and across the steel from
heel to point. Use only light pressure. Repeat
this operation on the other side of the steel
with the other side of the edge. Three or four
strokes on each side are enough.[7]

The process perfects the cutting edge and
can be done many times before a blade
actually requires sharpening.

One of the oldest methods of sharpening knives is by means of an oilstone
which, as the name indicates, is a stone
used with a few drops of light oil. One side
of the stone is coarser than the other. This
side is used first with the fine or smooth
side reserved for finishing.

Hand-operated or electric sharpeners
(Fig. 4-9) are a boon to the person who has
not taken the time to learn how to sharpen
a knife correctly. Disks or wheels of hard
steel or aluminum oxide are so spaced that

Figure 4-9. Can opener. Cutting wheel removes for
cleaning; knife sharpener on rear. (Sunbeam
Corporation)

the knife must go through or over them at
the right angle. However, it is still quite
important to exert the right amount of
pressure on the knife blade or it may be
over- or undersharpened. The knife should
not be heated during the sharpening process or the steel may lose its temper.
Knives should be clean before sharpening
and washed after sharpening. It is important not to force the blade against the
sharpening wheel. Any metal dust should
be wiped off the sharpener before storing
it. Watching a butcher or other experienced person is one step in learning the
correct procedure for sharpening knives,
but practice is essential.

Can Openers

Manufacturers have made much progress
in recent years in producing a number of
can openers that are safe, durable, and
easy to use (Fig. 4-10). The blade should
be made of a noncorrosive material. High-
carbon steel blades can be very sharp and
will remain so a long time. The can opener
should give a smooth, clean, and complete
cut. Most can openers cut through the lid
of the can; however at least one can

[7] Ibid., pp. 23, 24.

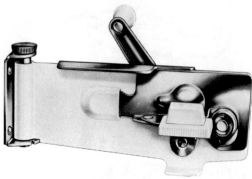

Figure 4-10. A wall-type can opener. (Swing-A-Way Manufacturing Company)

opener is designed to cut the rim of the can and not the lid. The lid may then be replaced on the can as both the edge of the can and of the lid are turned under so that they are dull. In either manner of opening, the edge of the can should be turned under so that the operator will not run the risk of being cut when removing food from the can. The leverage should be such that it is not difficult to start the can opening process. This also is facilitated by a sharp blade. A magnetic device to hold the can lid out of the food is a convenience and usually does not add much to the cost.

When mounting a can opener, care should be taken to mount it in such a way that large cans as well as small may be opened without difficulty.

Since the can opener blade must be clean in order to cut well, it must be washed at rather frequent intervals. Care should be taken to follow the manufacturer's directions for this. Some wall can openers are difficult to clean properly. One should look for a can opener that can be washed easily. Models that do not have all the gears encased in an attractive cover usually are easier to clean. Can openers should be kept sharp, clean, and in good working order. Wheels, inoperative because they are stuck with food, are essentially stationary blades. Dull, straight blades forced through metal form more and larger particles of metal fragments. These particles have not been shown as injurious to health, but are nevertheless undesirable in food.

A good can opener may be expensive, but, considering the number of cans many homemakers open, it is a worthwhile investment. Small portable can openers are essential for picnics and are suitable for vacation use at cottages, but it is preferable to have a good wall or electric can opener for the home.

A can opener that leaves a jagged edge should not be used. It should leave a turned-down edge without jagged pieces which might cut the hand. A knife should *never* be used as a can opener.

Electric can openers may be wall-mounted or counter-top models (Fig. 4-9). It is well to consider use of counter space, size cans to be opened, and possibility of wall mounting particularly in temporary living quarters before purchase. The opener should hold the can in place and not release it automatically when the can top is removed. The cutting wheel should be easily removed and replaced so that it can be cleaned often. This, like its non-electric counterpart, should leave a smooth, dull edge.

A switch that is manually controlled during all of the cutting operation helps to prevent a child from turning on the mechanism and then becoming unable to turn it off. As an added safety precaution in homes where there are small children be certain to disconnect the opener when it is not in use.

Shears

A pair of shears designed for kitchen use is a useful tool and may aid in maintaining knives in good condition. Kitchen shears, sometimes called poultry shears, are designed to cut up poultry and if used for this purpose they eliminate the dulling of knives on bones. Shears are also useful for dicing marshmallows, candied fruit, opening bottles, and for use as small pliers. One or both blades may be finely serrated.

Sometimes the blades are chrome plated; the quality of chrome plating is of course difficult to determine, and one can usually do best by buying from a reliable dealer and established company. Handles of the shears usually are painted.

Cutting Boards

Cutting boards come in a variety of sizes. They should be of a hard wood which is moisture resistant. The small round size is handy if only a small amount of food preparation is done, but a larger one should be chosen if food is prepared in quantity. It is quite possible to use one side of a dough board as a cutting board and the other for rolling out doughs. It is not good practice to use the sides interchangeably, since sooner or later the cutting side gets cuts and crevices in it, which are not only a nuisance to clean but also make rolling pie or cooky dough more difficult.

Cutting boards should be washed carefully and dried. They should not be allowed to stand in water. Some people mark one side of a cutting board for cutting onions only, because they think there is an intermingling of flavors if onions and other foods are chopped on the same side of the board.

Bowls

Mixing bowls should have slanting sides and a rounded bend at the bottom. It is much easier to mix or blend foods in a bowl with a rounded bottom than in one with a perfectly flat bottom. The size of the bowl should be chosen in relation to the amount of food that will be mixed in it. A small amount of food cannot be mixed conveniently in a large bowl, and if the bowl is too small, food will spill over the edges.

Some bowls can be used for mixing, refrigerating, and baking. It is possible to use smaller bowls as substitutes for casseroles if they are made of a heat-resistant material. For demonstration purposes glass bowls are highly useful, as the audience can see what is being mixed. They are rather heavy and tend to stay in place as foods are mixed. Foods can be stored in the refrigerator in them.

Bowls often are stored nested inside one another, but if storage space permits, it is much more convenient to store them singly.

Blenders

Blenders are of many varieties and at any one time there may be a number of new ones on the market, many of which make their appearance for a limited time only.

The pastry blender has been available for quite some time. It is useful not only for blending fat and flour for pastry but for mixing fat with other ingredients. The thinner the wires, the finer the blending. Wires should be of a material that will not rust and have a finish that will not wear off. They should be fastened securely to the handle, which should also be made of a material, or have a finish, that will not wear off or chip.

Another popular blender is the Foley fork, which can be used for making pie crust, blending gravy, mashing potatoes and carrots, and for other similar uses. It has stainless steel blades.

Beaters

There are many qualities of beaters on the market. Those of better quality will have a handle of moisture-resistant material and a finish that will not chip. The gear will be of cast metal or nylon and the blades of stainless steel. Quite thin cutting blades make a finer foam. The beater is easier to keep clean if the cutting blades do not extend to the gears, but they should not be too short or the beater will be very slow. The blades do not rest on the bottom of the bowl, but they should be as close to it as possible so that food on the bottom of the bowl will be beaten.

Rolling Pins

Rolling pins should be made of hard wood and have securely fastened handles. Some handles are attached to the pin in such a way that pin and handles move together. In others the pin turns while the handles are held firmly by the operator. Not all handles fit tightly, and care needs to be taken to keep them clean. Rolling pins should never be soaked in water. They should be washed, rinsed thoroughly, dried, and then air dried before storing.

MISCELLANEOUS

Spatulas

Spatulas should be made of a material resistant to staining. Since a sharp edge is not essential, and sometimes not desirable, the carbon content of the steel is not important. The blade should have a flexibility appropriate to its use. If it is too flexible, the spatula is of little use for lifting foods; if it is not flexible enough, it will not slip under a cake or a piece of pie. Spatulas used for blending should be somewhat more flexible than those used for lifting.

Spatulas come in various sizes. For removing cookies from a cooky sheet and for many other uses, the long or large-size spatula is desirable. For a multitude of other uses the very small spatula is handy. There are wide spatulas and narrow ones. Some of these are known under various names, such as meat lifters, pancake turners, or hamburger spatulas. For use as a turner it is desirable to have an off-set handle that makes it easier to get the spatula under the hamburger or hotcake. Perforated spatulas are useful for foods that need to be drained as they are lifted. The type of cooking one does determines to a great extent which should be chosen.

Wooden Spoons

The quality of wooden spoons may differ from one spoon to another. A hard wood with a smooth finish makes a spoon more useful and easier to care for. It is often very difficult to ascertain whether the wood is hard. White ash makes a very good spoon. Since the handle does not get hot, this type of spoon is particularly useful in such tasks as making jelly. Another advantage of a wooden spoon is that it does not scratch or mark pans and bowls. It should not be left standing in foods, as it is easily stained and will absorb flavors. Spoons are made with either deep or shallow bowls. Since their chief forte is stirring and beating, the shallow bowls are to be preferred. For tasting and for transferring food from one place to another, a metal spoon is a somewhat better choice.

Metal Spoons

Metal spoons are useful for stirring and mixing when the mix is not heavy. They are also useful for tasting. For large quantities of food the long-handled spoon with a solid or slotted bowl is useful. A stainless steel spoon is more desirable than a tin-plated one. All grades of stainless steel are not equally serviceable. A handle that does not conduct heat is easier to grasp than one made of metal. Painted wooden handles are generally unsatisfactory if the spoon is used frequently.

Graters

Graters should be chosen not only with the view of how well they will grate food but also how easily they can be cleaned. Plated ones usually show the effects of wear after a time. A grater made from too soft a metal will lose its shape with use. Some are made with punched holes like those made by driving a nail through a piece of metal. Others are made with drilled holes that resemble the holes left in paper when using a paper punch. The punched-hole graters give a mushy product, sometimes desirable when grated food is added to cakes. Drilled holes give a product that has definite form; this is more desirable when foods are to be used for salads.

Graters are available with various sizes of holes. Some have four sizes put together in one utensil; others come in sets of three or four and are preferred by some homemakers because only the size used needs to be washed. Handles on graters are usually attached by soldering and often wear loose.

Grinders

Grinders are made of cast iron or aluminum. If cast iron is used, they should be heavily coated with tin, as a thin coating may wear off quickly. A grinder should have a handle that is easily grasped and turned and made of wood or plastic rather than metal. The grinder should be so shaped that juices extracted run into the pan that receives the ground food rather than into the opening through which the handle is connected. In some modern kitchens it is almost impossible to find a suitable place for fastening a grinder. Sometimes breadboards inserted under the cabinet top are sturdy enough for this but often they are not. Grinders are also available with suction cups or rubber feet on the legs and with a wide base, and these do not need to be screwed in place for use. Grinders may come with an assortment of blades for various uses such as grinding meats, crackers, vegetables, and fruits. The blades should be of high-carbon steel for lasting cutting efficiency.

The grinder should be thoroughly washed and dried before it is put away.

An effort should be made to keep fruit pits and bones from being put through it, as this has a dulling effect on the blades.

Food Mills

Food mills should be of a metal that is sturdy and not easily bent out of shape. The metal need not be as heavy as that used for grinders, because food mills are used only for such tasks as ricing potatoes and pumpkin or separating the pulp of fruits and vegetables from the skin and seeds. There should be some type of support which will hold the food mill above the counter or bowl so that food once pressed through the mill will tend to fall off the perforated or screened surface.

Food mills take up quite a bit of storage space and should be chosen according to the amount of food one might be working with at a time. They should be so constructed that all parts are easy to wash. They should be simple to take apart and easy to put back together. It is a good idea to try this before buying one, as some appear to be much more simple in the store than they are when you get them home.

Strainers, Sifters, and Colanders

Strainers or sieves are made of tinned wire, stainless steel, or sometimes plastic. Since the mesh of the strainers varies, they should be chosen for the purpose for which they will be used. Fine mesh is desirable for tea strainers and flour sifters. For straining vegetables, macaroni, and similar foods, a larger mesh is quite satisfactory and also somewhat easier to wash. Strainers that have two support wires crossed underneath are stronger but may be difficult to wash. Some strainers are made so that the mesh basket slips out from the supporting cross wires and ring. While this makes cleaning easier, care should be taken in storage not to jam the strainer out of shape, because then it will

Figure 4-11. Five-cup aluminum sifter. (Foley Manufacturing Company)

not be easy to take apart and put back together again. Strainers should have a mechanical means, such as a projecting part, to support them on the side of a bowl, cup, or other utensil. Handles of strainers are usually of painted wood or plastic. Although the paint may eventually chip, a wooden handle is not too unsatisfactory because of the limited use most strainers get. A strainer is not a colander or food mill, and foods should not be pressed through it because this will result in the mesh pulling away from the ring support.

Because flour and water make a paste, sifters present a cleaning problem. They vary as to the gauge of the mesh and number of meshes. A relatively fine mesh is desirable, and the purpose of more than one mesh is to sift the flour two or three times on its way through the sifter. However, the major reason for sifting flour more than once is to incorporate air, and if the mesh layers are close together not much is accomplished. Such sifters are almost impossible to clean adequately.

However, sifters that are stored in the flour canister do not need to be washed each time they are used. Those not stored in canisters must be washed each time; in certain areas in the United States a used flour sifter left in a cupboard would be a welcome invitation to insects. Sometimes a vegetable brush will help in the washing process, and it is also easier to wash a sifter with a removable bottom (Fig. 4-11). Sifters should be scalded and thoroughly dried before they are put away.

Sifters usually come in two sizes, a small size that holds about 2 cups and a larger one that holds 4 to 5 cups. If only a small amount of food usually is baked, the small sifter is quite satisfactory and does not take up as much storage space. It would, however, exasperate the homemaker preparing food for a large number of people.

Sifters usually are operated by a lever device or a crank. Some sifters can be held and operated with one hand; others require the use of both hands. One's choice depends largely on habit. Either way the flour is not fanned over the work area as it is when the sifter must be shaken to work the flour through it.

A colander is used for draining foods and as a substitute for a food mill. The homemaker who has one may have specialized uses for it in her kitchen. Colanders should be made of a material not easily bent out of shape. In selecting one notice where the holes are placed. It should be so designed that all the food can be drained. It should have handle supports or a high enough base so that the bottom of the colander will not rest directly on the surface of the sink or another utensil.

A MINIMUM SET OF KITCHEN UTENSILS

The selection of a minimum set of utensils for the kitchen is a project that many beginning homemakers find most challenging and interesting. What constitutes such a set is open to question, because standards of living, income, habits, type of food prepared, kind of kitchen, and many other factors are involved. It is possible to cook a delicious meal over an open fire with the aid of a few sticks for cutlery and perhaps some tin cans for utensils, but most of us would not care to repeat this procedure day in and day out. Just what pieces of equipment one is willing to do without is pretty much an individual matter. To give a guide and *only* a guide to this selection, a list of utensils needed in many homes is given below. It should be remembered that the fewer utensils one has, the fewer there are to wash after the dinner is over! Also, all the articles listed need storage space, and since this is limited in most kitchens, care should be taken to choose only utensils and cutlery that will be used. Some utensils have dual or multiple uses; however, one should not be carried away on this point because it is quite possible to find oneself with nothing more than a complicated gadget—too time consuming to put together and much too much bother to clean. But within reason choices can be made for utensils in which food can be mixed, baked, and served; measured and mixed; and cooked, refrigerated and served.[8]

1 saucepan—1 quart
1 saucepan—2 quart
1 cast iron skillet—8 to 10 inches
1 chicken fryer or Dutch oven
1 griddle
1 glass or china casserole—1 quart
1 glass or china casserole—2 quart
1 cooling rack
1 set graduated measuring cups
1 quart glass measure
1 one-cup glass measure (liquid)

1 set graduated measuring spoons
1 small spatula
1 rotary beater
1 paring knife
1 French cook knife
1 utility knife
1 cheese slicer
1 cutting board
1 grater
1 blender
2 round baking pans—8 or 9 inches
1 loaf baking pan—10 by 15 inches
1 cooky sheet
1 muffin pan—6 cups
1 pie pan
1 set mixing bowls
1 coffee maker
1 teapot
1 set wooden spoons
1 wide spatula or pancake turner
1 can opener
1 bottle opener
1 tap opener
1 cooking fork
1 vegetable peeler
1 kitchen shears
1 rubber scraper
1 rolling pin
1 flour sifter
1 canister set
1 strainer

[8] Other lists of minimum sets of utensils can be found in "Tools for Food Preparation and Dishwashing," *Home and Garden Bulletin No. 3*, USDA, November 1951; in *Better Buymanship, Use and Care of Kitchen Utensils*, Household Finance Corporation (Reprint), 1947 no. 7; Margaret Davidson, "Brides Equip Their Kitchens," *Ladies Home Journal*, June 1970, vol. 87, no. 6.

EXPERIMENTS

The experiments in this group require no special testing equipment and could well be used by beginning classes in Household Equipment.

Experiment 1. Top-of-Range Utensils

Saucepans: Use identical pans approximately 6 inches in diameter. Have each student or group of students prepare a vanilla pudding from a prepared mix in each pan. Use similar sources of heat. (Differences in products will be due primarily to the way in which students manipulate sources of heat.) After puddings are made, pour into custard cups. Evaluate:

a. Quality of product
b. Amount of stirring needed
c. Heat setting used and amount of regulation required
d. Scorching or sticking in the pan

Repeat using saucepans of different (a) materials, (b) sizes. Compare the results with the first experiment.

Experiment 2. Measuring, Cutting, and Mixing Utensils

1. *Measuring cups:* Measure water into a U.S. Standard measuring cup. Be sure the cup is level full. Pour water, very carefully, into a measuring cup that is not a U.S. Standard; that is, a coffee cup, a dented measuring cup. Measure the amount of water necessary to fill the other cups or the amount by which the U.S. Standard measure exceeds the amount held by the other cup.
2. *Measuring spoons:* A similar experiment can be carried out using U.S. Standard measuring spoons and other spoons commonly used for measuring by many homemakers. Measure a dry food that can be leveled with a spatula and does not pack down. Sugar or cornmeal give good results.
3. *Knives:* Dice 1/2 cup celery with a paring knife. Repeat, using a French cook knife and cutting board. Compare ease and speed of accomplishing the task. Discuss relative efficiency for use of the two knives if larger amounts of food are prepared.
4. *Graters:* Grate a carrot on a grater with punched holes. Repeat, using a grater with drilled holes. Compare products. Which grater would be most satisfactory for grating carrots for a salad? For grating orange rind for a cake? Wash the graters. Which type is easier to clean after use for grating carrots?
5. *Beaters:* Evaluate ease of use of various beaters with bowls of various shapes by whipping a measured quantity of a high-sudsing detergent in a measured amount of water. Beat for a definite time at a uniform rate of speed. Different operators will not get the same volume if they do not use the same rate of beating. However, if all the operators use uniform rates for the same time the beaters should rate in the same order for volume of suds.
6. *Sifters:* Examine sifters of different types and sift 1/2 cup flour with each. Compare fineness of the flour sifted in each sifter by observing the volumes after sifting. Evaluate ease of handling, washing, and drying sifters.

Experiment 3. Oven Utensils

1. *Cake pans:* Experiment 1, Chapter 10, is suggested for studying characteristics of cake pans.
2. *Cooky sheets:* A similar type of experiment can be conducted using cooky sheets of the same size and different materials or of the same material and different sizes. Baking pans with high sides may also be used. Use a plain rolled sugar cooky or a plain butter cooky that can be put through a press. Evaluate:

   a. Quality of product.
   b. Effect of size of cooky sheet on heat distribution and browning of cookies
   c. Effect of shape of cooky sheet on browning cookies
   d. Effect of material of cooky sheet on browning of cookies

3. *Bread pans:* Use a one-hour recipe for bread. Bake bread in anodized and nonanodized aluminum and glass pans under similar conditions of time, temperature, location in oven, and type of oven. Compare browning.
4. *Pie pans:* Roll pie crust to fit the bottom of each pan. Take care to have the crust of uniform thickness in each pan and for all pans. Bake crusts under similar conditions as stated in Experiment 3–3. Use anodized, nonanodized, and disposable aluminum pans, darkened tin and glass pans. Evaluate browning crust and ease of handling and storing the pans.
5. *Muffin pans:* Bake muffins in Teflon-lined pans and in anodized aluminum pans. Standardize conditions of baking and time muffins are allowed to remain in pan after taking from the oven. Compare brownness and ease of cleaning the pans.

BUYING GUIDE

Consider the general factors for selection in Chapter 8. Then check the following list of questions for the type of item you wish to select. Remember it is not always necessary to buy the most expensive, the most sturdy, or the most deluxe to have a satisfactory purchase. Choose the item to meet your needs. However, a poorly made item often is a very expensive item if unsatisfactory results, excessive effort, much care, and lack of safety are present factors.

Saucepans, Skillets, Pressure Pans

1. What size do I need? What amount of food do I usually cook or is this pan to be used for a special purpose, such as candy-making or preserving?
2. Is the pan well balanced? Will it tip easily if only a little food is in it? (Try tapping the handle when cover is off pan.)

3. Is the pan perfectly flat on the bottom? Are the sides straight with the point of joining slightly curved?
4. Is the material of the pan a good conductor of heat?
5. What care will the pan require to keep it looking attractive? Does it water spot, darken easily, stain?
6. Does the cover fit tightly? Is the knob large enough to grasp easily without danger of burning the fingers?
7. Is the handle well insulated? (Metal rivets and screws get hot and should be located where the hand will not touch them during normal use.)
8. Is the insulated section of the handle securely fastened to the shank of the handle? Is the shank securely and smoothly fastened to the pan?
9. Does the pan stack easily? Is provision made for hanging? How and where will you store it?
10. Is the handle comfortable for you?
11. If a pressure pan, does it carry the UL seal of approval?

Oven Utensils

1. Does the material transfer heat readily if the utensil is to be used for short time cooking processes?
2. Is it light in weight but heavy enough gauge so that it isn't easily bent out of shape?
3. Is it easy to take hold of so that removal from the oven while hot will be facilitated?
4. Will the finish wear off? Can it be scoured off?
5. Is it the size I will use most?
6. Will the cooky sheet or roaster fit into the oven that I have?
7. How easy will it be to keep clean? Square corners, decorative lines, and uneven edges add to the cleaning problem.
8. Can I store it in an easily accessible place?

Can Openers

1. Are the handles covered with a plastic material?
2. Is the leverage good so that I can use it without too much effort?
3. Is the blade sharp so that it is easy to start the can opener?
4. Does it leave a smooth dulled edge with no jaggedness where the operation began or ended?
5. Will it open all sizes and shapes of cans?
6. Are the edges of the can left smooth and dull?
7. What provision is made for keeping the cutting blade clean?
8. Is the blade a hard steel that will stay sharp?
9. Does it hold the top away from the can so the top does not sink into the food?

(In addition, consider these points for electric can openers.)

10. Will it hold the can in position until the can is removed manually?
11. Does the opener only operate when the control is held in position?
12. Does it carry the UL seal?

Cutlery

1. What cutting operations do I perform most? What types of knives do I need?
2. Do I want this knife to last a long time or do I need one for cutting string or taking on picnics (where it may be lost)?
3. Is the handle securely attached by two or three medium-size rivets?
4. Is the blade high carbon steel with chromium plating, vanadium steel, or high carbon stainless steel?
5. Is the edge hollow ground? (This grind is suitable for most home use.)
6. Is the handle comfortable in my hand? Is it made of material that is moisture resistant and will not stain hands or towels?
7. Have I made provision for good storage for the cutlery that I buy?

Other Kitchen Tools

1. Is this a useful tool or a gadget?
2. Where will I store it?
3. Do measuring cups and spoons carry the U.S. Standard seal of approval?
4. Is the cup transparent if I want to use it for liquids?
5. Will material stain or rust? Will finish peel off?
6. Are there hard to clean cracks and crevices? Are edges and joins smooth and well finished?
7. Will I use this often? Is it a specialized tool that I really need even though I use it only occasionally?

CHAPTER 5
SMALL ELECTRIC APPLIANCES

The small electric appliances discussed in this chapter are those commonly used in kitchen or dining areas for food preparation.

GENERAL CONSIDERATIONS

Some general considerations are applicable for many small electric appliances. The finish on an appliance determines to a great extent the care required. One that fingermarks easily will need much polishing to keep it looking attractive. If the appliance or parts of it will receive much use, the finish should be one that will not wear off easily. Finishes are described in Chapter 3.

All electric appliances should carry the Underwriters' Laboratories seal of approval. If the appliance has been approved, the seal should be on the appliance, or it may be on the specification sheet which describes the appliance. A reliable dealer should be able to give this information to the consumer, but it is also relatively easy to locate on many small appliances. Sometimes it can be found by reading the literature that goes with the appliance.

Small electric appliances usually carry a one-year guarantee; a few carry a five-year guarantee. Often it is necessary to send a registration card with various data about the purchase to the company in order to have the guarantee put into effect. A guarantee is nearly always void if the appliance has been tampered with in any way. If an appliance needs service, it should be returned to the company or a company-approved service center. If in doubt about the correct procedure, write to the company, stating the problem with the appliance and ask for specific directions.

Storage for an appliance should be considered when selection is being made. If the storage place is inconvenient, it may become too much trouble to use the appliance, and very little satisfaction is then had from its ownership. Particularly, heavy appliances, such as the rotisserie and mixer, should be stored where they do not have to be moved each time they are used. Heat-controlled pans should not be stored in the oven, since oven heat may damage plastic parts of the appliance.

The terminals should be recessed or protected by a shield so that there is not the remotest possibility of touching a live terminal. Some other criteria to consider when evaluating quality are sturdy construction, no sharp edges, and parts that fit together smoothly. Drawers should slide in and out easily, covers fit without careful maneuvering, and legs on appliances should be even and of a simple style that will be easy to clean. Signal lights are of little value unless they are easily seen and they should respond simultaneously with the off and on cycling of electric current. Markings on controls are most useful if they can be easily read and understood. The markings should be durable so that they will last the lifetime of the appliance.

Some small appliances use electricity for power only, some for heat only, and a few use electricity for both power and heat.

MOTOR-DRIVEN APPLIANCES

Food Mixers

Food mixers can save a lot of energy for the homemaker if they are used properly and stored in a convenient location. Tasks such as whipping potatoes, mixing cake batters, and making frostings can be done more easily with a good food mixer. Mixers also permit easier standardization of some food preparation procedures than is possible with hand-operated beaters.

Mixers may be mounted in stands and provided with bowls or they may be portable or hand mixers. The power unit may be installed in a counter top (Fig. 5-1). Other appliances may be used in the same installation. Some mixers mounted in stands

Figure 5-1. Power unit installed under counter, with mixer, knife sharpener, and juicer attachments. Attachments not pictured: blender, meat grinder, salad shredder, and ice crusher. (NuTone Division of Scovill Manufacturing Company.)

may be removed from the stand. Mounted motor heads generally weigh about 3 pounds more than portable mixers. This is not particularly important if the appliance is not used as a portable mixer and can be used in the place where it is stored. Some portable mixers are also provided with stands.

Mixers mounted in stands are usually sold with two bowls, a large size of 3- or 4-quart capacity and a smaller one of 1- or 2-quart capacity. Some mixers have only one bowl of 4- or 5-quart capacity. The bowl to be used depends on the amount of food to be mixed. The mixer will do a better job if enough but not too much food is in the bowl. Bowls are made of translucent and clear Pyrex, stainless steel, and aluminum.

Beaters are similar on many mixers. Often there are two separate beaters much like those of a rotary hand beater. But mixers may have the two beaters fastened to a center shaft which is tightened in place in the mixer head with a thumbscrew. This design of beater sometimes is considered rather difficult to clean but easy to assemble. For most mixers, the beaters rotate in one place in the bowl, and the bowl is also supposed to rotate. Mixers are also available with a bowl that does not rotate. Bowls are designed to rotate because of the friction developed between the beater and bowl. The batter serves to make better contact between the two. In some mixers a small plastic button is on the bottom of one beater so that good contact will be made even if the mixture is thin. The position of some beaters in the bowl can be changed while the food is being mixed by changing the position of the turntable (and bowl) in relation to the beaters or by changing the position of the beaters. In the latter case the position of the turntable is fixed. Turntables in both types rotate.

A beater with a planetary motion mixes food in all parts of a stationary bowl. As the beater rotates in one direction, the shaft to which it is connected rotates in the other. This beater is larger and somewhat different in shape than those discussed above.

The speed control on the mixer should be easily read and convenient to use. Most speed controls also turn the motor on and off. The control should not be too easy to turn or it might be turned on accidentally. Proper placement of the control can also contribute to safe use. The speed with which the beaters rotate varies at the different speed settings, but it is important that it remain somewhat constant at any one setting. Beaters should not slow down excessively as the mixture thickens or speed up excessively as it becomes thinner. A governor-controlled motor helps to maintain a fairly constant speed as do some solid-state controls.

The mixer should be well balanced. When the mixer head is tilted back it should not tip the mixer. For convenience it should stay in this position and not tilt back into place so easily that it can be done unintentionally. It is not a good idea to use the mixer close to the edge of a work surface as it could be knocked over when the mixer head is tilted back.

Most beaters are relatively easy to wash. It may be convenient, however, to have an extra set of beaters or one extra beater. If large amounts of food are prepared regularly or if the attachments are used often, a mixer with a good reputation for heavy-duty operation might be a good choice.

The motor of a mixer is a device for transforming electrical energy into mechanical energy. Universal motors, designed to be used on both alternating and direct current, usually are used in mixers and blenders. When a motor is operating the necessary conditions for generating voltage exist—closed conductor, magnetic

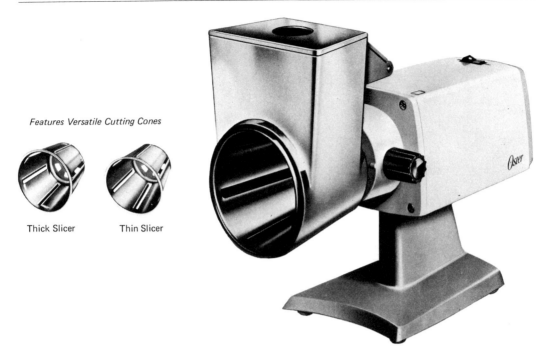

Features Versatile Cutting Cones

Thick Slicer   Thin Slicer

Figure 5-2. Food grinder with salad-maker attachment. (Oster Corporation)

field, and relative motion between conductor and magnetic field. The voltage generated is in the opposite direction to the voltage applied to the motor and it is called back electromotive force. The difference between the applied voltage and the back electromotive force is the voltage that determines the current. The wires of the armature are designed to carry this current. If the motor is slow to start or is doing an extra load of work the back electromotive force is small; as a result wires in the motor may overheat. Some mixer motors are so designed as to eliminate radio interference. Rating of motors is done in horsepower and is the output or ability to do work. The input in watts is greater than the output.

Manufacturers' directions should be followed for oiling and lubrication. All mixers get warm when the motor runs for a period of time but a mixer should not be used until it gets hot. Either running it for long periods of time or beating too heavy a mixture may cause the motor to overheat. If it heats excessively when used only a short time, the motor should be checked for electrical safety.

The mixer head should never be put in water. It should, however, be wiped off after use with a damp cloth. Likewise, dust (flour, etc.) should be kept out of the motor. Placement of the air vents at the rear of the mixer head helps to avoid particles of food being drawn into the motor. The rubber-covered cord should be kept clean by wiping occasionally with a damp cloth. This should be done when the cord is not connected to the source of electricity.

The electric food grinder is a motor-driven unit to which may be attached a grinder, slicer-shredder (Fig. 5-2), can opener, knife sharpener or ice crusher, or it may also be sold as a separate appliance. The motor and gear mechanism are suitable for heavy duty tasks.

A food grinder should be so designed that a bowl can fit close to it. The ground food should fall into the bowl or pan and never on the table. The grinder head should be locked into place so it cannot loosen during use. A hard wood pusher to force food into the hopper is a protection for the user's fingers. All parts of the grinder should be rustproof, easy to clean, and easy to assemble.

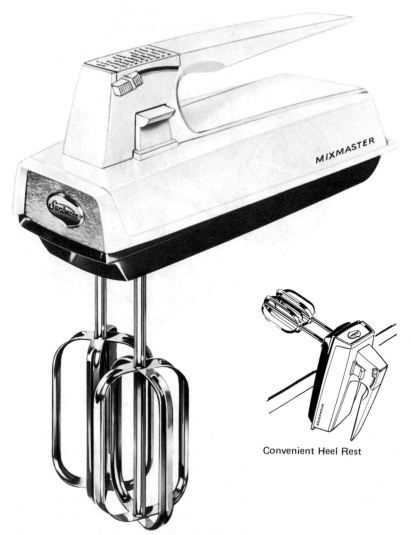

Convenient Heel Rest

Figure 5-3. Portable mixer has automatic thumb-tip beater ejector, heel rest, and governor-controlled motor. (Sunbeam Appliance Company)

Portable Mixers

Portable mixers consist of motor head and beaters (Fig. 5-3). The wall hanger, bracket, or stand is an accessory. Most portable mixers are light in weight—between 1 1/2 and 3 1/4 pounds. They are small in size and can be put into ordinary kitchen storage drawers easily. Care should be taken not to bend the blades in drawer storage. Most of them can be operated on either alternating or direct current. They do a good job for most mixing tasks. It is relatively easy to mix all the ingredients in a bowl because the beaters can be moved around. They are not recommended for mixing large amounts of food or for extra-thick mixtures such as stiff cooky dough.

The "on and off" switch and the speed control should be clearly labeled. Ordinarily, fewer speeds are marked on these mixers than on those of standard size.

A heel rest and a storage hanger make it easier to use and store the beater. If the motor head with beaters in position can be stored near the place where the portable mixer is used most often, the mixer will be especially convenient to use. If there is no heel rest, it is difficult to put the beater down without letting the beaters rest on the counter. The heel rest should be designed so that food drips off the bottom end of the beaters and does not flow toward the mixer head. One mixer has clips on the side for storage of the beaters.

A beater ejector is convenient. Smooth beater blades are easier to clean than those that are not smooth. Care should always be taken to lay the mixer down so that the beaters are near the back of the counter. This precaution will tend to eliminate the danger of a child turning a mixer on and getting a finger in the beaters. Where there are small children, it is a good idea to leave the mixer disconnected when not in use.

Blenders

Blenders have specialized uses in food preparation. In some homes, they are used regularly.

Figure 5-4. Blender has 14 speeds, cord storage compartment, controlled-cycle blending. (Oster Corporation)

Blenders consist of blades that cut or mix the food; a glass, metal, or plastic container for food; a cover for this container; and a base that houses a motor unit, an "on and off" switch, and a speed control (Fig. 5-4). Blades in a blender, usually made of surgical stainless steel, rotate at speeds of 16,000 to 18,000 revolutions per minute (rpm). Blenders may have a number of speeds, but those actually used by homemakers may be considerably fewer than the number provided on the appliance. Since there are usually four cutting blades, the number of cuts per minute is determined by multiplying the rpm by four.

Containers that are smooth inside are somewhat easier to clean. It is especially important that the top opening be sufficiently large for easy cleaning. Plastic containers may not stand the heat of a mechanical dishwasher. Ordinary fruit jars can be used as the container on some blenders. The regular containers hold about 40 ounces, but of course should not be filled with food before blending. Accurate seating of the container on the motor is important and there should be no possibility of starting the motor with the container inaccurately seated.

The blades are usually extremely sharp and care should be taken to wash and put them away after each use. They should be kept out of the reach of children. If they are permanently attached to the container, wash by putting water and detergent into the container and running the motor for a few seconds. It is a little difficult to remove such foods as cheese spreads, chopped dates, etc., from a container in which the blades are attached permanently.

Blenders are used to mix beverages, prepare vegetables for creamed soups, prepare mayonnaise, blend fruits and vegetables to a purée-like consistency, grate, and do other similar operations. It is an easy way to soften cream cheese and blend it with other ingredients. A few seconds in a blender and cottage cheese is transformed into a smooth creamy product. Some direction booklets indicate that the final mixing for cakes and muffins should be done by hand or with an electric mixer. The blending of liquid ingredients, shortening, and eggs can be done in the blender. Blenders are usually not recommended for whipping egg whites, mashing potatoes, or extracting pure juice from solids.

The motor and blades should be completely stopped before using a rubber scraper to push foods down into the container. The fingers should never be used to push foods down as these blades are *sharp.*

Blending is done in seconds, rarely ever in minutes. It is essential that directions be followed and blending times be clocked at least until one is familiar with the appliance. Foods may be overblended and the motor may overheat if used too long at one time. Controlled-cycle blending helps to avoid overblending of foods.

HEATING APPLIANCES

Heating appliances for cooking should be constructed so that feet and handles are well insulated. The handles should be so designed that pockets of hot air are not held under them. An opening between the appliance and the outer part of the handle will let the hot air rise away from the handle. The feet should be insulated so that heat from them will not mar the surface on which they rest. The feet should hold the body of the appliance far enough away from a table or counter surface so that heat radiating from the bottom of the appliance does not blister or mar the table finish. The feet should be smooth so that the appliance can be moved on a table without scratching it.

Heat-controlled appliances depend upon a thermostat to control the temperature. The accuracy of the thermostat on a small electric appliance depends primarily on the manufacturer's standards. Some appliances may have quite a wide range of actual temperatures from that set on the dial, and others will maintain temperatures within ±5°F. Both hydraulic and bimetallic thermostats are used. Once a recipe and temperature combination is found to be satisfactory, it is easy to duplicate results.

Wattage of small heating appliances is generally *not* small. Before buying an appliance, it is a good procedure to check the wattage by reading the nameplate. It is quite possible that it will be so high that only one appliance can be used on a circuit at one time. Voltage requirements should also be checked on the nameplate. Most appliances operate satisfactorily if the voltage supplied in the home is within ±5° F volts of that stated on the nameplate.

Open-coil heating elements are often a hazard even when recessed in ceramic forms. An open-coil unit supported by ceramic bridges or tunnels is especially vulnerable to jarring. Once the coil touches or almost touches the metal of the appliance it becomes a shock hazard. The safest heating units are tubular enclosed coils of the type used on electric ranges.

Many small appliances have indicator lights to show when the appliance is heating and/or when it is ready to use. These lights are somewhat easier to see if the light is from a small bulb. If a section of the heating coil is the source of light, the "light" comes on and goes off gradually as the coil heats and cools off.

For some appliances the heat-control dial and indicator light are detachable and interchangeable with other appliances manufactured by the same company.

Mini-brew basket permits 2-3 cup brewing.

Completely immersible for easier cleaning.

Figure 5-5. Coffee maker is completely immersible. (General Electric Company)

Only one heat control needs to be purchased, unless it is planned to use the appliances simultaneously.

Some heating appliances can be completely immersed in water for cleaning (Fig. 5-5); others can be immersed up to the indicator light or thermostatic dial control; still others must not be immersed in water at all. The thermostatic dial control, indicator lights, and heating units must not get wet. An immersible appliance has a sealed-in heating unit and a control plug that houses thermostat and indicator light. The manufacturer's directions for a particular appliance should be followed carefully. Models change from year to year; also, not all the appliances by any one manufacturer will necessarily be cared for in the same manner. Cold water should not be put into or on hot griddles, saucepans, or frypans, as this may cause the metal to warp.

Appliances that are finished in chromium should be wiped with a damp, soapy cloth, then dried and polished with a dry cloth. Spatterings of fat that may have burned on can be cleaned off with whiting and water. Harsh abrasives will scratch the shiny finish and should almost never be used. If a griddle, interior of a saucepan, frypan, or rotisserie needs scouring, it should be done only with a very fine abrasive such as a fine steel wool soap pad.

Many new heating appliances will give off a slight odor or smoke a little when first heated. This is nothing about which to be alarmed. Manufacturers' directions for first use of an appliance should be followed carefully. This first use can, in

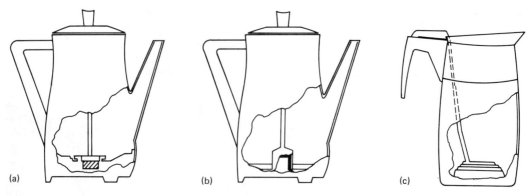

Figure 5-6. Three types of heating units used in coffee makers: (a) well; (b) projection; (c) removable.

some cases, determine to a great extent the satisfaction that comes with later use.

Coffee Makers

Electric coffee makers are of the percolator type. It is usually necessary to practice a few times in order to determine the exact amount of coffee to use and the setting that is best for an individual coffee maker. What constitutes "good" coffee varies with personal taste, and there are numerous factors that influence the flavor of coffee.

Coffee makers are made of aluminum—anodized, polished, or with a chromium finish—of copper, plated with nickel and chromium, of stainless steel, of Pyroceram, of glass, and of plastic. The top of the coffee maker should be large enough to put one's hand into in order to wash it with ease.

The heating unit may be completely separate from the percolator, it may be in a raised projection in the bottom of the pot, it may be a well-type heating element, or it may be a flat unit beneath the the bottom of the pot (Fig. 5-6). The heating unit should always be covered with liquid when it is connected to electricity.

Percolators may be automatic or non-automatic. In the nonautomatic type, the coffee is perked until the operator disconnects the percolator from the source of electricity. This percolator does not maintain a keep-warm temperature. In non-electric percolators the heat is from an exterior source. In the automatic type the coffee is perked until the desired strength is attained, at which time the electricity is automatically shut off or cut down to maintain a keep-warm setting. Most modern percolators have the keep-warm setting. In some of these, however, this setting is reached only after the percolating cycle; each time this type of coffee maker is connected, the perking cycle is started. In other percolators, the keep-warm setting can be used at any time.

Some percolators that reheat without reperking have a heat control that can be set at reheat and any setting between mild and strong. Both thermostat switches (Fig. 5-7) are closed when the percolator is cold and the control is set at one of the perk settings. The two 400-watt units are energized but the pilot light and keep-hot unit are shorted out while coffee is perking. When the temperature of the

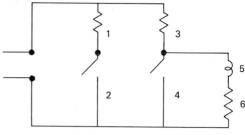

1. 400-watt booster unit
2. Booster thermostat
3. 400-watt pump unit
4. Pump thermostat
5. Pilot light
6. 55-watt keeps hot unit

Figure 5-7. Wiring diagram for percolator that has separate reheat setting. (General Electric Company)

brew reaches the desired point the booster thermostat opens and that 400-watt unit is out of the circuit. At a slightly higher temperature the pump thermostat opens and the keep-hot element and pilot light are then in series with the pump unit. With the additional resistance in the circuit the total heat generated by the two units will just keep the brew hot. Percolation stops at this time.

A cam, operated by the control lever, controls the temperatures at which the booster and pump circuits open. When the control lever is set at reheat the pump thermostat is open. Percolating is stopped when the pump circuit is opened and the quicker this takes place the milder is the coffee brew. For strong coffee the booster unit is cut out at a lower temperature and the pump unit at a higher temperature which extends the time of percolation. At the strong setting the booster thermostat switch cuts out the unit between 120° and 150° F. The heater thermostat switch opens between 185° and 201° F. The water does not reach the boiling point.[1]

In a percolator without a manually operated heat control switch the brewing process also is started with cold water. Percolation takes place at approximately 187° F. When the coffee has perked the unit is cut off (Fig. 5-8). The serving temperature, 185° F, is maintained by cycling of the heater unit. Water does not boil with this type of control.[2]

When coffee is not served immediately, the grounds should be removed from the coffee maker to keep the brew at an optimum flavor.

Some percolators are said to make as few as two cups of coffee or as many as nine or ten. Others do not make less than four cups successfully. The cup marks on

[1] *General Electric Service Manual*, Form PS6–4, July 1961.
[2] West Bend Service Notes.

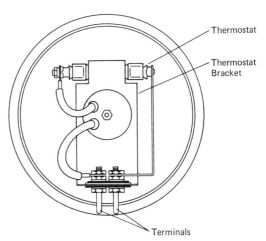

Figure 5-8. Bottom view of a coffee maker without a manually operated heat control switch. Follow path of electricity from terminal to heating unit to thermostat and back to terminal. (The West Bend Company)

the coffee maker often are not accurate measures in terms of a standard measuring cup, but indicate coffee-cup measures. Many manufacturers have standardized capacities in terms of a 5-ounce cup of finished brew. If the amount of water corresponding to the different markings in the percolator is determined once, it is not necessary to measure the water each time coffee is made.

The time required to brew coffee of a certain strength varies considerably with different coffee makers. However, even those that take the longest time will make a full pot of coffee in less than 20 minutes. For any one model the strength selected, as well as the amount of coffee made, influences the length of time required. Some coffee makers make more noise than others.

It makes for ease in using if the coffee basket, stem, and cover can be removed at one time. It is also convenient if the stem and basket will stand upright when removed.

The cover on a coffee maker should not tilt forward or fall off when pouring. Even a cover which merely *acts* as if it is going to fall off is a nuisance to use! The spout should pour easily and be so constructed that it does not drip. Spouts on many coffee makers are not too easy to clean, and for that reason often are ignored by homemakers. A small brush should be used to keep the spout as clean as the other parts of the percolator. The handle should be easy to grasp, made of insulating material, of such a design that the knuckles or fingers do not touch the coffee maker, and securely attached.

Only cold water should be used in percolators that start perking almost immediately upon connection to electricity. If hot water is used, the time for perking will be considerably shortened and the flavor of the coffee will be weak. In coffee makers that heat all the water before perking starts, the temperature of the water used is not so important. Such coffee makers are similar to percolators used with an exterior source of heat. However, many authorities state that the best coffee is made by using freshly drawn cold water, regardless of the type of coffee maker used. Some percolators must be cooled before the perking process can be repeated; others will brew more coffee immediately.

It is generally recommended that the interior of a coffee maker be washed with hot water and a detergent and rinsed with clear water after each use. Periodically, the interior of all but aluminum coffee makers can be cleaned by boiling a solution of soda and water in the percolator. Aluminum baskets, stems, and covers should not be cleaned with the soda solution. Other cleaners also are recommended by individual manufacturers. The interior of a coffee maker that is finished by plating should not be scoured. A coffee maker should be stored with the cover off, or at least not on tightly. If this is not feasible, air the coffee maker after it has been washed and dried before putting it away.

The vacuum-type coffee maker has two bowls—a bottom one that holds water and a top one in which coffee and filter are placed. When the water is heated, steam is formed and the air in the bottom of the coffee maker is forced out. As more steam is formed, pressure is exerted on the water, and the water is forced into the upper bowl. The water will remain in the upper bowl as long as the heat applied to the lower bowl is sufficient to maintain steam. When the heat is reduced the steam in the lower bowl condenses, and a vacuum is formed. Immediately, air pressure forces the water, which is now coffee brew, back into the lower bowl. Its strength is adjusted by changing the amount of coffee in relation to the amount of water used, and the time the water is held in contact with the coffee which depends upon the time the coffee maker is allowed to remain on the heating source.

The upper bowl is easier to remove from the coffee maker if it has an insulated handle. The lower bowl must have a handle in order to use it as a coffee server. The cover usually is designed to fit both bowls.

Filters may be of stainless steel with very fine openings, of cloth, or a glass rod. The cloth filter must be kept clean or it may give an "off" flavor to the coffee.

Large-capacity percolators are a boon to the homemaker who entertains large groups of people. They usually make 30 cups of coffee in about 30 minutes. Larger sizes are also available. The minumum amount of coffee brewed is usually 12 cups. The faucet should have a secure "off" position so that there is no drip. At the same time the faucet should operate easily and be so designed that a cup and saucer or coffee mug will fit under it without tipping. These coffee makers are made of aluminum or stainless steel and usually have an electrical rating of 1,000 watts or more. A keep-warm setting without reperking is highly desirable. A signal light

Figure 5-9. Buffet cooker/server for fondues, casseroles, soups, entrees—a variety of foods. Aluminum base, porcelain exterior, interior Teflon II coated. (Oster Corporation)

to tell when the coffee is ready to serve is a convenience, but usually one can determine this by listening. Most of the perks can be easily heard during percolation. The capacity gradation marks on the body of the coffee maker should be clearly marked. Provision should be made for easy cleaning of the glass gauge used on some models. A detachable cord makes it somewhat easier to wash the coffee maker as it is not necessary to be concerned about the cord getting wet.

It is easier to make a large amount of coffee in an electric percolator than in a nonelectric dripolator. Pouring a large quantity of boiling water over the coffee grounds in a dripolator is quite difficult for some users. Water is used at tap-water temperature in the percolator and no extra heating unit is required. Small sizes, that is 3- to 10-cup dripolators, are quite easy to use.

Cooker-Fryers, Saucepans, Dutch Ovens, and Fondue Cookers

Cooker-fryers, saucepans, and Dutch ovens can be used for a number of different cooking operations. Foods can be steamed, stewed, and blanched as well as deep-fat fried. They can also be used as bun warmers, for making soups, pot roasts, and other specialties (Fig. 5-9).

Cooker-fryers usually come equipped with a cover and a deep-fat frying basket with a detachable handle. The heating unit and controls are housed in the same shell as the container for the food. These appliances should be well balanced, have good insulation on the bottom, and have sturdy, securely fastened, insulated handles. A good heat-conducting material should be used for the container so that the foods will cook evenly. A thermostatic control maintains the temperature at a specified setting. It is especially important that this control be placed where food cannot drip on it when being removed from the utensil. The cooker-fryer should have an indicator light that shows when the fat is at the correct temperature. NEMA Standards state that the capacity of a deep-fat fryer should be at least 2 1/2 times the volume of cooking oil recommended by the manufacturer.[3] The quan-

[3] DA9–1964, *NEMA Standards Publication*, Household Automatic Electric Fry Pans, Griddles, Saucepans, Dutch Ovens, and Deep Fat Fryers, p. 3, National Electrical Manufacturers Association.

tity of food used will influence the time needed for the selected temperature to be reached. Temperature setting may need to be adjusted for varying quantities of food.

Overshoot temperature on preheat should not be excessive, but a certain amount is desirable. When cold foods are added to a preheated medium the temperature falls and then tends to cycle fairly close to the temperature set on the control. The range of the cycling temperatures depends upon the design of the thermostat. When cooking several batches of food, such as doughnuts, ample time should be allowed between batches for the temperature of the fat to recover.

There is quite a difference in the design of fry baskets, and some are much easier to clean than others. Usually this can be observed rather quickly even without using them. The handle, though detachable, should be secure when it is attached. A place to fasten the fry basket on the edge of the cooker-fryer to enable fat to drain from the food into the cooker is a convenience. This fry basket also makes a convenient blanching basket. For deep-fat frying, a large surface area for the fat is somewhat more important than a very great depth for the fat. A cooker with a fairly large diameter will enable more foods to be fried at one time.

Hot fat should be allowed to cool slightly before it is emptied into a storage container. If the cooker has a drain spout, care must be taken that it is securely fastened during the use of the cooker and completely cleaned after the fat is drained out. Crumbs from cooked food can clog the drain spout. This spout should have a valve that cannot be opened simply by turning, as this might be done accidentally while the fat is very hot. If the fat is drained out by a spout, two counter-top levels in the kitchen must be available for this. Depending upon construction of the

sink and length of the spout, the sink well can sometimes be used as the lower level.

A fondue cooker should be well-balanced so that it cannot be unintentionally tipped. Fondue cooking is designed for informal occasions and the appliance should be foolproof especially when designed for use by groups of people at one time. A deflector to minimize spattering can also serve as a support for the fondue forks. The cooker should be well insulated on the bottom so that the table on which it is placed is not marred by excessive heat (Fig. 5-10).

Many of the foods used for fondues are heat sensitive. Chocolate scorches rather easily, fats can reach a smoking point quickly, and cheeses need controlled heat for best results. Therefore, a fondue cooker that has a sensitive thermostat is highly desirable. Most thermostats probably control the heat within narrower limits than humans can control the non-electric heat sources used for some fondue pans. In addition, some fondue cookers

Figure 5-10. Fondue cooker—aluminum base, Teflon lined. (National Presto Industries, Inc.)

have a second bowl which is used for water in the same manner as a conventional double boiler has a container for water.

Some of these appliances can be used for steaming foods. If hard water is used in aluminum utensils the addition of cream of tartar (1 teaspoon to 2 cups water) will minimize darkening of the interior of the utensil.

Corn Poppers

Although other appliances can be used to pop corn, not all will work well without stirring. Some do not reach a temperature high enough to pop corn successfully.

Corn poppers usually consist of three parts—a base that houses the heating unit, a pan that holds the fat and the corn, and a cover. A see-through cover has some advantages but it is not a necessity. An open coil set in ceramic is the most common type of heating unit. It has a relatively low wattage, about 450 watts, and can be used on a general purpose circuit. An automatic popper has a heat control, making it possible to keep the popped corn warm for serving. It also adds an element of safety since the possibility of the fat overheating is lessened.

If the popper pan is kept clean and directions are followed, consistently good results can be obtained. It is, of course, necessary to use popcorn with a moisture content such that the corn will explode.

Salt left on aluminum may accelerate pitting. As soon as the pan has cooled, wash it carefully with a detergent and rinse well. If all the oil is not removed use a soap-filled steel wool pad.

Frypans and Griddles

It is difficult to define some of the small electrical appliances. Frypans can be used for roasting, baking, serving, braising, pan-broiling, frying, and as griddles for toasting sandwiches. A rack, similar to oven racks, that fits the frypan makes it easier to bake or steam some foods. Some appliances, formerly sold as griddles, have been redesigned and had a cover added; they are now also sold for roasting and baking. Most of these, as well as frypans, are also promoted as servers for buffet and patio meals. Other griddles are combined with trays to serve as food warmers. Some, with broiler units built into a cover, serve as a broiler as well as a frypan. Some frypans can be immersed in water up to a point on the handle where the dial of the thermostatic control and the connections for the cord are located. Although this type is not on today's market there are many, no doubt, still in use. If this control should accidentally get wet, the pan should not be used until the control has had ample opportunity to dry thoroughly. This might require several days. Most frypans have a detachable heat control (Fig. 5-11). These pans can be completely immersed in water, and so cleaning is similar to cleaning any other pan. It is, perhaps, a little more difficult to clean the exterior of a frypan than a nonautomatic skillet because it is necessary to clean around the legs. The heating unit usually is not flush with the bottom of the pan and adds to the cleaning needed. However, because it is heat controlled there is less scorching and burning of foods, and so the interior is likely to be easier to clean. Temperatures on frypans and griddles are maintained over a fairly wide range. Lows are approximately 150 $\pm$ 10° F and highs are about 425-450° F.

Frypans are made of cast aluminum, sheet aluminum, stainless steel, and a combination of stainless steel and aluminum. Exterior finishes may be porcelain; the interior often is coated with Teflon II. Electric frypans and griddles, like their nonelectric counterparts, are somewhat more satisfactory if the metal of the pan transfers heat evenly.

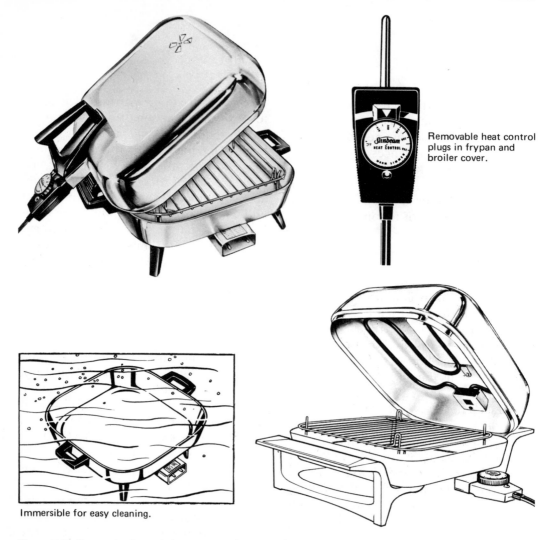

Removable heat control plugs in frypan and broiler cover.

Immersible for easy cleaning.

Figure 5-11. Frypan, broiler unit in cover. (Sunbeam Appliance Company)

They are either square or round in shape and vary in depth and size. Choose a size suitable for the amount of food usually cooked.

Covers on frypans vary a great deal. The cover is often of the same material as the pan but usually a thinner gauge and approximately 2 inches high. Other covers may be 5 or 6 inches high making it possible to cook larger roasts and fowl and deeper casseroles. Some frypans have covers with steam vents with a small movable lid; on others the steam vent is controlled by adjusting the handle on the cover. Most of the covers do not have steam vents. It

is possible to get a frypan with a cover that can be positioned on the pan at different angles. The chief advantage of this is that counter space does not need to be available for the cover when it is removed. Frypans and griddles use 1,200–1,500 watts; they should be connected only to an appliance circuit.

Griddles should be made of a heavy-gauge material that is a good heat conductor. Some models have a heat control, (Fig. 5-12) others do not. Except for portability, a griddle without a heat control has little advantage over one used on the unit or burner of a range. If the griddle has a

Figure 5-12. Automatic griddle with 8-gauge stamped aluminum, Fired-On, No-Stick surface, and phenolic handles. (The West Bend Company)

cover, it should be easy to move about. The griddle should not be too large to be manipulated easily with the handle or handles on it.

Some workable method of removing excess fat from cooking foods at high temperatures adds to the usefulness of a griddle. In cooking some foods, the griddle will need to have a coating of fat spread over it; for other foods this will not be necessary. Until familiarity with a griddle is gained, it is a good procedure to follow recipes in the instruction booklet. A griddle should be wiped with paper toweling after it has cooled enough to be safe to do so; then it should be washed in hot soapy water, rinsed, and dried. Some manufacturers recommend the regular use of soap-filled steel wool pads to eliminate the possibility of a build-up of fat on the griddle. It is rather difficult to wash a griddle unless it has a detachable heat control so that the griddle may be immersed in water.

Roasters

Roasters can serve as a second oven for baking or roasting foods. Whole meals can be prepared in them, and they are also convenient for preparing large amounts of food for a crowd. Roasters are not as heavily insulated as most range ovens, and the covers are often not insulated, which may allow them to get uncomfortably hot during baking. The handle of the cover should be of insulated material. Steam vents in the cover serve the same purpose as the oven vent in the electric range oven.

Roaster ovens use less wattage than range ovens and can be connected to an appliance circuit outlet. However, they often take somewhat longer to preheat, and so the cost of heating per unit area may not differ significantly.

Roasters are thermostatically controlled and unless the manufacturer's instructions state otherwise, the temperatures used for baking should be the same as for baking in the range oven. A light that indicates when the roaster oven is preheated is almost standard equipment. An inset pan is also standard equipment. It can be removed from the roaster and washed the same as any other large pan.

Roasters can be purchased with automatic timers, enabling the user to put the food in the roaster and go away knowing the food will be cooked at the time the user has specified.

Toasters

The automatic toaster is so widely sold that many people may have forgotten or perhaps never have known that there ever was a nonautomatic toaster on the market. It cost much less and was usable for many years, but the quality of the toast did depend a great deal upon the user.

It is not necessary to time the toasting

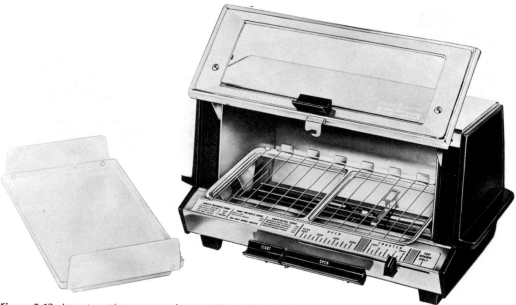

Figure 5-13. An automatic toaster and oven. (General Electric Company)

of bread in an automatic toaster (Fig. 5-13). Theoretically each slice of toast will be evenly browned, but there are a number of factors, which may make this a bit improbable. The browning of the toast is controlled in two ways. In one method the control is set according to degree of brownness desired; that is, the darker the toast, the greater the time. If the voltage supply is constant and within the range for which the toaster is designed to operate, evenly browned toast will result. However, if the voltage is higher or lower than that specified on the nameplate, the time required to make toast will be changed. Also, the same setting will not produce the same degree of brownness on fresh bread and bread several days old. Toasters do not seem to be as well designed as might be expected from the number of years they have been on the market. Laboratory use of toasters indicates that obtaining a toaster that gives uniformly browned toast even under standardized conditions is more nearly the exception than the rule.

The second method of controlling is by radiant heat from the toast. As the toast browns, more heat is given off, and this activates the thermostat switch that shuts off the electricity and thereby releases the lever or spring that raises the toast from the toast well.

Some toasters have a voltage adjustment (usually on the bottom of the toaster) that makes it possible to get the desired brownness of toast. These are quite useful if the voltage supplied is consistently different from that required; but if the voltage fluctuates, they are generally not so useful.

All toasters should have a means of raising the toast from the well manually. This is necessary to remove the toast if the automatic device does not work and to remove toast for inspection as it is being browned. A fork or knife should not be used to dislodge a piece of toast since it is possible thus to injure the heating element and also to receive a bad shock if the toaster is connected to electricity. The latter can happen whenever the toaster is connected to the source of power, whether the switch on the toaster is on or off since most toasters have a single-pole switch. If a toaster has a double-pole switch this possibility does not exist *if* the toaster is off.

Toasters usually are insulated only at the handles and around the base of the well. The metal may get very hot, particularly if several repeat runs of the toasting cycle are made. Toasters should be out of the reach of small children, but this is generally true of most electric appliances.

There should be a crumb tray that is easy to empty and stays closed at other times. It should be cleaned weekly in most homes. A collection of crumbs can interfere with the making of the toast as well as with the pop-up mechanism.

Toasters are available with relatively wide slots for thick slices of bread. Some toasters are designed to raise very small slices of bread far enough out of the well so that one can take the toast out easily. Others have to be turned upside down to remove small slices of toast. Toasters are available that make one, two, three, or four slices of toast at one time. However, not all toasters will toast all the slices to the same degree of brownness or produce uniform browning over the entire slice. A properly designed toaster should make uniform toast, the color desired, for any possible combination of slices. One toaster has not only the oven well but an oven drawer for toasting sweet rolls and English muffins. Some broilers are sold as toasters and vice versa.

The wires which protect the heating unit from the toast should be small in diameter to aid in even browning of the toast. If the wires are large, light-colored lines are seen on the toast.

If a light brown crunchy toast is desired, a toaster should be chosen that does not claim to be extremely speedy. The faster toast is made, the less the bread is toasted all the way through. For those who like soft toast, speed is the answer, providing, of course, that one starts with soft bread!

Waffle Bakers

Waffle bakers, one of the oldest small electrical appliances, are now sold under various names. Current models are heat controlled and most have indicator lights.

Grids are nearly always made of cast aluminum. Often they are coated with Teflon II. The size and closeness of the knobs on the grids will influence the kind of waffle that is baked. Knobs that are large and fairly close together produce a crisp waffle. If the knobs are rather small and far apart, there is more room for batter, and a softer waffle will result. Grids should not be scoured or washed. Instead, after each use brush out any crumbs with a soft brush. The baker should be left open after use until the grids have cooled. It should be closed when stored to keep the grids free from dust. Grids tend to darken with use, but this is not harmful to foods.

There should be an overflow rim around the grids to catch extra batter in case the amount needed is misjudged. This rim keeps the batter from running down the outside of the baker, which not only looks unattractive but may soil the tablecloth. If batter does get on the outside of the baker, it will be easier to remove when the appliance is cool.

The expansion hinge between the two parts of the baker should be sufficiently flexible to allow the waffles to rise as they are baked. If the baker has grids for grilling, the hinges should allow plenty of room for making grilled sandwiches without mashing them.

The thermostatic control is set according to preference for a light or dark waffle. Different types of thermostats are used. In one type the radiant heat from the baking waffle is directed toward the thermostatic control. In another the heat from the lower grid heats a strip of metal that actuates a bimetallic thermostat. In a third the difference in expansion of the aluminum grid and a Pyrex rod actuates the thermostat.

The wires that carry the electricity to the top grid should be well protected from wear and guarded so that the user cannot touch them.

Waffle bakers may have several sets of grids—one set or more for baking waffles of different designs and one for grilling foods. In some bakers the grids for grilling are permanently attached and the waffle grids are removable. This eases the storage problem but it adds others. The grill will need washing, whereas the waffle grids seldom, if ever, should be washed. If the grill is not cleaned well, the bottoms of the waffle grids get dirty. Waffle grids may not heat evenly in bakers with permanently attached grills. In other models, the grills are simply reversed or turned upside down. Overheating and sticking can be a problem in this type.

Ovens, Broilers, and Rotisseries

It is possible to buy an oven, a broiler, or a rotisserie as a single purpose appliance and also in combination with one another. For example, a portable oven may contain a broiling unit or an appliance sold as a rotisserie may be designed for use both as an oven and as a broiler. It is a good idea to consider before buying not only the safety of these appliances but also the ease of cleaning. Ovens and rotisseries should be thermostatically controlled, although this is not essential for broilers. In a study of portable electric cooking appliances it was found that a longer time was required to bake foods in portable ovens than in range ovens.[4]

These appliances vary in wattage from a relatively low wattage of about 700 watts for a small broiler to 1,500 for a rotisserie-oven large enough to bake a two-layer cake. As a general rule, most of these appliances use enough wattage so that they should be connected only to an appliance circuit. When appliances have the capacity

[4] Faith Churchill and Lenore S. Thye, "Portable Electric Cooking Appliances," *Journal of Home Economics*, April 1963, vol. 55, pp. 261–267.

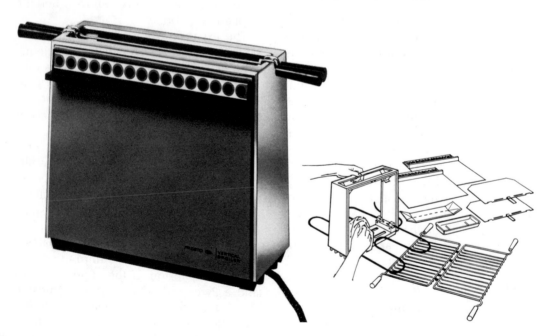

Figure 5-14. Vertical broiler. Broiler rack, doors, and drip pan may be washed in dishwasher. (National Presto Industries, Inc.)

(a)

(b)

(c)

to cook large roasts or fowl or to bake, the insulation on handles, the whole exterior, and especially the bottom, is very important. The heat output over the long period of use can cause these parts to become uncomfortably hot. These appliances should be placed where no one is apt to touch them accidentally; otherwise a burn is quite possible.

In most portable broilers the heat source is located so it will be above the food (Fig. 5-11). However, a broiler which is specifically designed to cook luncheon meats or bacon without fat spattering on the counter or range surfaces has the heat source below the food. Another broiler (Fig. 5-14) has the heat located in two areas so that foods are cooked on both sides simultaneously.

Rotisseries are designed to broil foods on all sides without the user turning the food. The spits are designed to rotate horizontally or vertically by means of a motor. The food to be broiled is centered on a spit and secured in place by means of prongs. The food needs to be balanced if it is to rotate smoothly.

If one anticipates viewing the rotisserie in action, a model should be chosen that is protected with glass, or so designed that the area around it does not get too hot or splattered with particles of fat. If a rotisserie is designed to be spatter free it might be well to ask for a demonstration. If the rotisserie is completely closed, it is more like a baking unit than a broiling unit. However, many models spatter grease if not closed.

The ease in cleaning a rotisserie, oven, or broiler depends in great part upon the ease in disassembling the various parts to be cleaned. It also follows that the various parts should be easily reassembled without danger of incorrect assembly. Although these appliances are not designed to be completely immersed in water for

Figure 5-15. Food warmers range from portable to built in. [(a) Hamilton-Beach Division of Scovill; (b) The West Bend Company; (c) Corning Glass Works]

cleaning, many of the component parts can be washed in a dishwasher.

Small portable broilers are especially useful for the single homemaker or homemaker with a small family. These broilers have a low-wattage rating. Sometimes the heating unit can be removed from the broiler cover, making it easier to keep the top part of the broiler clean. The unit should be removed from the cover only when disconnected from the source of electricity.

Food Warmers

Food warmers are usually of low wattages and are designed to keep foods at serving temperature once they have been cooked (Fig. 5-15, p. 107). There are a number of these which have a variety of uses. One is designed to be built in and used with a heat-proof counter. The heat-proof counter may be portable or part of the permanent counter installation. This warmer uses infrared heat and is designed to keep foods at serving temperature without flavor or moisture loss. The small portable keep-warm devices can be used to keep foods warm from a buffet service or at the table. In essence these devices make it possible to use nonelectric utensils in the same manner that an electric appliance is used to maintain temperatures of food once they are cooked. The snack warmer serves to freshen snacks and improve their taste.

EXPERIMENTS

Experiments 1 and 4 can be done without any special testing equipment, but Experiments 2, 3, 5, and 6 require some special equipment.

Experiment 1. Use of Heating Appliances

Read the instruction book carefully. Identify all the parts mentioned. Determine the voltage requirements and wattage. To what type of circuit should this appliance be connected?

Gather materials for the food preparation involved. Foods that might be prepared include hotcakes, French toast, fried eggs, and bacon in the frypan; plain waffles or French toast in the waffle baker; prepared pudding mixes, carrots, or frozen green vegetables in the saucepan; hotcakes on the griddle; refrigerator biscuits in the roaster; potato chips and biscuit doughnuts in the cooker-fryer; coffee in the coffee maker; toast in the toaster or broiler; wieners or shish kebab in the rotisserie; and plain refrigerator cookies in the oven.

Follow directions and prepare the food.

Were the instructions complete and easy to understand? If not, make suggestions for changes. Did the appliance perform satisfactorily in the preparation of the food? Were the handles insulated adequately? Was the signal light easy to see? Did the table underneath the appliance remain cool? Was the appliance easily cleaned? Did the thermostatic control or signal light get dirty when food was removed from or added to the appliance?

A suggested variation of this experiment is to repeat the food preparation while the appliance is still warm. What are the differences, if any, in this second run? Compare toast made in timed and radiant-heat-control toasters.

Experiment 2. Effect of Varying Voltages on the Operation of a Heating Appliance

Be sure the appliance is cool before beginning this experiment. Determine voltage required by the appliance from nameplate data. Connect a variac, ammeter, and wattmeter in a circuit with the appliance to be tested. Set variac for the correct voltage. Preheat the appliance. If it is an appliance such as a coffee maker that should not be heated without water in it, put in a measured amount of water at a measured temperature. Check the time it takes to preheat the appliance or in the case of a coffee maker the time it takes to start perking.

Repeat at a voltage setting of 10 volts lower than nameplate requirements. What is the effect of low voltage on time required? On the current supplied to the appliance? What would it cost to preheat the appliance at your local electric rate?

Experiment 3. Electrical Characteristics of Household Mixer Motors

Connect a wattmeter, voltmeter, ammeter, and variac in a circuit with an electric mixer. Observe starting current and wattage, as well as steady values at different speed settings. Are current and wattage readings the same at all speed settings? Place 1 1/3 to 1 1/2 cups of flour and 1/3 cup water in the mixer bowl and mix. Observe wattage, voltage, and amperage at 30-second intervals from start until batter becomes quite stiff. If necessary, add more flour to make a very stiff batter.

How do meter readings vary as batter becomes stiffer?

Can you explain from your current and wattage readings why a mixer gets hot when a stiff batter is mixed in it?

Experiment 4. Capacity Check of Appliances

Frypan: Measure quantity of water necessary to fill level full; to within 1 inch of rim. Determine number of eggs, slices of bacon and/or hotcakes (1/4-cup batter) that can be cooked satisfactorily at one time. Consider not only the appearance of the food but the space required to turn the eggs or hotcakes.

Coffee maker: Do capacity marks indicate 5-ounce cups, 5 1/2-ounce cups, or 8-ounce cups? Use U.S. standard measuring cup to check measurements. Make the smallest amount of coffee for which the coffee maker is designed, the largest amount, and an intermediate amount. Does the coffee maker perform better for one quantity of coffee than another? How much time is required for each?

Blender: Measure total capacity of container; capacity to the full mark. With motor running add water to container until the blender is full—to the point of overflow. Will the container hold more at one speed than at another? What is the largest quantity of food that can be put into the blender without danger of overflow?

Experiment 5. Heat Controls with Various Loads

Frypan: Thermostat settings required to simmer water may vary. Each frypan should be checked before it is used for cooking foods. Procedures for this are given in some instruction books. If not available, the following will give a close approximation of the simmer set on the frypan. Pour 2 to 3 cups of water into cold frypan. Set thermostat for 250° F and allow water to boil for approximately one minute. Slowly turn the thermostat down until signal light goes off. This setting is considered to be the simmer temperature. At this setting water should bubble gently when light is on. Repeat test to check results.

Cooker-fryer: Set temperature to that recommended for frying doughnuts. When signal light has gone off check temperature with thermometer. Add prepared biscuits cut in fourths. Check temperature immediately after this addition, when signal light goes on, when signal light goes off, just before removing the food from the fat, and after food has been removed. Compare temperatures with those recommended for cooking doughnuts. Evaluate the quality of the product.

Repeat test using plain doughnuts. Compare results.

Connect cooker-fryer in circuit with ammeter or wattmeter. Measure actual preheat time, time to recover temperature when load is added, and time that unit is heating while food is cooking. Check these against heating periods as indicated by the signal light.

Coffee maker: Follow directions and make a full pot of coffee. Record the temperature of water used and time to start perking. Record the time to complete perking. Read temperatures as soon as percolation is completed and every 3 minutes for a 30-minute keep-warm period.

Use 85° F water; otherwise follow directions and make a full pot of coffee. Take same measurements as before. Compare results and finished products.

Experiment 6. Performance tests

Frypan and saucepan: With paper towel spread unsalted vegetable oil smoothly over bottom of pan and about 1 inch up the side. Sprinkle with 1 tablespoon flour, shake until pan is lightly and evenly floured. Discard flour that does not cling to the pan. Standardize heating conditions. Use a medium high heat and preheat pan. Disconnect when signal light goes off. Use different size pans of the same material and same gauge; use pans of the similar size of different materials. Evaluate evenness of browning, time required to preheat, time required for browning at end of preheat period, and when pan has cooled. Repeat tests using low-heat and high-heat settings.

Frypan and griddle: Bake a capacity load of three 4-inch hotcakes. Turn cakes only once. Evaluate evenness of browning.

Repeat with one hotcake near the heat control and a similar size cake on the opposite side of the frypan or griddle.

Position thermocouples or thermometers on table top with masking tape before starting cooking tests for heating appliances. Read temperatures while appliance is cool, after it is preheated, after load has been added and signal light has gone off, after it has been in continuous use 30 minutes and after 2 hours (250° F with full load).

BUYING GUIDE

Check through the following list. Because many of the answers depend upon the situation of the individual, it is not always possible to state a "right" answer.

1. How heavy is it? Would a lighter one be better or is weight important for satisfactory performance? Is it sturdy, well balanced, and resistant to tipping? Be sure to check the evenness of the legs—do not assume a wobbly appliance is due to a table that is not level. Appliances that tip easily are unsafe.
2. Is the appliance well insulated—especially the bottom and the handles? Are the handles large enough to enable the user to move the appliance easily?
3. What finish is used? Will it be easy to maintain? Is it resistant to scratches? Does it need polishing with a dry cloth after it is wiped with a damp cloth? Does it require a special cleaner? Are there decorative lines to collect soil? Are the corners or edges sharp, the surface rough? Are there hard-to-clean crevices? If the interior is to be washed manually, is the opening large enough to accommodate movement of the hand?
4. Is it easy to disassemble, clean, and reassemble? Are the parts sturdy, rust resistant, and machined well so that they fit together smoothly? Are the parts easily identified?
5. How many volts are required to operate it? How many watts are developed? Does it operate on AC or DC or both? Will this wattage overload the circuit where it will be used? Is an outlet located in a handy position for use of the appliance? If not, will this appliance perform satisfactorily if an extension cord is used?

Most appliances will not unless a heavy-duty extension cord is used. This sometimes involves buying a custom-constructed cord. Use of an extension cord often indicates a potentially overloaded circuit.

6. Do both the appliance and the cord carry the UL seal of approval? The UL seal is the best assurance available to the average consumer that the appliance is safe to use under conditions stated on the nameplate.

7. Does the appliance have an "on and off" switch or must it be controlled by disconnecting the outlet plug? This is not only an inconvenience but is sometimes unsafe if the user is apt to have wet hands. Is the "on and off" switch thoroughly dependable? Most small appliances should always be disconnected when not in use.

8. Are the controls clearly marked (especially important to the homemaker who wears bifocals!) and are they easily understood? Will the control markings wear off? A control without clear markings decreases the value of the appliance. Is the control easy to manipulate; is it placed where food does not drip on it, where it will not be turned on and off accidentally, where it will not get too hot to handle?

9. Is the heat control detachable? Can the appliance be safely immersed in water? Is it important for the expected use that it be immersible?

10. Is the power or heat supplied enough to do well the particular task or tasks for which the appliance is designed?

11. What safety features are incorporated?

12. Does the manufacturer have a reputation for making good appliances of the type under consideration? Does the dealer have a reputation for fair dealing? Reasonable prices are necessary if a dealer and manufacturer are to stand back of products sold.

13. What kind of guarantee is included? Have you read it? Do you understand it? If not, ask questions.

14. Where can the appliance be serviced? Is it highly probable that it would be necessary to ship it to a distant point if service were needed?

15. Are directions for use explicit and easy to read? Are the pictures helpful? Is it clear to which model the instruction book refers? Always read and follow instructions before using a new appliance.

16. Will the possession of this appliance decrease labor or be otherwise useful? Homemakers sometimes discover they have been greatly influenced by advertising and have purchased an appliance, although perfectly good, for which they have little use. If money and space are important think twice before a decision

is reached. A corn popper is a decided convenience for the family that likes popcorn; it is not much use in other homes.

17. Will the special features be useful? Is the usefulness outweighed by the added cleaning and maintenance problems?

18. Is this an attractive appliance? Will its use bring pleasure by its appearance and by its performance? An appliance should be the slave and not the master.

19. Is there a convenient place in the home to store it so that it *will* be used?

CHAPTER 6
KITCHEN DESIGN

One architect noted in his book *The House and the Art of Its Design* that "it would seem to be true that the space and equipment required for cooking by the urban-surburban upper middle class family can be pretty well pinned down."[1] Steidl and Bratton in their book *Work in the Home* give a definition of functional design for workplaces that can be applied to the home kitchen. The definition for workplaces can be stated so: an arrangement of the important parts to serve a special purpose that meets the requirements of the worker and the work.[2] The definition is applied to the kitchen area by accepting that the special purpose of the kitchen is the preparation of food for a family and its guests. At the present time, cleanup of food preparation utensils and soiled tableware is generally also considered part of the special purpose of a family kitchen, although in the future, cleanup might be planned for another area.

The special purposes do not preclude use for additional purposes and, where practical, kitchen areas should meet individual family requirements. These may include a fireplace, special space for teenage entertaining, special lighting over a kitchen table because some members of the family like to read there, or anything else important to the family.

Adequate counter space where needed, convenient and adequate storage space for articles used in the kitchen, appropriate amounts

[1] Robert Woods Kennedy, *The House and the Art of Its Design*. New York: Reinhold, 1953, p. 208.
[2] Rose E. Steidl and Esther Crew Bratton, *Work in the Home*. New York: Wiley, 1968, p. 265ff.

of activity (clearance) space for the worker(s) in the kitchen, and appliances appropriate for the family's living style and the available kitchen space—all arranged to form an assembly that permits food preparation to proceed smoothly—are likely to contribute to *satisfaction* as well as ease in food preparation. Thus the requirements of the worker(s) are met in a broader sense than solely saving steps or energy. At the same time, if some individual family requirements such as suggested above are met, the family that uses the kitchen enjoys it; in addition the resale value of the residence is likely to be better than it would be if the kitchen had not been designed as a functional space for food preparation.

Home kitchen arrangements or assemblies *are not* similar to a factory production assembly in which parts are delivered at one end and added in a "straight line" process until the finished product comes out the other. Foods brought into the kitchen do not all go into the refrigerator, to the sink, to a mix area, to the range, to the table—in that order. Some foods are stored without refrigeration; some meats are wiped with damp paper near the sink, then wrapped for freezer storage; some berries and other fruits are "picked over," then placed in the refrigerator.

Also germane to kitchen design is the recognition that a functional kitchen considered as an arrangement for food preparation and cleanup may use an entire room or part of a room. When the kitchen is part of a room, the other part may fill a need special to the family such as a children's play area, family room, or informal dining area. In planning a dual or multipurpose room or area, it is advantageous to think *big*. For example, if informal eating space is to be provided as well as kitchen "work space," think of enough space for the work to be done in the kitchen *and* enough space for a comfortable dining area (p. 125).

WORK CENTERS

During food preparation and serving much tripping usually takes place between sink, range, refrigerator, counters adjacent to these appliances, and the dining table—not necessarily in the order listed. Previously prepared salads, for example, may be served directly from the refrigerator.

The concept that generally is accepted as most useful for organizing within the kitchen is that of *work centers*. Although these will be described separately one or more usually connect with one another, except when individual preference or architectural features such as passageways dictated that one or more of the centers be an island.

Major appliances in modern United States kitchens vary in number from three to seven, except for an occasional kitchen that has only two because it was planned that cooking would be done with portable appliances only. The usual major appliances are, of course, sink, range, and refrigerator. In addition, the kitchen may have a dishwasher, separate freezer, barbecue, and, in place of a range, a built-in cooktop and a built-in oven.

The major *work centers* in the kitchen are:

1. Sink or sink-dishwasher center
2. Range-serve or cooktop-serve center
3. Refrigerator center
4. Mix center

If a separate cooktop and built-in oven are used, the cooktop rather than the oven is associated with serving. In fact some workers in kitchen planning suggest that the built-in oven can be placed almost anywhere in the kitchen.

A kitchen with separate cooktop and

oven is likely to require more space than one with free-standing range for equal convenience of storage.[3] However, the built-in cooking appliances have special merit for some handicapped persons and are the preference of some other home-makers.

The mix center often will be close to an appliance—range, refrigerator, or sink—but it should be planned as a complete center (see p. 117).

Irrespective of the number of separate centers planned, *storage of frequently used articles at place of first use* is accepted as a basic principle. The logic is that the user does not need to think of where articles are and carry them to where they will be used, but has them there. This is especially helpful when several items frequently are used together in one process. (An analogy might be storing tooth-brush and toothpaste near each other.)

Organization within centers is planned so that work normally will proceed from right to left, since for right-handed persons this method of working usually involves smoother and fewer motions. Also, the centers are planned to be part of a total kitchen arrangement with reasonable placement of centers with respect to each other.

Sink Center

This is the most often used center; more trips are made to and from this center than to and from any other one.[4,5] This center

contains the sink and, if one is provided, the dishwasher. Provision is made here for disposing of food waste and sometimes waste paper. Storage space is provided for the following articles: dishwashing supplies; cooking ware requiring addition of water in use, such as teakettles and coffee makers; vegetables and fruits that do not require refrigeration but do need washing or peeling, such as potatoes and oranges; foods for which water is needed at the start of preparation, such as canned soups, dried beans, and so on; measuring cups (for water); and cutlery and accessories for cutting, peeling, and straining.

Saucepans are used with water and without water. If practical to do so, storing them near the sink is reasonable according to research at the University of Minnesota.

Dinnerware is stored in the sink center, the serve center, the dining area, or all three areas. Storage of dinnerware in the sink center in some cases contradicts the principle of storage at place of first use, but has some practical advantages—for example, putting away the clean dishes requires less walking. When the sink center is close to the serve center, dishes in which food is served will be conveniently located for serving.

A minimum of 24 to 36 inches of counter to the right of the sink for soiled dishes and 18 to 30 inches to the left of the sink has been suggested when dishes are washed by hand from right to left.[6]

Laboratory work at the University of Minnesota suggests that 24 inches at the right of the sink is adequate when a dish-washer, built in or free standing, is next to the sink and an adequate counter, 36 inches or more, is available at the left of

[3] Florence Ehrenkranz, "Effect of Free-standing Cooking Appliances on Wall Space and Cabinet Frontage in Corridor and L Kitchens," *Journal of Home Economics*, May 1961, pp. 364–370.

[4] Rose E. Steidl, *Trips Between Centers in Kitchens for 100 Meals*, Cornell University Agricultural Experiment Station Bulletin 971, 1962, p. 8.

[5] Florence Ehrenkranz, "Functional Convenience of Kitchens with Different Sink-Dishwasher Locations," *Journal of Home Economics*, November 1965, pp. 711–716.

[6] Wanslow, *Kitchen Planning Guide*. Small Homes Council—Building Research Council, University of Illinois, August 1965, pp. 34–35. (The smaller figures are suggested for houses less than 1,000 square feet.)

the sink. (The 36 inches or more at the left of the sink was needed for setting out soiled tableware removed from the dining table for the locations of dining table, sink, and dishwasher used. For another "sequence" of the two appliances and dining table, 36 inches or more at the right of the sink would have served as well for the *unstacked* soiled tableware.)

Range-Serve Center

The range-serve center is the second most frequently used kitchen center.[7] Most trips *between* centers in meal preparation are between sink and cooktop.

The range is of course the major appliance in the center. Associated with the range is storage space for skillets and lids for saucepans, stirring spoons, ladles, turners, etc.; seasonings and shortening used directly at the range; and foods used with boiling water, for example, macaroni or noodles.

Counter in the amount of 15 to 24 inches or more adjacent or near the range is suggested for setting out dinner plates and serving bowls. Associated with the serving function is storage space for trays, some cutlery, linens such as placemats and other articles used at the eating table—for example, salt and pepper shakers and paper napkins.

Refrigerator Center

Basically this center has the refrigerator and counter frontage of 15 to 18 inches on the door-opening side of the refrigerator. In addition, storage space may be provided for juice glasses, cans of fruit and juice that will be refrigerated before use, and refrigerator dishes. The need for counter space on the correct (door-opening) side for setting out articles to be put into or taken out of the refrigerator is especially appreciated by persons who have worked in kitchens with no adjacent counter or with a counter on the incorrect (door-hinge) side of the refrigerator. Although a counter is needed for articles to be put into the refrigerator, it does not follow, as some older and indeed some current literature states, that for all kitchens the refrigerator should be close to the door through which supplies are brought into the house because as stated earlier, even articles that are to be refrigerated—and not all food supplies are—may have wrappings removed or be washed or have some other type of handling *before* they are refrigerated.

Mix Center

Instead of an appliance, this center is distinguished by a "long length," 36 to 48 inches or so, of uninterrupted counter and specialized storage. The shorter length of counter perhaps is more reasonable for families that do little assembling of ingredients for casserole dishes, baked foods, and other mixed products.

Some kitchens have a mixer cabinet and file storage space for baking pans. Often, storage space is provided for commercial mixes, sugar, flour, shortening, spices, and other foods used in mixing operations; utensils such as bowls, baking pans, and casseroles; and accessory equipment such as sifter, mixer, beater, grinder, rolling pin, measuring cups and spoons, and spoons and forks for mixing.

Duplication of small equipment such as measuring cups and spoons at sink and mix is reasonable.

While the mix center need not be "attached" to an appliance, it usually is best located between the three major appliances or within the *work triangle* (p. 124).

Steidl and Bratton[8] observe that: "a

---

[7] Steidl, op. cit., p. 10.

[8] Steidl and Bratton, op. cit. p. 308.

counter level lower than the present standard of 36 inches would better accommodate the worker at the mix center."

ARRANGEMENT OF WORK CENTERS

Different arrangements (assemblies) of the kitchen centers are used. Perspective views of some of the kitchen types are shown in Figures 6-1 through 6-5. "Complete" kitchens with factory-made *custom-type* cabinets are shown in Figures 6-6 through 6-8.

The *one-wall kitchen*, not shown in perspective, has the three major kitchen appliances—sink, refrigerator, and range—on one wall. If this type has adequate cabinet space for storage of articles used in the kitchen and adequate counter space for mixing and other operations done on kitchen counters, it is a long kitchen and therefore likely to involve excessive walking. If it does not have adequate cabinet and counter space, it can be short. Short one-wall units might be used because space is at a premium and/or because they can be purchased in a single package. (See p. 123.)

Parallel-wall or corridor kitchens have two of the major kitchen appliances on one wall and the third on an opposite wall. This arrangement formerly was somewhat downgraded because the doors or doorways at opposite ends of the corridor permit traffic (cross traffic) through the kitchen and this traffic might interfere with movement of worker(s) during food preparation. This is indeed a possible limitation but it is not a necessary limitation of the corridor. (See p. 138.) On the other hand, the corridor has some distinct advantages—the sink wall can be long with good counter and storage space, the distance between sink and range or cooktop can be short, while at the same time adequate storage and counter space can be provided near the range.

The *L kitchen* has two major appliances on one wall and the third on an adjacent wall, all three appliances being "connected" by a counter. The *broken* or *interrupted L* also has two appliances on one wall and the third on an adjacent wall but the counter is *discontinuous* between two of the appliances.

The *U kitchen* has one major appliance on each of three adjacent walls and the appliances are connected by a counter. In the *broken U*, the counter is discontinuous between two of the appliances.

The L, broken L, broken U or corridor kitchens are likely to be the easiest arrangements to fit into a given space with *adequate cabinet* and *counter space*. The L, broken L, and adequate-width corridor are particularly convenient when two persons work in the kitchen at the same time.

In a small U kitchen, a worker can "shuffle" or sidestep between appliances rather than make complete body turns. This may lead to waste motion in that the worker glides between major appliances and/or counters without first completing tasks that could be finished at one location. However, the small U may be particularly liked by a homemaker who wants to be able to reach different counters from one location. If two persons work in the kitchen at the same time, the U needs to be somewhat large.

The *complete island kitchen* has the three major appliances separated from each other. Some kitchens in older homes are like this because of openings in all four walls. Some new kitchens are like this also—with one island usually in the center of the room. To be most satisfactory, each island must have adequate counter and storage space near the major appliance.

The best arrangement is determined in part by the space available. Research does *not* point up one arrangement as the one to strive for at all costs.

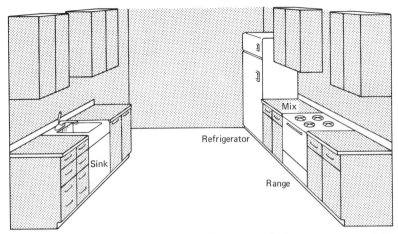

Figure 6-1. Parallel-wall kitchen. Range, refrigerator, and mix area on one wall. Refrigerator door hinged at left. Base cabinet at right of sink planned for pots and pans. (Adapted from *Southern Cooperative Series Bulletin No. 58,* Southern Regional Housing Research Technical Committee, published by Georgia Agricultural Experiment Station, 1958, p. 12)

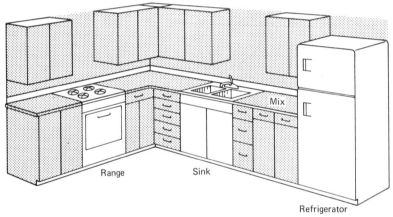

Figure 6-2. L kitchen with refrigerator, sink, and mix area on one wall. (Adapted from *Southern Cooperative Series Bulletin No. 58,* p. 15)

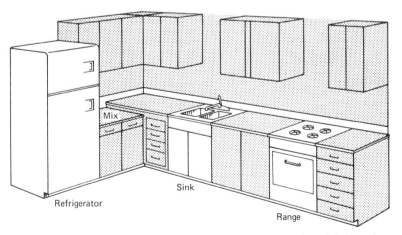

Figure 6-3. L kitchen with refrigerator and mix area on one leg of L. Note lower height of mix counter. (Adapted from *Southern Cooperative Series Bulletin No. 58,* p. 21)

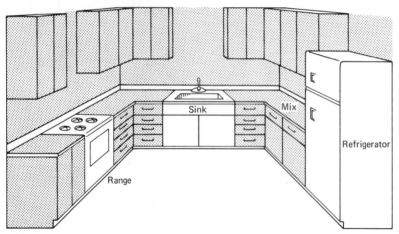

Figure 6-4. Traditional U-type kitchen with sink at base of U. (Adapted from *Southern Cooperative Series Bulletin No. 58*, p. 23)

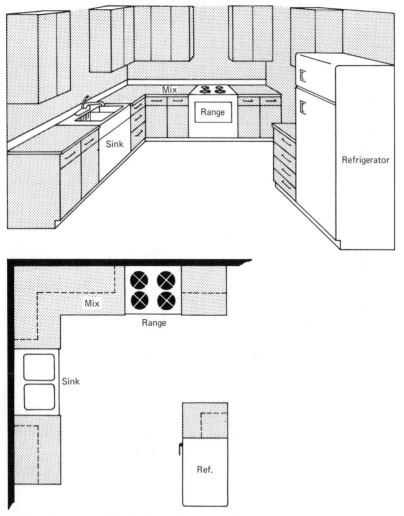

Figure 6-5. Interrupted-U kitchen with range at base of U. (Adapted from *Southern Cooperative Series Bulletin No. 58*, p. 29)

Figure 6-6. L kitchen with 30-inch range and refrigerator on long wall of L. Rotating-shelf base cabinet is used in base corner and diagonal fixed cabinet in wall corner. Dining table utilizes "clear" space in L. Floor is linoleum that can be installed directly over smooth suspended old flooring. (Photograph courtesy of Armstrong Cork Company)

Figure 6-7. U kitchen with pass-through space over range "wall". Kitchen working area is shown as part of a dual-purpose room. Very little standing space is provided in corners of sink wall to right of sink and left of dishwasher. Attractive wood cabinets add warmth to whole room. (Coppes, Inc.)

Figure 6-8. Kitchen with Pyroceram cook top in a center island and eating counter of table height. The base cabinet to left of dishwasher has drawers of unequal depth. (Coppes, Inc.)

Factory-made kitchens consist of sink, range or cooktop, and refrigerator assembled in a single unit. These factory-made units are available with various special features. Figures 6-9 and 6-10 show two of the several models offered by one manufacturer. The model shown in Figure 6-9 is 39 inches wide and 81 inches high. It is available either with two gas burners as shown or electric surface units. An oven is not provided. The undercounter electric refrigerator has a net capacity of 6 cubic feet. The standard wiring when two electric surface units are used is 115/230 volts, three-wire, single-phase AC. Each of the two surface units is 1,250 watts and the grounding-type appliance outlet is wired for 1,440 watts. The kitchen may also be "special-wired" for 115 volts; in this case

Figure 6-9. Factory-made kitchen 39 inches wide. Six-cubic-foot refrigerator has a freezer of 30-pound capacity. Shelves in refrigerator are roll-out type on nylon roller bearings. (Dwyer Products Corporation)

Figure 6-10. Factor-made kitchen 84 inches wide. Ten-cubic-foot refrigerator has roll-out shelves, full-width freezer, and full-width drawer storage at bottom. Oven in this model is 16 inches wide, 19 inches deep, and 15 1/2 inches high. Upper cabinets have porcelain door fronts and a net storage volume of 19.3 cubic feet. (Dwyer Products Corporation)

interlocking switches are provided that limit the electric load to 1,250 watts. When the kitchen is wired for 115 volts, the two surface units will operate simultaneously on "low" and "medium" heats and either can be operated singly on "high."

The model shown in Figure 6-10 is 84 inches wide and 87 inches high. Besides an oven, this model has a 10-cubic-foot refrigerator with full-width freezer and utility storage drawer. When wired for 115/230 volts, the total connected load is 8,990 watts. This type is also available with a gas range.

Work Triangle

The *work triangle* is a triangle on the kitchen floor that connects the center bottom fronts of the refrigerator, sink, and cooktop or range. The perimeter of this triangle is the locus within which *most* food preparation takes place.

ACTIVITY SPACES (CLEARANCES) AND EATING AREAS

Space needs for the people who use kitchens relate to activities in the room. Guides based on activities have been developed for clearances between appliances and/or cabinets and for eating areas.

Clearances in Kitchen Work Area

Test observations on 250 subjects indicated that one person working in the kitchen needs 36 inches in front of (perpendicular to) a base cabinet, a refrigerator, and a built-in wall oven; 38 inches in front of a conventional range; 42 inches in front of a front-opening dishwasher. When two workers use appliances or base cabinets in the kitchen, besides the 36–42 inches needed by one worker, 16 inches of "edging" clearance or 26 inches of walking clearance should be allowed for the other worker.[9]

The need for standing room between the edge of a sink, dishwasher, or range and the corner of a counter for kitchen assemblies that "turn corners" often is overlooked by planners anxious to get

[9] Helen E. McCullough, Kathryn Philson, Ruth H. Smith, Anna L. Wood, and Avis Woolrich, *Space Standards for Household Activities*, University of Illinois Agricultural Experiment Station Bulletin 686 (in cooperation with Alabama Agricultural Experiment Station, College of Home Economics, Pennsylvania State University, Washington Agricultural Experiment Station, U.S. Department of Agriculture), May 1962.

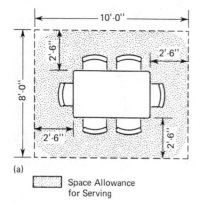

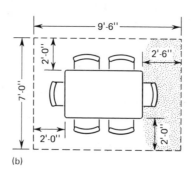

(a)   Space Allowance for Serving   (b)

Figure 6-11. Minimum space requirements for family meals for six persons: (a) serving four sides; (b) serving one end. (Adapted from *Southern Cooperative Series Bulletin No. 58*, pp. 54–55)

Table 6-1. Space Requirements for Individual Place Settings and Clearances Around Table

Item	Minimum	Liberal
Space for individual place settings (cover)		
Width, side to side	24 in.	29 in.
Depth	12 in.	15 in.
Clearances—table edge to wall		
Getting up	24 in.	30 in.
Serving	30 in.	36 in.

Source: Southern Regional Housing Research Technical Committee, *Planning Guides for Southern Rural Homes*, Southern Cooperative Series Bulletin No. 58, published by Georgia Agricultural Experiment Station, 1958, p. 53.

"everything" into too small a space. Standing room between edges of appliances and fronts of counters that turn corners is illustrated in Figure 1-15 as follows: 2 feet between north edge of range and front of northwest corner of counter on sink wall; 2 feet 6 inches between west edge of sink and front of northwest corner of counter on range wall; 3 feet 6 inches between east edge of dishwasher and front of northeast corner of counter on refrigerator wall. Less standing room than shown is adequate, but less than 9 to 12 inches is inconvenient.

Eating Areas

The space standards study specifies 44 inches for serving a seated person (20 inches sitting space plus 24 inches walking space).[10] The recommendations for space requirements by workers in the southeastern area of the United States are given in Table 6-1.

Minimum dining space recommendations for serving meals to six family members are shown in Figure 6-11. The planning guides recommend more liberal dining space when company meals are to be served.

SMALL HOMES COUNCIL RECOMMENDATIONS FOR KITCHEN DESIGN

In addition to the Kitchen Planning Guide prepared by Wanslow, the University of Illinois Small Homes Council-Building Research Council in 1965 prepared a circular to provide "standards" for designing and judging kitchen plans using conventional equipment.[11] These vary according to size of house. Minimum standards are described for houses with a floor area, exclusive of garage, attic, or cellar, of less than 1,000 square feet; medium standards for houses with a floor area between 1,000 and 1,400 square feet; and liberal standards for houses with a floor area over 1,400 square feet.

The recommendations for total base-cabinet frontage not counting cabinet under the sink, drawers in ranges, and corner base cabinets with stationary shelves are: liberal 10 feet; medium 8 feet; minimum 6 feet. Wall-cabinet storage recommendations relate to amount of dinnerware stored in the kitchen. If dinnerware for four people is stored, wall-cabinet frontage should be the same as base-cabinet frontage and this amount is in addition to cabinets over ranges, refrigerators, built-in

[10] Ibid, p. 5.

[11] Kapple, *Kitchen Planning Standards*, University of Illinois Bulletin Circular Series C5.32, March 1965.

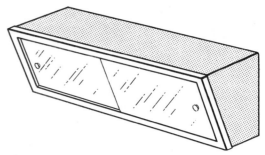

Figure 6-12. Slanting metal midway cabinet.

ovens, sinks, and corner wall cabinets with stationary shelves.

Recommendations on counter frontage in different centers compare with those of other workers. "Combined counters" are of particular interest. "Whenever two or more counters (right of sink and left of refrigerator for example) are combined, the multiple-use counter should equal the longest counter in the group plus one foot. However, the combination of multiple-use counters should not reduce the total base cabinet frontage."

The recommendations also call attention to desirable and undesirable features affecting the kitchen as a working area. Examples of desirable features are: adjacent work centers, adequate clearances between base cabinets or appliances opposite each other, length of work triangle preferably less than 23 feet, good traffic pattern (traffic from front or "rear" door does not cross work triangle or an alternate route is also provided outside of the work triangle). Examples of undesirable features are: a range below a window, two work centers separated by a "tall" appliance (refrigerator or built-in oven), a portable dishwasher stored where it interferes with use of other appliances or cabinets.

KITCHEN CABINETS

Mass-produced, somewhat standardized kitchen cabinets are sold by kitchen-planning firms or centers, department stores, appliance stores, mail-order houses, lumber yards (wood only). They include the following types: base cabinets that sit on the floor; wall cabinets that hang on the wall; midway types, installed usually between the base and wall cabinets but sometimes in place of wall cabinets; and 13-inch deep 7-foot high cabinets (Figs. 6-12 through 6-15).

Custom-built kitchen cabinets are made

"on the job" by contractors and "to order" by manufacturers for a specific installation. Types are limited partly by functions served by kitchen cabinets, partly by cost considerations, and partly by lack of imagination.

The tops of base cabinets serve as kitchen counters. Midway cabinets can hold a surprising amount—canisters of flour and other foods, canned foods, cake mixes, measuring cups, and more—and can fulfill particularly well the function of storage at point of first use. Tall shallow cabinets either have one or two shelves only and are used for cleaning articles, or have many shelves for foods and utensils and thus serve essentially as double-decker wall cabinets, except that they may be located out of the work centers.

Materials used are wood, steel, steel with wood doors and fronts, and plastic

Figure 6-13. Wood midway cabinet installed over sink with sliding glass doors. (Coppes, Inc.)

Figure 6-14. Four-drawer base cabinet with drawers of unequal height and compartmented top drawer. (Mutschler Brothers Company)

(melamine) laminates usually with wood doors and hardwood frames. Factory-made wood and steel cabinets are available in a broad range of prices according to material and finish and/or construction characteristics. Steel with wood doors and fronts ordinarily would not be in the very lowest price range. And at present the cabinets that have a frame of pressed wood (particle board or hardwood) and all visible surfaces (insides, sides, doors, stiles, drawer fronts) covered with melamine plastic are custom made for specific installations and are at the high end of the price range. Advantages of the plastic laminate on wood cabinets are their relative ease of cleaning, their availability in all the

colors, and the patterns and texture effects of the laminates manufactured under different trade names (Formica, Consoweld, Textolite, etc.).

Tight-grained wood such as maple and some birch and more open wood such as oak and pines of different varieties are used for doors, fronts (including drawer fronts), and exposed sides of wood cabinets. For a warm "feel," the wood is likely to be finished with a stain. In some cases the wood is finished with synthetic enamel paint and, depending on colors used in the kitchen, this may be dramatic. Often the wood is painted with a conventional semigloss paint.

Good quality all-steel cabinets have a coating of rust-preventative material and a baked-on enamel finish.

Functional Criteria for Kitchen Cabinets

The places where cabinets will be installed, as well as construction characteristics, should determine which cabinets are most desirable. Generally base, wall, and midway cabinets will be used in conjunction with one of the three major kitchen appliances (refrigerator, range, sink) to prepare, cook, and serve foods. Kitchen cabinets located near a major appliance should have convenient storage for articles used with that appliance. Cabinets used in the mix center should have

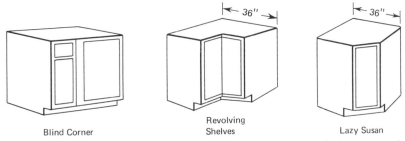

Blind Corner

Revolving Shelves

Lazy Susan

Figure 6-15. Types of base corner cabinets. Revolving-shelf type has shelves mounted on two panels that form front corner and is opened by pushing on one of the panels. "Lazy Susan" has a door. (Curtis Companies, Inc.)

storage space that is useful in this center.

Since the tops of base cabinets usually are the counters near the major appliances, enough base cabinets should be provided to meet counter requirements near the appliances.

Four criteria are useful in deciding which cabinets should be installed in different locations. One criterion is ease of seeing articles that logically will be stored in the cabinet. Another is ease of grasping articles stored in the cabinet. A third is suitability for convenient storage of most frequently used articles in the top part of base cabinets and bottom part of wall cabinets. For convenient storage, like articles, such as plates of the same size, are stacked and unlike articles, such as plates or bowls of different sizes, are separated. The fourth criterion is suitability for storage of less frequently used articles in the bottom part of base cabinets and the top part of wall cabinets. Examples of less frequently used articles might be large roasters and extra or "company" dishes.

Dimensions

Standard base cabinets usually are 34 1/2 to 34 3/4 inches high, and the counter with its backing is approximately 1 1/4 to 1 1/2 inches thick, thus making the height to working top of counter 36 inches. Factory-made cabinets that are 29 1/2 inches high sometimes are used for special purposes (such as for a mix area). Other heights might be available from some manufacturers, even for "standard" cabinets.

Depth (front to back) of base cabinets is usually 24 or 24 1/4 inches, and depth of counter including overhang is usually 24 to 25 1/2 inches.

Widths of cupboard-type base cabinets range from 9 to 48 inches and wider in steps of 3 inches, although any one manufacturer will generally not supply every width. The 9- and 12-inch widths are for specialized use, such as tray storage or limited utensil storage. Widths of standard drawer cabinets range from 15 to 30 inches. The widest cabinet that has features wanted and will fit into a space supplies the maximum storage. For example, a 48-inch cabinet with two side-by side top drawers plus the usual shelf and floor is likely to be a better choice than two 24-inch cabinets or one 15-inch cabinet and one 33-inch cabinet, unless special considerations are involved such as need for a 24-inch multiple-drawer cabinet.

Base and wall unit fillers (wood or metal strips to match the cabinets) up to 6 inches wide are used to make cabinets match kitchen dimensions. Corner base unit fillers take care of the problem caused by counter overhang and door pulls at corners.

Standard wall cabinets also are made on the 3-inch module, with 12 inches likely to be the narrowest one. Usually they are 30 inches high, but heights of 15, 18, 21, 24, 33, and 36 inches are available. The shorter models are used over a refrigerator or sink and sometimes over a range hood. Depth to the front of the door exclusive of handles generally is 13 inches. A greater depth, 13 1/2 or 14 inches, has advantages and is available from some manufacturers.

Metal midway cabinets 24 or 30 inches wide, 9 or 10 inches high, about 7 inches deep at the bottom, slanting to 9 to 11 inches deep at the top have been available (Fig. 6-12). Wood midways are custom made in any width and are more likely to have a constant depth than to slant forward (Fig. 6-13).

Doors, Drawers, and Shelves

Sliding glass or translucent plastic doors are usual for midway cabinets. Single doors hinged on the right or left are used with base cabinets up to 21 inches wide and wall cabinets up to 18 inches wide. Base cabinets that are 24 inches wide may have one or two doors; wider ones have two.

Wall cabinets 21 inches or wider usually have two doors.

Doors need not have exposed handles or pulls; instead they may have a grasping place on one edge for the fingers. When pulls are provided, they are most convenient when mounted on the bottom third of doors of wall cabinets and the top third of doors of base cabinets.

The doors of good quality steel cabinets have acoustical insulation to minimize noise due to vibration when they are closed.

Full-height (34 1/2 inches high) drawer base cabinets usually have three or four drawers; less than three or more than four drawers have been available on "standard factory-made" cabinets and, of course, can be provided on custom-built ones. For maximum usefulness, drawers should have inside clear heights appropriate for articles to be stored in them. Research in several institutions indicates that kitchen drawer cabinets with four drawers of unequal heights are very useful.

Better quality cabinets use drawer slides to give smooth rolling drawers. These consist of rollers made of nylon or other plastic and treated or plastic channels and guide strips.

Some drawers are needed and standard base cabinets with drawer at top is one way of getting them. Cupboard cabinets usually but not always have one drawer. Sometimes more functional storage is achieved by omitting a drawer in a particular location. Corner cabinets of the "lazy Susan" type or the rotating-shelf type (Fig. 6-15) are likely not to have a useful drawer. Also a drawer in a 12-inch cabinet supplies minimal clear space for storage.

Wood base and wall cabinets usually have wood floors and shelves. Metal base and wall cabinets have a solid metal floor and shelf or an open wire shelf. The latter sometimes is covered with Plastisol (a plastic material). The wire shelf is liked for its "see-through" quality, especially by some short people.

The shelf in base cupboards usually is at a fixed height, but some manufacturers supply supports so that the shelf can be installed at different heights. In research on kitchen space at the University of Minnesota it was found that one full depth and one partial depth shelf were useful, provided the heights of both were adjustable and the drawer above the shelves was not more than about 4 1/2 inches high.

A base cabinet shelf may be designed to pull out. This adds to the cost and the pull-out type is useful in some cases and not others. For example, one would not store a frequently used lightweight utensil on a pull-out shelf and pull the shelf forward *after* opening the door. A frequently used article should be stored as conveniently as practical and this might be near the front of the shelf so that pulling would be unnecessary. On the other hand, kitchen and table linens stored in kitchen cabinets are conveniently stored on pull-out units. Also, if a drawer is not provided, a pull-out shelf is convenient at the bottom of a mixer cabinet (cabinet with typewriter-like support that springs up).

Good quality wall cabinets 30 inches or taller usually have two shelves whose heights are adjustable. Research-based planning guides recommend two adjustable shelves for wall storage of food supplies and two or three for kitchen storage of everyday dishes.[12]

Corner Cabinets

Often the most reasonable use of a base corner in an L or U kitchen is to leave it "dead," that is, not to have cabinets extend into it. Even though storage is lost in the corner, work surface is not if an L- or U-shaped counter covers the corner on adjacent walls.

One method of using a base corner for storage is a 45-inch wide "blind" cabinet on one of the two adjacent walls. A door and a drawer are provided for 18 or 21 inches and access to the corner is through the door under the drawer. (The door should be hinged on the corner side.) A second method is to use a cabinet with rotating shelves. Good access to stored articles is provided when the rotating-shelf cabinet uses 36 or 37 inches on each wall. That is, convenient use of the approximately 24-by-24 inch corner involves using 12 or 13 extra inches of frontage on each of two walls. A third method of using the corner is to have access to it from an opening outside the kitchen. The space logically then is used for storage of nonkitchen articles.

A wall corner may have a lazy Susan cabinet that uses approximately 24 inches on each wall; a fixed-diagonal shelf cabinet also using 24 inches on each wall; a "blind" type that uses 12 or 13 inches on one wall and 27 inches on the wall with the access door; or a 13-by-13 inch dead space.

CRITERIA FOR SURFACE COVERINGS AND CEILINGS

Kitchen floor, counter and wall coverings, and ceiling should be easy to maintain, durable, pleasing in appearance, and reasonable in cost, including installation cost. In addition, specific functional characteristics may be stated for the coverings and ceiling. Some of the materials noted below have been used for many years, others have not. In our changing world we expect new materials to be introduced. An informed consumer will seek evidence that the materials meet the functional and aesthetic requirements for their proposed use.

Resilient Floor Coverings

Specific functional characteristics desired for floor coverings include grease resistance, alkali resistance, resilience,[13] good retention of color, washability with water and detergent or organic solvent, and for some materials, ability to hold a wax-type finish.

Different types of *resilient* coverings for kitchen floors that meet other functional characteristics are available. In current use are vinyl, vinyl-abestos, "seamless resilient" vinyl, inlaid linoleum, asphalt, rubber, cork (Figs. 6-16 through 6-19). Except for the seamless vinyl, gauges and sizes of the resilient floor coverings are standard.

Vinyl and inlaid linoleum are available in sheet or tile form. Six-foot sheet widths are usual. An advantage of the sheet form that often is overlooked is that it can be laid in unusual *curved* designs. Tile sizes for vinyl, linoleum, vinyl-asbestos, asphalt, rubber, and cork are usually 9 by 9 inches or 12 by 12 inches, though some other sizes are available for some materials

The Construction Lending Guide gives a table in which functional characteristics of the resilient floor materials, except seamless vinyl, are given relative numerical ratings.[14] For materials which may be used

[13] Resilience relates to indentation resistance, quietness, and underfoot comfort. The last factor is the instantaneous yielding of the covering to the impact of the foot and the return of the covering to its original position. The 1955 Small Homes Council Circular Series F4.6 on Floor Materials refers to National Bureau of Standards tests that show that for all practical purposes affecting human fatigue, there is no more "give" under the heel or foot in wood, asphalt tile, rubber tile, or inlaid linoleum than there is in concrete.

[14] Schmidt, Lewis, and Olin, *Construction Lending Guide.* New York: American Savings and Loan Institute and McGraw-Hill, 1966, pp. 220–227.

Figure 6-16. Sheet vinyl floor covering for kitchen and dining area. (Photograph courtesy of Armstrong Cork Company)

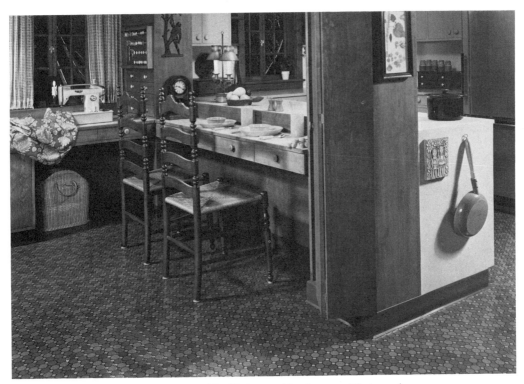

Figure 6-17. Embossed inlaid linoleum used for kitchen and family room. (Photograph courtesy of Armstrong Cork Company)

Figure 6-18. (a) Vinyl asbestos tiles in different patterns. (Photograph courtesy of Armstrong Cork Company) (b) Kitchen floor of vinyl asbestos tile. (Photograph courtesy of Armstrong Cork Company)

Figure 6-19. Floor of vinyl wear surface bonded to undercushion of foam latex. (Photograph courtesy of Armstrong Cork Company)

below grade, on grade, or suspended (above grade including a surface over a ventilated crawl space) solid vinyl tile, asbestos-backed vinyl tile, and vinyl sheet have the highest rating for resistance to grease and alkalis and for durability; next to the highest rating for ease of maintenance; intermediate ratings for resistance to indentation and for quietness and resilience. Vinyl-asbestos tile has next to the highest rating for durability, ease of maintenance, resistance to grease and alkalis and next to the lowest rating for resilience and quietness. Asphalt tile, except for its resistance to alkalis and its durability, has intermediate or low ratings. The other material that can be used below grade, on grade, or suspended is rubber tile. This is given next to the highest rating for durability, quietness, and resilience and intermediate ratings for other characteristics.

Linoleum sheet and tile are considered for use above grade or suspended only. They are ranked highest, 1, in resistance to grease. Linoleum sheet is ranked next to highest, 2, in ease of maintenance and linoleum tile is ranked 3; both are ranked 3 in durability; and both have lower ratings for the other characteristics.

In addition to the backing on the material, a lining felt is placed on the floor before linoleum or vinyl sheet is installed (Fig. 6-20). This makes the floor "quieter"; also, "picking up" the linoleum or vinyl sheet is easier when new material is installed. Good installation practice uses an underlay of *hardboard* or plywood. The hardboard or plywood is nailed with special type nails to the wood subfloor in new installations or to the wood finish flooring in an existing home. Hardwood sometimes is not used in existing houses when the top or finish flooring is firm and smooth.

Cork tile and vinyl cork tile are both considered useful for on grade or suspended installation. Their numerical ratings are different; vinyl cork tile has a relative rating of 1 (highest) for resistance to grease and cork tile 4 on a scale of 1 to

7, but cork tile rates 1 for quietness and resilience while vinyl cork tile rates 3.

The "seamless-resilient" covering is a poured covering "consisting of a two-component polyurethane liquid binder plus colored chips which are broadcast into the system during application forming the colored decorative portion of the finished installation."[15]

Consumer Bulletin in a preliminary report observes that this new type of surfacing material is perhaps more accurately described as a "permanent" floor covering.[16] "It can be installed over almost any smooth floor which is clean and free from water and grease." Disadvantages noted in the *Consumer Bulletin* report are "that its hard surface transmits impact sounds readily . . . and it is noisy." Advantages are relative ease of maintenance and, for some users, its slightly irregular surface.

Kitchen Carpeting

The kitchen carpeting used at present is made of polypropylene—a man-made olefin fiber. Hercules, Inc., introduced the

[15] "Torginal Seamless Resilient Flooring," A.I.A. File No. 23-G.

[16] "Poured on Canned Floors," *Consumer Bulletin*, February 1968, p. 4.

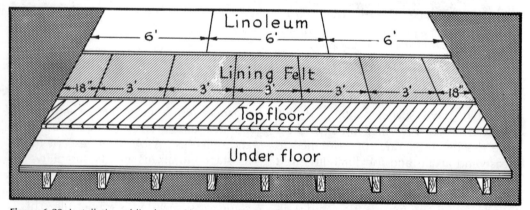

Figure 6-20. Installation of linoleum. (Armstrong Cork Company)

Figure 6-21. Kitchen with indoor/outdoor carpet of "Alfresco" by Magee Carpet Company. Surface is a low, dense, nonabsorbent pile and the secondary backing is embossed foam. Surface pile and backing are mildew resistant. (Herculon Olefin Fiber)

fiber under the name Herculon for *outdoor* carpeting. Currently, several companies manufacture indoor-outdoor carpets and carpeting for kitchen use only. Brand names for kitchen carpeting include Leisure Turf, Alfresco, Omnibus, Vectra (Figs. 6-21 and 6-22).

A report from the Good Housekeeping Institute in July 1969 states that it found "a significant variation in performance among the many indoor-outdoor carpets."[17] The report notes further that:

1. Mildew and odor will develop unless the backing is of synthetic material or a special rubber that won't absorb mois-

[17] "Institute Reports," *Good Housekeeping*, July 1969, pp. 6–7.

Figure 6-22. Carpeting for dining and kitchen area are "Omnibus" by Mohawk. (Herculon Olefin Fiber)

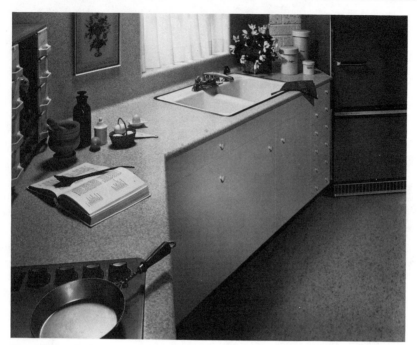

Figure 6-23. Vinyl counter with roll-type or no-drip edge and wall flashing. (Photograph by makers of Armstrong counter Corlon)

ture. (To avoid mildew the backing must not retain a liquid spill.)

2. The carpeting will be relatively easy to clean if the surface is dense and one-level so that crumbs and other dirt are not too difficult to dislodge. (A suitable surface is feltlike or tufted with tightly packed, low-level loops.)
3. The carpeting will be resistant to most stains if spills are handled promptly.

Counter Coverings

Specific functional characteristics for counters include resistance to staining by materials such as lemons that are regularly used in kitchens, washability with water and detergent, resistance to penetration of water if the counter is near the sink, enough resilience to minimize noise and breakage of dishes and glassware, reasonable resistance to heat, and good color retention.

Materials widely used in the United States now are laminated plastic (Formica, Textolite, Consoweld, Micarta, and others), vinyl, and inlaid linoleum. Materials less commonly used are stainless steel, ceramic tile, wood, and marble.

Of the materials listed, linoleum is the most likely to deteriorate from water splashing. However, for any material used near the sink, seams and edges need to be tight and waterproof or water will get under the material and loosen the adhesive bond.

Generally a back splash 3 or 4 inches high is used, but some homeowners like a back splash or *flashing* that extends from counter top to bottom of wall cabinet.

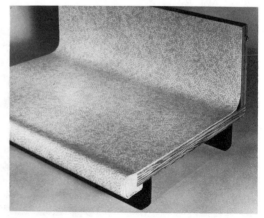

Figure 6-24. Cross section of vinyl counter. (Photograph by makers of Armstrong counter Corlon)

Counters of plastic laminate and vinyl either have a no-seam coved back splash and raised "no-drip" edges and sides (Figs. 6-23 and 6-24), or a stainless steel molding at the seam between flat part and back splash and at the edges and sides.

A counter or counter inset of stainless steel or Pyroceram adjacent to the range is useful for holding hot utensils. The wood counter or inset sometimes recommended for use near the sink is likely to offer more maintenance problems than a separate wood cutting board.

Wall Coverings

Kitchen wall coverings should be resistant to moisture and grease and should reflect light somewhat diffusely. Semigloss washable enamels and paints are much used—the enamel on woodwork and the paint on walls.

Wallpaper may be used to provide color and pattern interest. From a maintenance viewpoint, wallpaper that can be washed with soap and water is likely to be more satisfactory than that which is cleaned dry. Plastic-coated wallpaper is described as washable and sunfast, and stains do not penetrate the surface.

Plastic laminates, steel, and tiles—aluminum with backed-on enamel, ceramic, plastic—are sometimes used. Generally, the steel, tiles, or plastic laminates are installed as a flashing near the sink and/or cooktop.

Still other "coverings" or wall materials are possible and used—pegboard, plastic finish hardboard, brick or stone masonry, and glass blocks.

Ceilings

Traditionally, kitchen ceilings have been flat and covered with a paint that had a high reflectance, of the order of 60 to 90 percent. It seems reasonable to require that beamed ceilings, when used, should be finished so that some of the light directed at them will be reflected back into the room.

Suspended ceilings are used to give the effect of a totally illuminated (lighted) ceiling, to lower old high ceilings, and/or to cover unfinished ceilings. They usually consist of ceiling panels installed in a suspended framework. Translucent panels are used for a "lighted" ceiling and acoustical panels for an acoustical ceiling.

More commonly, acoustical coverings installed directly on the existing ceiling are used. Acoustical coverings are desirable where the kitchen is open to a dining area, family area, or other living room area. The rationale of this is obvious if the food waste disposer, range hood fan, dishwasher, or other appliance is noisy, although low noise level should be one selection criterion for home equipment. It is, of course, still true that an acoustical ceiling in the kitchen makes the rest of the house quieter when the homemaker or other members of the family are making noise in the kitchen.

Acoustical tiles of mineral fiber prefinished with white paint are available. Supports for ceiling lighting fixtures extend through the tiles. If range hoods and duct work involve the ceiling, the tiles would be cut to fit snugly around the pipes or ducts.

THE KITCHEN AS A WHOLE

This section summarizes some characteristics of the kitchen considered as a unit in a family residence.

Location and Size

A convenient location has easy access to the front door for callers and to the service door and garage for delivery of supplies, removal of waste, and getting into the yard. Preferably the passage to the remainder of the house from the outside does not cross the work triangle or, if it does, an alternate route also is provided.

Kitchens in the United States commonly are at the back or toward the center of single family residences. However, many very good house plans for houses with an area less than 1,200 to 1,400 square feet have the kitchen near the front of the house.

Kitchen sizes suggested by architects might be as follows: For a minimum area (800 square feet) two-bedroom house, the kitchen might be 8 by 10 feet; for a 994 square foot, two-bedroom house, the kitchen might be 9 by 12 with a minimum width of 8 feet; and for a 1,470 square foot, three-bedroom house, the kitchen might be 9 by 14 with a minimum width of 8 feet.[18] These, of course, are nominal room dimensions and are guidelines. The shape, special considerations and relationship to other plan elements cannot be codified but are just as important as the dimensions.

Figure 1-15 in Chapter 1 used a 9-by-14-foot U plan with sink-dishwasher center on the long wall and entrances opposite each other on the short walls. In another house plan the same dimensions, 9 by 14 feet, would have a corridor kitchen *without* traffic across the passageway if the house plan were such that the two entries to the kitchen were at the ends of the two parallel 14-foot walls.

Also, if the house plan were such that a 9-by-14-foot L kitchen could be achieved with one entry on the 14-foot wall, again an excellent kitchen in terms of counter and storage space requirement and floor arrangement might be possible.

A careful analysis of house plans often reveals that the U kitchen supplied is too small. An adequate kitchen in terms of

[18] Schmidt, Lewis, and Olin, op. cit., vol. II, design section, p. 47.

Figure 6-25. Hood fan for wall installation. Model is available in widths of 30 inches, 36 inches, 42 inches, and 48 inches, and in several finishes. The hood shown is made of stainless steel. It has twin blower power units with automatic reset thermal-overload protected motor. Manufacturer's rating is 300 cfm (cubic feet per minute) with a sound level range from high to low of 4.3 to 1.9 sones. Twin aluminum mesh grease filters are provided. Speed and sound controls are solid state. (NuTone Division of Scovill Manufacturing Company)

counter and storage space, and a good plan, would be an interrupted L or U or a corridor kitchen rather than an uninterrupted U.

Light and Ventilation

Windows usually are liked in the working area and/or dining area part of the room both for light and ventilation. The relevant Federal Housing Administration (FHA) standards do not require windows in the kitchen for daylight, but do require artificial light distributed so as to provide effective illumination of work area and dining area.[19] Also, ventilation shall be pro-

[19] U.S. Department of Housing and Urban Development, Federal Housing Administration, *Minimum Property Standards for One and Two Living Units,* November 1966, sec. 603.

Figure 6-27. Roof-mounted hood fan. A roof-mounted fan eliminates noise of a fan in kitchen. Model shown has a cfm rating of 1,000. (NuTone Division of Scovill Manufacturing Company)

vided either by natural means in amounts required for habitable rooms, or by mechanical means. Natural ventilation is provided through openable windows, skylights, or other suitable openings in exterior walls or roofs. Net area "shall be not less than 4 percent of the floor area of the room or space."

Mechanical ventilation is achieved by a range hood or by a ceiling or wall fan.[20] If a fan is used, it should have sufficient capacity in cubic feet of air per minute (cfm) to provide a minimum of 15 air changes per hour in the space occupied by the kitchen.

$$cfm = \frac{cu \text{ ft of room volume} \times 15}{60}$$

If a range hood is used, it shall be at least as wide as the range, at least 17 inches deep (front to back), and the bottom of the hood shall not be more than 30 inches above the range (Figs. 6-25 through 6-27).

[20] Ibid., sec. 1002.

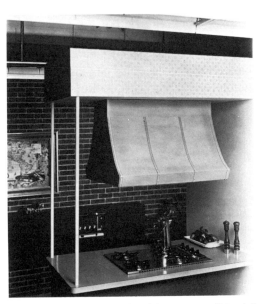

Figure 6-26. Canopy-type hood for use over "islands" and pass-through counters. (NuTone Division of Scovill Manufacturing Company)

The range hood fan (or blower) shall have a minimum capacity of 40 cfm per lineal foot of hood length except that when installed over a range located in an island or peninsula, minimum capacity shall be 50 cfm per lineal foot of hood length. Further, if a range hood is installed in a kitchen not provided with adequate natural ventilation, the hood fan capacity shall be calculated from the range width and from the requirement of 15 air changes per hour and the larger value supplied.

The calculations for the cfm capacity of the range hood fan of a 12-by-14-foot kitchen, with an 8-foot ceiling and a 30-inch (2 1/2 foot) wide range located on a wall are as follows:

cfm (from range width) = 40 × 2.5 = 100

cfm (from 15 air changes per hour)

$$= \frac{12 \times 14 \times 8 \times 15}{60} = 336$$

The hood fan capacity to be supplied therefore is 336 cfm.

Compliance with the requirement is shown by a Home Ventilating Institute or manufacturer's label, showing capacity under specified test conditions (Fig. 6-28).

Figure 6-28. Label of the Home Ventilating Institute. Label may state number of cfm moved by fan. (Home Ventilating Institute)

Safety and Relevance to Family Composition and Living Style

Safety, as far as possible, is built in. A range or cooktop is not located near a window curtain that can blow against a hot utensil or open flame. Wall cabinets have a counter, base cabinet, or in some instances an appliance such as a dishwasher below them. Doors on base and wall cabinets have good catches. Clearances are adequate. Doors of cabinets or appliances do not open into passageways. Rounded rather than sharp corners are used to the extent possible. Wax used on the floor or the flooring material itself is of a nonslip variety. A step stool and space for it are provided when some cabinet shelves are too high to be reached conveniently from the floor. Appliances selected for homes where small children will be in the kitchen have controls not readily manipulated by small children. Proper storage is provided for knives. Cords on electric appliances are replaced when worn. Appliances that "leak" electric current and cause a shock are serviced.

Appliances and materials are selected with ease of cleaning as an important consideration.

The appliances selected are appropriate for the family's living style *and* for the kitchen space.

The electric wiring is adequate, that is, the homemaker can use all the appliances she wants to, safely, at design voltage, at one time, and at locations that are convenient.

Total Appearance

Equipment; floor, counter, and wall coverings; ceiling; and accessories should make an interesting and somewhat unified whole and a pleasant place in which to do work and carry on other appropriate family activities.

The color scheme logically will take into account physical aspects of the space—location, size, exposure, amount of light.[21] Good kitchen accessories are well-designed ones that add accents of color and are personal joys to the individual who uses the kitchen most. The style of the kitchen is compatible with that of the remainder of the home. Especially in a windowless kitchen, a luminous ceiling (one that is suspended and has translucent panels) will add warmth if the light sources are well chosen.

Kitchens "Around Which" Houses Have Been Planned

The Clothing and Housing Research Division of the Agricultural Research Service of the U.S. Department of Agriculture designed and tested a kitchen with dining area that is part of a kitchen-workroom and two kitchens without dining area, one of which is "oriented to a family room with dining area."[22]

The first design is a kitchen-workroom 17 1/2 by 18 feet. Kitchen and workroom are separated by a storage island, with wall refrigerator (not currently manufactured), counter and base cabinets on the kitchen side, and planning desk and shelves on workroom side. The dining area in the kitchen has space for six persons. The workroom includes, in addition to desk and shelves in the island, laundry facilities, a food-storage pantry, a freezer, and closets for cleaning and ironing supplies and equipment.

A serving cart and a posture chair are

[21] Lorraine Allen, "Kitchen Planning Combines Efficiency and Beauty," *What's New in Home Economics*, March 1970, pp. 41–44.

[22] Mildred S. Howard, Lenore Sater Thye, and Genevieve K. Tayloe, *The Beltsville Kitchen-Workroom with Energy-Saving Features*, Home and Garden Bulletin No. 60, U.S. Department of Agriculture, 1958; *Beltsville Energy-Saving Kitchen Design No. 2*, Leaflet No. 463, U.S. Department of Agriculture, 1959; Mildred Howard, Genevieve Tayloe, and Russell Parker, *Beltsville Energy-Saving Kitchen Design No. 3*, Leaflet No. 518, U.S. Department of Agriculture, 1963.

planned as part of the kitchen equipment. The most frequently used supplies and utensils are stored between 28 and 64 inches from the floor. Undercounter knee space is provided for sitting at the mix counter and at the shallow well of the double-well sink.

Counters, dining table, and serving cart are covered with laminated plastic.

The electric wall oven is placed so that the bottom of the interior is 32 inches from the floor and the most-used rack positions are between 35 and 40 inches from the floor.

Daylight and ventilation are provided by two large windows—a broad one over the sink and counters on either side of the sink and a picture window in the dining area. In addition, a ventilating fan is located in the ceiling over the range.

Work areas are lighted by fluorescent ceiling fixtures, and the dining table is lighted with a pull-down incandescent fixture.

The second and third designs are kitchen arrangements without workroom. Design 2 is somewhat unusual, though not unique, in that two house plans have been developed "around" this kitchen by USDA architects.

One variation (plan A) of the second design is shown in Figure 6-29 (a) and (b). The center island of plan A has a free-standing refrigerator, cooktop, drawers, half-circle revolving shelves, and space for a waste paper basket. In plan B's center island the cooktop is replaced by an oven.

The sink wall for design 2 is 16 feet long. Midway-type cabinets with sliding doors at the right of the sink are used for sink supplies and other storage. The mix area is designed as a sit-down working area, with wall cabinet starting at counter level. Flour and sugar canisters are designed with vertical handles and caster-equipped platforms.

Three arrangements of design 3 have

been developed.[23] One is shown in Figure 6-30. This design uses a free-standing range. The dishwasher is on a wall at right angles to the sink. The island consists either of refrigerator with adjacent counter and cabinet or range with counter at both ends and cabinet for "pot and pan" storage. The distinctive feature of this kitchen is *slant-front, wall-hung* cabinets—one for the mix center and a narrower one for the range center. The bottom shelf of these cabinets is 5 1/2 inches deep and the top shelf is 9 1/2 inches deep. The cabinets are hung 4 inches above the counter rather than 14 or 15 inches above. Thus the wall cabinets facilitate height of reach and extent of reachover—the user reaches over less to get an article from the top shelf than one from a lower shelf.

Modules

Modules that comprise a complete kitchen, laundry, bath and the plumbing, heating and air conditioning facilities for an entire house were described in the early 1970s.[24] The modules would be factory assembled with appliances, cabinets, and fixtures and shipped to the site to become part of the house construction.

KITCHENS FOR THE HANDICAPPED

Rehabilitation personnel and home economists concerned for the handicapped have developed kitchen plans for wheelchair patients and for heart patients. In addition, many workers in kitchen planning and design have stressed, perhaps overstressed, saving physical energy.

Kitchens that will serve for a homemaker (man or woman) in a wheelchair have special space and design requirements. Research on these requirements has been

[23] Howard, Tayloe, and Parker, op. cit., pp. 2, 3.
[24] Maidee Kerr Spencer, "The Heart of the House," *American Home*, May 1971.

(a)

Figure 6-29. (a) Beltsville "energy-saving kitchen," variation A, design no. 2. (U.S. Department of Agriculture Leaflet No. 463) (b) Center island of variation A. (U.S. Department of Agriculture)

reported by McCullough and Farnham.[25] The Institute of Physical Medicine and Rehabilitation of the New York University Medical Center has a well-known on-going program for assisting severely handi-capped homemakers in relearning and in developing aids for them.[26]

The book by May, et al., is a "classic" reference for home economists for its bib-liography and other materials.[27]

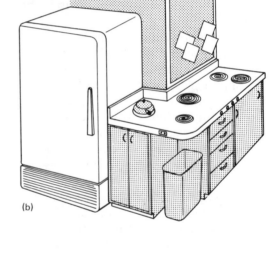

(b)

[25] Helen E. McCullough and Mary B. Farnham, *Space and Design Requirements for Wheelchair Kitchens*, Bulletin 661, University of Illinois Agricultural Experiment Station, June 1960.

McCullough and Farnham, *Kitchens for Women in Wheelchairs*, University of Illinois Extension Circular 841, November 1961.

[26] New York University Medical Center, Institute of Rehabilitation Medicine, *Kitchen in a Low-Income Housing Project Adapted for a Paraplegic Home-maker*, February 1969.

Edward W. Lowman and Howard A. Rusk, *The Help-ing Hand—Self-Help Devices*, Institute of Physical Medicine and Rehabilitation and the Arthritis Self-help Device Office, April 1963.

Virginia Hart Wheeler, *Planning Kitchens for Handi-capped Homemakers*, Institute of Physical Medicine and Rehabilitation. (Not dated but available for purchase at the Center.)

[27] E. E. May, N. R. Waggoner, and E. M. Boettke, *Homemaking for the Handicapped*. New York: Dodd, Mead, 1966.

Kitchens for persons in wheelchairs pro-vide sit-down possibilities for work at sink and mix areas *but* also significant are such functional characteristics as well-organized work centers. For *non-wheelchair* persons, an observation based on a review of re-search findings is made by Steidl and Bratton: "Whether or not the [physical] energy cost for sitting to work is even slightly less than for standing to work seems to depend on how you sit. In any case, the differences are so slight and the total energy cost for tasks that can be done sitting down is so low that in a practical way there is no difference."[28]

[28] Steidl and Bratton, op. cit., p. 153.

Figure 6-30. Beltsville "energy-saving kitchen," design no. 3. (U.S. Department of Agriculture)

RANGE HOOD BUYING GUIDE

Selection Points for Ducted and Nonducted Hoods

1. Is the UL seal on the hood housing? Is it on the motor only?
2. Dimensions: What is the width relative to the range over which it will be installed? What are the other dimensions?
3. What is the capacity of fan or blower in cubic feet per minute? Is this value certified by Home Ventilating Institute and/or by manufacturer?
4. Will it be easy to clean the hood (housing), interior surfaces, filters (if provided), and other parts that need to be cleaned?
5. Will it be easy to lubricate the motor if lubrication is necessary?
6. What about the material and design of housing? How will the hood "fit into the kitchen"?
7. Is there a means provided to minimize the noise of the blower? What is the *loudness* level in *sones*? Preferably this is below 7.
8. Is there a guarantee?

Special Selection Points for Ducted Hoods

1. Is the hood designed to be ducted appropriately for the house— that is, horizontally through an exterior wall or vertically through a roof?

2. What about special cfm capacity consideration for kitchens that are not ventilated by windows, and so on? Does the cfm rating meet or exceed the FHA requirement?
3. Special quietness consideration for ducted hoods: Is it practical to mount blower and hood on an exterior wall or roof?

KITCHEN-PLANNING EXERCISES

Comments: It is helpful to draw a plan view *and* elevations. A direction, such as N, should be marked on the plan view and the elevations should be marked N, S, E, or W. The plan shows the widths and depths (front to back) of appliances and wall cabinets in the kitchen. Elevations show heights of cabinets, appliances, windows, and so on.

A scale of 1/2 inch = 1 foot is convenient and paper ruled 8 lines to the inch also is convenient. In addition, overall dimensions, such as 10 by 12 feet, should be marked on the plan view.

The doors to the kitchen should be marked to show space such as family room or hall to outside or whatever is appropriate.

Some cautions:

1. Get dimensions of appliances including sinks from specification sheets or other reliable sources.
2. For corners, note that a good corner base rotating cabinet requires 36 inches on each of the two walls making up the corner. A lazy Susan wall corner cabinet requires 24 inches on each of the two walls making up the corner.
3. A corner sink uses more space than a sink on one wall. For example, one manufacturer requires for a single-bowl sink 39 inches on one wall and 29 inches on the adjacent wall. For a double-bowl sink, the same manufacturer requires 42 inches on one wall and 32 inches on the adjacent wall.
4. A 12-inch or narrower base cabinet is expensive and suitable for limited and specialized storage *only*.
5. Remember to allow at least 12 inches standing room on each side of the range, sink, and dishwasher, if a dishwasher is included. Also, a refrigerator should not be placed in a corner next to base cabinets at right angles to it. The same is true for a freezer in the kitchen.
6. It is not necessary to record what will be stored where. It is necessary or it is desirable to provide enough storage space that usual kitchen articles can be stored in appropriate centers.
7. Have the mix center inside the work triangle; that is, arrange the mix counter and storage between two of the major appliances—

sink and range or cooktop, sink and refrigerator, or range and refrigerator. If the counter of a built-in or free-standing dishwasher is part of the mix center, extra storage space may be needed in the mix area to compensate for that lost to the dishwasher.

Exercise 1

Draw elevations for the kitchen plan shown in Figure 1-15.

Exercise 2

Draw a plan and elevations for the 9-by-14-foot corridor discussed on page 138.

Exercise 3

Draw a plan and elevations for a 9-by-14-foot L as suggested on page 138.

Exercise 4

Get dimensions and location (*within the house plan*) of a U plan for a kitchen in a "stock" or development house and draw a kitchen plan and elevations for the U. Evaluate the plan.

Exercise 5

Use the suggested minimum dimensions of 9 by 14 feet and "locate" a kitchen in a three-bedroom house. Draw a plan only and evaluate the plan.

Do the same for kitchen dimensions of your own choice.

Exercise 6

Draw a sketch to scale of your mother's kitchen, your own kitchen, or that of an understanding friend. Take an inventory of utensils, accessories, small appliances, dishes, linen stored in the kitchen, and foods exclusive of those in the refrigerator or freezer. Change the original plan, *on paper*, as necessary to provide more convenient storage for the articles in the inventory. Indicate where the articles in the inventory will be stored in the new plan.

Suggestion: The locations of the range and the refrigerator often can be changed at moderate cost. But it may be quite expensive to move a sink or dishwasher more than 4 inches or so because the drains on sinks are vented through an outside stack.

CHAPTER 7
FOOD WASTE DISPOSERS, GAS-FIRED INCINERATORS, AND TRASH COMPACTORS

Disposal of household food waste and rubbish by householders in a manner that contributes as little as possible to environmental pollution is desirable and in some communities already mandatory. Different methods and types of equipment are available. The *food waste disposer* is used with running cold water and is installed in the kitchen sink. The particles into which the waste is ground are carried by the running water into the city sewerage system or into a private septic tank. (Contractors are likely to recommend that a disposer not be installed in a house that uses a cesspool.) Some cities do not permit connection of food waste disposers to the city sewerage system, presumably because the system is not adequate to handle the extra waste water. On the other hand, some new city developments reportedly require that new residences have a food waste disposer. A prospective purchaser of a disposer or a new home will check.

Recommendations on size of septic tank into which a food waste disposer discharges can often be obtained from the State Health Department or from the State Extension Service. A 500- or 600-gallon tank often is suggested for new homes. Actually, the size needed depends on the number of persons in the household and on factors associated with some of the water-bearing equipment, such as number of bathtubs, showers, and toilets, and provision for disposal of backwash water and brine effluent from mechanical water softeners, as well as disposal of liquid waste from food waste disposers. In new construction, disposers can be counted as another family member when determining size of septic tank needed.

Gas-fired incinerators, like food waste disposers, handle most types of household food waste and in addition will burn paper. A gas incinerator must be vented to a flue or chimney and therefore might be installed anywhere in the home that is close to a chimney. Frequently, the incinerator is installed close to the furnace or water heater for homes that have a furnace and/or water heater connected to a chimney. Models that meet the September 1969 requirements of the American Gas Assocation (see the section on gas-fired incinerators in this chapter) incinerate to a specified level in terms of uncarbonized material; this level appears to be accepted as satisfactory by most communities.

The *trash compactor* is reported to compress waste to about one-fourth the original bulk. The bag of compacted waste then is disposed of as other bulk waste— in a sanitary land fill for example. (Since the waste may include metal and other difficult-to-burn materials, ultimate disposal by burning may not be practical.)

FOOD WASTE DISPOSERS

Current models of food waste disposers handle vegetable parings, fruit rinds, fruit pits, fibrous material such as corn husks, carrot tops and celery, eggshells, seafood shells, fats and greases, coffee grounds, bones approximately as large as chop bones, and other food wastes. Instead of storing these food wastes in a container for later removal from the kitchen, they are put down the drain as they occur in food preparation and after each meal. Food waste disposers are not designed to handle paper, tin cans, glass bottles, crockery, cloth tea bags, string, aluminum foil or other metal, or large bones.

CONSTRUCTION: TYPES AND PRINCIPAL PARTS

A disposer is a motor-driven shredding device in which food wastes are cut into small particles. As stated earlier, all models are designed to operate with cold water running through them. Besides flushing waste through the disposer into the drain line, the cold water solidifies fats and greases, so that they can be shredded and washed down the drain. Unsolidified fats would coat the inner walls and shredding mechanism of the disposer.

The top of the disposer is mounted in the sink, replacing the flange in the sink opening and the strainer. On some models a flexible cushion insulates the disposer from the sink and thus decreases transmission of vibration and noise to the sink. The motor is located in the housing in the lower half of the disposer. The shredding action takes place above a water seal that protects the motor. The drain outlet from the disposer is located just above the water seal.

Types

The available types of disposers are the continuous-feed type and the top-control (batch-feed) type (Figs. 7-1 through 7-4).

The continuous-feed type can be used in two ways. One is to scrape or feed waste into the disposer as it operates. The other

Figure 7-1. Exterior views of three models of continuous-feed food waste disposers. Hush model on left is described as having a quick hush mount that permits easy installation; jam-free turbine-quality hardened stainless steel swivel impellers; hardened stainless steel turntable; precision-machined, file-hard tool steel grind ring; surgical steel undercutter blade. Hush master on right has in addition complete sound conditioning and deep removable splash guard. Super hush in center is described as having dual-sound conditioning and a dynamically balanced hardened stainless steel turntable with heavy flywheel, as well as characteristics of hush master. (Waste King Universal)

(a)   (b)

Figure 7-2. Interior of disposer mechanism shown in Figure 7-1. (a) Turntable, impellers, grind ring (on interior of housing), upper part of motor housing. (b) Pictorial showing what waste may be added. (Waste King Universal)

Figure 7-3. Exterior views of four models of top-control food waste disposers. (The Maytag Company)

is to add waste before the disposer is turned on. The top-control type must be loaded before it is used because the starting switch is in the top control part of the disposer. Waste is fed in and the top control is positioned to automatically close the motor circuit and thereby start operation.

When the top-control disposer is not in use, its top serves as a sink stopper. Since the top of the continuous-feed type does not serve as a stopper, a separate sink stopper or plastic disk is needed.

Dimensions and Capacities

Waste disposers are cylindrical or rectangular in shape and have overall dimensions from approximately 7 inches wide by 13 inches long to 11 inches wide by 16 inches long. (The width is to the outside of the

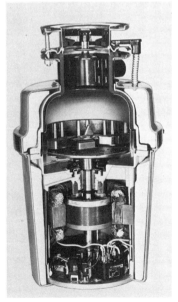

Figure 7-4. Interiors of disposer mechanisms shown in Figure 7-3. Features include swivel-mounted impeller arms and a positive-pressure water seal. Polyurethane impeller cushions (left photo) absorb noise from impeller arms as they rebound from hitting hard objects. Spring-loaded water seal (below grinding chamber in right photo) is designed to protect against water leaks to motor section. (The Maytag Company)

drain pipe.) The specifications for some models include a volume capacity or liquid-plus-waste capacity. One-quart and 2-quart capacity models are available.

Shredding Mechanisms

The shredding mechanisms in the two types of disposers are similar in action. Both utilize blades and/or shredders, usually mounted on the inside of the hopper on a part designated as the shredder ring, and impellers mounted on a turntable (flywheel) located on one end of the motor shaft. As the turntable rotates, food waste is forced against the cutting and/or shredding edges. Some manufacturers use fixed impellers or lugs on the turntable; others use hammermill impellers (small hammers attached so they can swivel and thus be "jam-free"). In disposers with rigid impellers food waste can more easily become lodged between the impeller and the cutting teeth and then the action of the disposer is stopped.

In some disposers food waste cannot be discharged into the drain line until water is introduced to carry it away.

The shell or outside housing of the disposer may be made of propylene; the hopper is lined with or made of stainless steel or other corrosion-resistant material; and in the better models the space between hopper and shell is filled with a sound insulating material. Certainly it is wise to choose disposers that are sound insulated.

Electrical Characteristics of the Motor

The motor of the better models is usually a 1/2 horsepower capacitor or induction type designed for use with 110- to 120-volt, 60-cycle alternating current. Some use a 1/3 horsepower motor and the least expensive or "competitive" models may use 1/4 horsepower.

Most models have a thermal overload protective device built in. If provided, the reset device that starts the motor again, after the cause of a stoppage has been corrected, is commonly a button that is pushed by the user.

Controls

Besides the disposer reset mechanism, an on-off switch is provided for continuous-feed models. This may be installed on a wall or elsewhere. The top control models have the on-off switch *inside* the disposer where the top contacts it. It is possible to design a single switch for motor and water.

INSTALLATION

Local electrical and plumbing codes should be followed when installing a disposer. For the electrical part, manufacturers recommend an individual-equipment circuit fused with a 15-ampere fuse for a disposer only, or an individual equipment circuit fused with a 20-ampere fuse for a disposer and electric dishwasher.

Manufacturers' specific recommendations on plumbing vary according to whether the disposer is to be used with a one-bowl or well sink, one well of a two-well sink, or a combination sink and dishwasher. In all cases, if the local plumbing code permits, any grease trap in the kitchen sink waste line should be removed before installing the disposer.

In general, a P trap or a double trap is used on the outlet side of the disposer if the outlet drain goes into a wall, and an S trap is used if it goes into the floor (Fig. 7-5). (A special trap is sometimes required by a community or a state.) A trap with special fittings that lengthen the S sometimes is suggested to serve for the disposer and the drain of the second well of a two-well sink. However, separate traps are better.

Separate drains and traps may be used for the disposer and the dishwasher. A less expensive and less desirable installation is to have the dishwasher drain through the disposer; current models of disposers usually have a drain connection for a dishwasher. If this type of installation is used, an air-gap or antisiphon assembly should be included in the drain line between dishwasher and disposer to prevent possible back siphonage of waste water into the dishwasher. Also the disposer should be empty when the dishwasher is in operation; otherwise waste in the disposer may interfere with rapid flow of water from the dishwasher.

USE

For the top-control model, drop or scrape waste into the hopper, turn on the cold water, and position the top control to turn on the switch. A grinding sound indicates that waste is being shredded. When the grinding sound ceases, wait a few seconds, turn off the switch, and then turn off the cold water.

For continuous feed into a continuous-feed model, turn on the cold water, turn on the switch, and drop or scrape waste into the hopper. Scrape waste directly into the unit while it is operating. When the waste is ground, turn off the switch and turn off the cold water.

The rate at which cold water should flow varies for different models; 2 gallons per minute or approximately 1 quart in 8 seconds is typical, but some models use more.

Do not stuff or pack waste tightly into the hopper. In addition, if the manufacturer's booklet so specifies, cut fibrous waste such as carrot tops into short lengths. A full load of waste may be disposed of in 5

Figure 7-5. Representative plumbing installations for a food waste disposer. (General Electric Company)

to 30 seconds; harder materials take longer. If convenient to do so, mix soft waste with hard.

Under normal conditions, disposers require no regular attention. No drain-cleaning chemicals should ever be used, because the corrosive action of drain cleaners might damage the interior of the disposer. To clean the appliance, fill the sink with cold water and, with the cold water running, operate the disposer until the sink is empty.

An obstruction may cause a disposer to jam or stall. Know the recommendation of the manufacturer for correct procedure in case the disposer jams. In general, if the disposer jams due to an obstruction such as a fork or a paring knife, reach first for the cold water faucet and next for the switch. Then reach into the disposer for the fork or knife. In other words, protect your hands first by having the disposer *positively off* before you retrieve the fork or knife.

GAS-FIRED INCINERATORS

As indicated earlier, gas disposers are designed to burn household waste. Different types of waste (not food waste only) are defined. The American National Standard on incinerators defines two household types:

Waste Material, Type 1. Rubbish, consisting of dry combustible waste with a moisture content not exceeding 25 percent by weight. It would be made up of material such as paper, rags, scrap wood, cartons, floor sweepings. The 25 percent moisture represents normal moisture content in dry waste material.

Waste Material, Type 2. Refuse consisting of an approximate even mixture of rubbish and garbage by weight. This type of waste is common to apartment and residential occupancy.[1]

The standard applies to models used in fixed installations with natural, manufactured, mixed, and liquified petroleum (LP gas) and for models installed in mobile homes or travel trailers when used with LP gas. Models for installation inside a house are connected to a Fire Underwriters' Class A chimney flue that serves a central gas-fired heating system or water heater. Those for mobile homes may utilize a special vent terminal that extends through the roof.

An incinerator used daily and/or one that has a constantly burning pilot aids in keeping dry a chimney connected to a gas-fired appliance.

CONSTRUCTION CHARACTERISTICS AND OPERATING COMPONENTS

A gas incinerator is usually approximately 36 inches high, exclusive of flue connection and 18 to 20 inches square. The interior of the combustion chamber is low-carbon steel, ceramic, firebrick, or other corrosion resistant material. The exterior surface of the appliance, except for lid or door, is likely to be baked enamel or porcelain. Thermal insulation may be used between combustion chamber and exterior or a firebrick lining may be used in the combustion chamber. The waste capacity of the combustion chamber varies between approximately 1.4 and 2 bushels.

Waste, wrapped in paper if damp, is

[1] American National Standard, formerly United States of America Standard, *Domestic Gas-Fired Incinerators,* Z21.6–1969, September 1969, part III, Definitions. (The sponsor for this standard is the American Gas Association.)

dumped directly into the combustion chamber. (Some materials—glass, tin cans, foil—will not burn in a domestic incinerator.) In models that meet the current standard, a secondary compartment is provided in addition to the main combustion chamber. As the waste burns, odor and smoke which circulate due to convection and air entering the appliance are consumed by the secondary flame, thus giving a smokeless-odorless type incinerator.

A grate at the bottom of the combustion chamber is rotated by a handle on the exterior. Below the grate is the removable compartment into which the ashes fall when the exterior handle is operated. "Dry" waste such as a newspaper gives the greatest amount of ash (Figs. 7-6 and 7-7).

The standard cited earlier specifies the following performance requirement for incinerating effectiveness.[2] "An incinerator shall be capable of completely incinerating domestic wastes as typified by a mixture of 40 percent dry combustibles and 60 percent food refuse. This provision shall be deemed met if the uncarbonized material at the end of the test does not exceed 6 ounces per bushel of incinerator capacity."

The Btu's per hour input rating may relate to bushel capacity and the heat

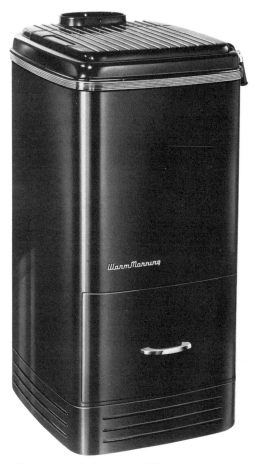

Figure 7-6. Exterior view of 2-bushel-capacity gas-fired incinerator. (Locke Stove Company)

requirement in the secondary or afterburner section. One two-bushel model has an input rating of 50,000 Btu's. A 1.4-bushel model by the same manufacturer has an input rating of 45,000 Btu's.

TRASH COMPACTORS

The appliance was introduced to the public in 1970. It measures about 15 inches wide by 25 inches deep and is 34 3/4 inches high to fit under a standard 36-inch high kitchen counter. Trash is dropped into the bag of the pull-out compartment or waste drawer; the compartment then is closed and a spray of deodorizer is injected

automatically into the drawer. To compress the refuse the operator turns a key and pushes a button. This causes a ram powered by a 1/3 horsepower motor to compress the refuse under very high pressure (2,000 pounds for some models). The compressed material is held in the deodorized bag which is polyethylene lined and coated for strength and moisture resistance (Figs. 7-8 and 7-9).

[2] Ibid., sec. 2.10.

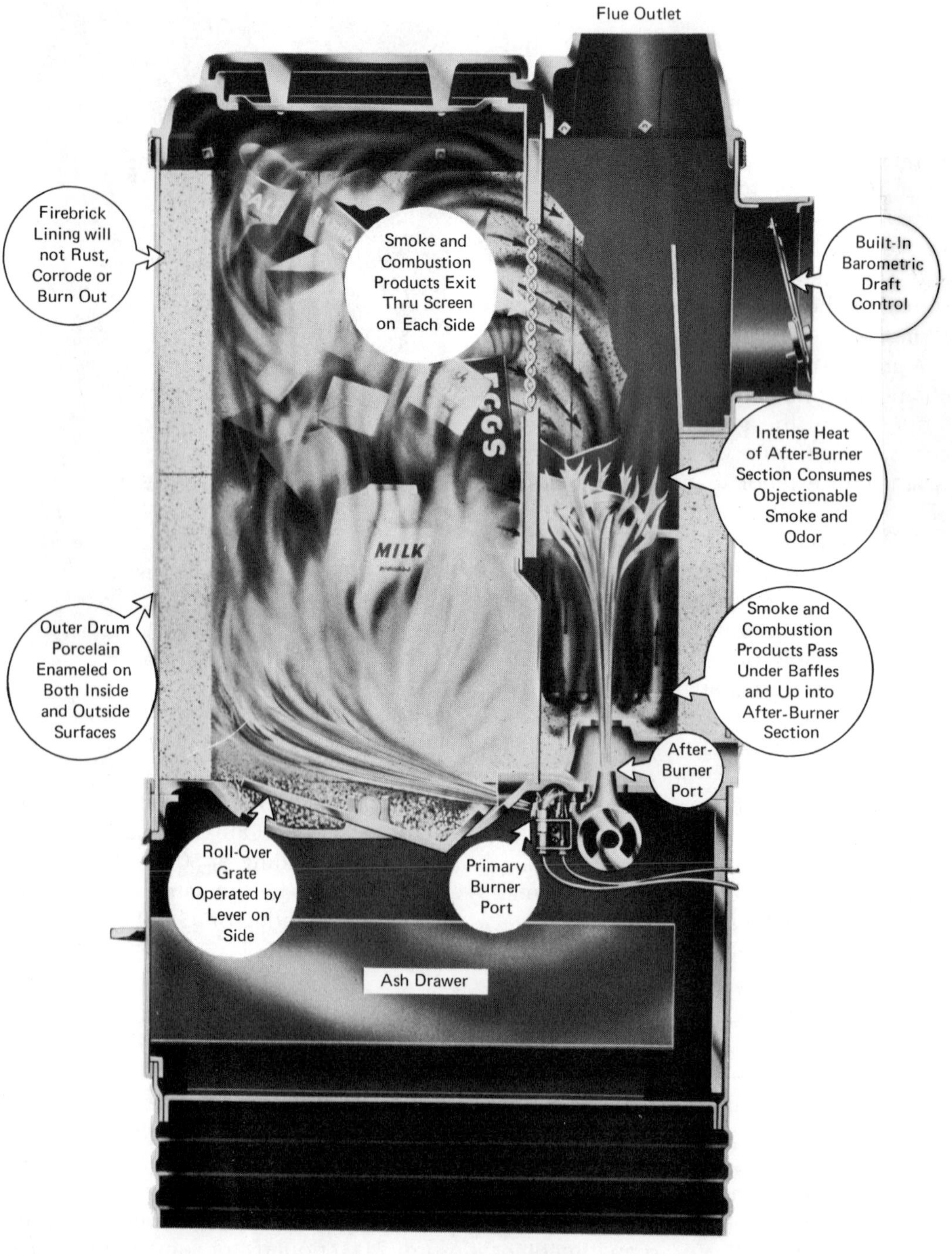

Flue Outlet

Firebrick Lining will not Rust, Corrode or Burn Out

Smoke and Combustion Products Exit Thru Screen on Each Side

Built-In Barometric Draft Control

Intense Heat of After-Burner Section Consumes Objectionable Smoke and Odor

Outer Drum Porcelain Enameled on Both Inside and Outside Surfaces

Smoke and Combustion Products Pass Under Baffles and Up into After-Burner Section

After-Burner Port

Roll-Over Grate Operated by Lever on Side

Primary Burner Port

Ash Drawer

Figure 7-7. Cutaway of incinerator shown in Figure 7-6. Dimensions are: height, 36 1/2 inches to top of feed door; width, 18 1/2 inches; depth, 21 1/2 inches. Main burner input is 45,000 Btu's per hour. AGA certified under latest requirements for smokeless-ordorless operation for use with natural mixed and LP gas. Factory equipped for use with type of gas specified. (Locke Stove Company)

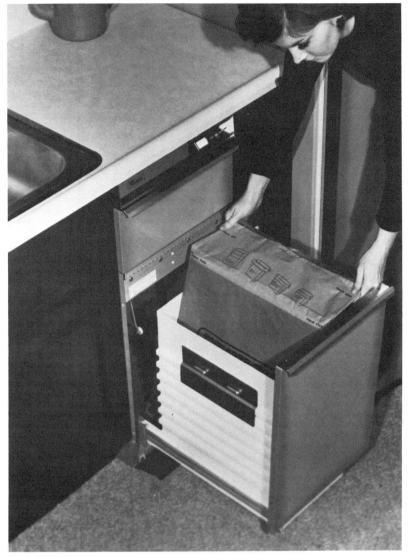

Figure 7-8. Removing bag of compressed waste from a trash compactor installed under the counter adjacent to the sink. (Whirlpool Corporation)

Figure 7-9. Controls for compactor. Total cycle time is 60 seconds. Compactor will not operate unless the safety key is in place and on. (Whirlpool Corporation)

When the bag is filled, the operator lifts it out of the compactor. An interlock shuts off the motor whenever the refuse compartment is pulled out. The bulk of the trash added to the appliance is said to be reduced by a ratio of 4 to 1.

Consumers' Research[3] does not recommend use of the machine for bottles or garbage. Consumers' Research does not indicate, on the other hand, that the appliance should not be used for cans or other refuse suggested by the manufacturer.

[3] *Consumer Bulletin Annual, 1972,* published in September 1971 by Consumers' Research, Inc.

Students in household equipment classes at the University of Minnesota who have compactors in sorority houses, homes in suburbs where garbage collection is expensive and/or not easy to arrange, and in vacation homes have reported considerable satisfaction with the appliance. Satisfaction would be expected too in a city home where much bulky refuse is a normal part of living since city collection of refuse may be once a week. (Outdoor burning may not be allowed.)

BUYING GUIDE

FOOD WASTE DISPOSERS

For this as for many other appliances, information is needed on dimensions, electrical characteristics, UL seal, construction features, and materials used. Additional information is needed on operating characteristics of this particular appliance. Finally, information on probable installation cost is relevant because for some homes, plumbing and electric wiring costs of installation may equal one-half or more of the cost of the appliance itself.

General

1. Is there a UL seal?
2. What are the dimensions in inches: length, diameter, horizontal distance from housing needed for waste drain?
3. Can the disposer be installed with its waste drain and trap connected to a wall drain, a floor drain, either a wall or floor drain?
4. Is the waste drain movable?
   (A rotating waste drain may make installation easier.)

Electrical Characteristics

1. What is the motor rating in horsepower? Amperes at rated volts?
2. Is there motor overload protection? If a manual reset button is provided, is the button likely to be accessible when the appliance is installed?
3. Is a waterflow-interlock switch available as an accessory? Is one built into the appliance?
4. What is the manufacturer's estimate of kilowatt-hours used per month?

Construction, Materials, Operating Characteristics

1. What type is the disposer—batch feed, continuous feed?
2. What is the capacity of disposer in quarts? (For batch-feed and continuous-feed type, capacity is a measure of amount of food waste that the appliance can handle at one time.)
3. If the appliance is a continuous-feed type, is a flexible shield

provided for the top opening to eliminate splashing of water? Does the shield *look* functional?

4. What materials are used for the hopper, turntable, shredder?
5. Are impellers on turntable mounted so they can move?
6. List special provisions that are designed to aid in decreasing noise of operation: insulation in housing, flexible or cushioned mounting at sink, at drain.
7. List any special features not previously noted—for example, is the turntable rotation reversed each time the disposer starts?
8. What does the warranty promise?

Installation

1. Will the dealer who sells the appliance arrange for installation?
2. What is an average installation cost in your community?

GAS-FIRED INCINERATORS

The first questions may seem negative but really they are part of the information needed by a consumer.

1. If you are planning on indoor installation, does your home have a Class A chimney and is space for the incinerator available near it, or do you have information on what will be involved in venting the incinerator in your home?
2. What are the special requirements of your local community and your state? Do all gas-fired incinerators sold in your community meet the requirements? (A negative answer limits the models you will consider.)
3. Of the different resources available to you for disposal of waste, is the incinerator the best choice? Does the convenience of the incinerator for immediate or daily disposal of most wastes "balance" accumulation of waste for garbage collection weekly, possibly twice weekly, in some communities? In making this comparison you may wish to consider cost of garbage collection versus average cost of operating an incinerator.
4. What are the exterior dimensions in inches?
5. What is the capacity in bushels and nominal pounds per hour?
6. What is the certifying agency marking on the incinerator—American Gas Association, Canadian Gas Association, or another?
7. What are the construction and operating features: interior of firebrick, steel, or other material? the exterior?
8. Is it designed to consume odors and smoke, as well as garbage and paper trash? Is this accomplished by two burning zones within the appliance, a dual-purpose burner, or other? (Even if the local community or state does not require the certified "smokeless-odorless type," neighbors and homeowner and family are likely to appreciate this characteristic.)

9. What additional features are provided, and are these important to you?
10. What does the warranty state?

TRASH COMPACTORS

1. Is there a UL seal?
2. Are its dimensions such that it will fit into a convenient location in or near the kitchen in your home? Is an electric outlet of a small-appliance circuit available at that location?
3. Does the information on use supplied by the manufacturer indicate that it will be suitable for your household?
4. What special arrangement if any do you need to make for disposing of the compressed waste?
5. What information does the manufacturer supply on average operating cost including the cost of bags and deodorizer?
6. What does the warranty promise?

CHAPTER 8
SELECTION OF
MAJOR APPLIANCES

Needs and Wants

Information Needed for Good Selection

Sources of Information

Buymanship Guides and Procedures/*Buymanship Guides; Buymanship Procedures*

Check Lists/*Check List Before Purchase; Check List After Purchase*

About a quarter of a billion major appliances, such as washers, dryers, refrigerators, ranges, air conditioners, dishwashers, disposers, and refuse compactors, are in use in the 64 million homes in the United States.[1] Homemakers invest nearly as much in appliances as they do in cars and, like cars, appliances need servicing and wear out.

As well as performing satisfactorily at reasonable cost (initial, operating, servicing), a well-selected appliance is appropriate for the family's life style. An appropriate appliance might be obtained by an impulse purchase due to a bargain price, a glowing advertisement, the chance remark of a relative or acquaintance, or another isolated cause. But one is more likely to select an appliance that is appropriate if the family's needs and wants are determined and information on the appliance is acquired, synthesized, and evaluated.

NEEDS AND WANTS

Some aspects of major appliance needs and wants are clear; others must be weighed. A beginning family moving into an unfurnished house needs an appliance on and in which to cook food. Appropriate size and features are much less self-evident. A family with a 15-year-old refrigerator that requires substantial servicing such as installation of a new unit has a spectrum of considerations to be weighed: should the refrigerator be serviced, should it be replaced at this time, should it be replaced with one that has the same, more, or less freezer space, should a no-frost type be purchased, should a side-by-side model be obtained? and so on.

Hopefully, decisions are not always of the emergency type. A family with a 15-year-old refrigerator that does not need repairs at the moment should realize that the replacement decision is ahead.

Appliances also fulfill wants related to easier, safer, more pleasant

---

[1]A communication from Guenther Baumgart, President, Association of Home Appliance Manufacturers, titled "A 1971 Design for Helping Consumers Help Themselves," April 1971.

living. Examples are making more space available by replacing an appliance with one that needs less space, making the home more comfortable by air conditioning, freeing family members from hand dishwashing by a dishwasher, and making the laundry area safer by installing a dryer that will not operate until a start switch is actuated.

A thoughtful consumer also considers other points. Will the appliance add an excessive amount of heat or noise in the home? Would it be better to save the money or use it for another purpose? Will the appliance use the space that is needed for something else? Will it further complicate or clutter the life style of the homemaker or other family members? For example, with a large chest freezer, would the homemaker think it is important to buy and freeze large quantities of meat with the hope of saving money?

If the decision is for making the purchase or for exploring the purchase, the consumer is ready for considering *kind* of information needed for good selection, sources of information, selection procedures, and buying guides.

INFORMATION NEEDED FOR GOOD SELECTION

Five "vital" areas in which consumers need information are suggested by Morris Kaplan of Consumers Union and listed by Sylvia Porter.[2]

Performance: How well does the product do the job for which it is purchased? How effectively does a washer wash? How effectively does a refrigerator refrigerate?

Operating costs: Manufacturers stress purchase price. But operating costs are just as important. An example is given of re-

frigerator A which retails at a higher price than refrigerator B but over a 15-year life span costs less than B because its monthly operating cost is less.

Repairability: How long will a product be likely to operate trouble free, and what is the probable cost of repair over the anticipated lifetime of the product?

Safety: Safety information is needed, especially about hidden hazards and dangers associated with misuse.

Instructions: How should a product be used to make it perform best, preserve its life, and avoid accidents? In what ways is a product most likely to fail or become hazardous?

Answers in all these areas are not currently available for individual consumers. But many sources of information exist.

SOURCES OF INFORMATION

The list given here is alphabetical. Consumers find different sources more or less useful.

Association of Home Appliance Manufacturers' literature. This includes standards, lists of certified models for selected appliances, proceedings of the annual conference, various special publications.

Books on household equipment.

Catalogues and other consumer literature of mail order houses.

Changing Times (published by the Kiplinger Washington Editors, Inc.).

Consumer Bulletin (published by Consumers' Research, Inc.).

Consumer Reports (published by Consumers Union).

Federal publications. *Consumer Product Information*, available from the Consumer Product Information Distribution Center, Washington, D.C., is an index of selected federal publications of consumer interest.

[2] Sylvia Porter, *Minneapolis Tribune*, p. A10, June 24, 1971.

Friends who have used the same or a similar appliance. Ask what they like and do not like about a particular model and why. The "why" is important. A woman might say, for example, that she does not like the dishwasher she has because it does not get dishes clean. But you might learn from chatting with her that the dishwasher is 12 years old, that she does not use hot enough water, that she expects more than she should. Also you might find that a family likes an appliance chiefly for a reason that is unimportant to you. For example, a family likes a freezer because the father "bags" an elk each year.

Home economics publications. *Journal of Home Economics*—articles, abstracts, advertisements. Other publications of the American Home Economics Association—for example, *Handbook of Household Equipment Terminology*. Monthly magazines of different publishers concerned with home economics.

Manufacturers literature — specification sheets, value-line folders, and others. A retail dealer may or may not be willing to give a prospective purchaser a specification sheet. Value-line folders with information on features of several models are often available at the store.

Other manufacturers' literature includes publications of their home economics or consumer units, user's booklets, service manuals, fact sheets. Some of these materials are available to persistent seekers.

Shelter magazine articles.

State Extension Service pamphlets.

Utility companies' booklets, folders, pamphlets.

Your Equipment Dollar (published by the Consumer Education Department of Household Finance Corporation).

BUYMANSHIP GUIDES AND PROCEDURES

Specific characteristics to look for relate to specific appliances. Good buymanship procedures are similar for many major appliances.

Buymanship Guides

Guides are given for major appliances at the end of the relevant chapters. These guides were current at the time the manuscript of this book was prepared. The thoughtful reader will realize that equipment changes: new features and new appliances are introduced; some features and some appliances go off the market—examples are combination washer-dryers, and ironers.

The guides may also apply to new homes with "contractor-supplied equipment" because when asked, contractors are likely to be willing to let an insistent home buyer specify models. In some cases you will pay more than you would if you accepted the contractor's choices. The contractor may be using "builder's models" and/or the contractor as a large purchaser may get a more favorable price than an individual homeowner would for the same equipment. However, especially when cost of the major appliances is part of the mortgage cost, it is reasonable to get equipment that best meets the family's needs. (When appliances are part of the mortgage package, one may be paying for them long after they have been replaced.)

Buymanship Procedures

The following advice has been given to university students and others in a large metropolitan area for many years and has been found useful. Visit stores that sell the appliance. When you go to the store have in mind the main points in which you are interested. Look at appliances on display.

Try to see a demonstration or a sales presentation on features. If a manufacturer's representative is in the store, chat with her or him and ask for printed materials—specifications and the user's booklet if possible.

Try to visualize the appliance in your home: How will it fit into the family's life style? Remember that if you have really informed yourself of the possibilities, you might not get all the features you think you want in the price range you are willing to pay. Weigh the combination of features most important for your family.

Be attentive and interested in what the salesman says. Get information on safety seals; warranty; servicing arrangements; cost, including installation cost if any; trade-in allowance, if appropriate; approximate operating cost.

At home, review and synthesize the material you have collected. Confer with other members of your household. Consider the check list in the following section.

One *management* approach to buying major household appliances stresses several things.[3]

Factors involved in intelligent consumer decision-making: Shop around; comparing prices can save money. Shop defensively; guard against misleading advertising and selling, business incompetence (clerks not well informed), consumer incompetency (you are not well informed).

Minimize cost of searching for the good choices: Use consumer references listed in the previous section. Use the telephone and yellow pages.

Profit from seller's discrimination: Your profession—home economics, clergy, architect, and so on—may entitle you to a discount.

[3] Edna K. Jordahl, *13 Questions When Buying Household Equipment*, Extension Folder 252, Agricultural Extension Service, University of Minnesota (slightly modified).

An additional management point is to pay cash, buy on installment, buy on credit, according to what best serves your needs, wants, and resources. The statement that it is cheapest to pay cash is not relevant if one does not have the cash when a demonstration piece of equipment is available (at a substantial saving) or when a store has a sale on a model that is appropriate for you.

CHECK LISTS

Use a check list before and after purchase of a major appliance.

Check List Before Purchase

1. Safety and other seals. In the United States the Underwriters' Laboratories seal is the important safety seal for electric appliances and the American Gas Association seal is important for gas-burning appliances.

   The AHAM (Association of Home Appliance Manufacturers) seal is found on air conditioners, refrigerators, freezers, and other appliances. The seal certifies the accuracy of such characteristics as cooling capacity in Btu's per hour, refrigerating volume, and so on. Consumers' Research reports,[4] that in the case of air conditioners, the AHAM program is effective and well controlled. Actual testing for the program is done by a well-qualified independent laboratory organization. On the other hand, the net refrigerated volume and net shelf area given on AHAM certified consumer tags on the appliances were overstated. (Students in advanced Household Equipment classes have also found these parameters difficult to verify in some instances.)

[4] *Consumer Bulletin*, June 1971, p. 38.

The Good Housekeeping seal is described in the August 1971 issue of that magazine as follows:

We satisfy ourselves, by evaluation and samples and other pertinent data, that the products and services advertised in *Good Housekeeping* are good ones, and that the claims made for them in our magazine are truthful. If any product or service advertised in *Good Housekeeping* proves to be defective, it will, upon request of the consumer and verification of the complaint, be replaced or the price which the consumer paid for it will be refunded. . . . We cannot be responsible for faulty installation or service by dealers or independent contractors.

2. Effectiveness of the appliance for the primary purpose for which you will purchase it. For example, if you are about to purchase a washer, consider what *unbiased* information you have on how clean different models wash the types of laundry loads your family has. Think also about requirements such as softened water and enough hot water that are important for effective performance of the appliance in your home.
3. Ease of use of appliance and clearness of instructions on controls, safety devices, etc.
4. Space needed by the appliance relative to the space you plan for it.
5. Ease of cleaning and maintaining the appliance.
6. Whether the design of the appliance is satisfactory to you.
7. *Unbiased* information on durability of the appliance or your judgment on durability.
8. Servicing organization for the appliance.
9. Understandability of the warranty. (For practical purposes, the words warranty and guarantee as applied to

appliances are identical.) Note that the manufacturer's warranty does not cover claims made by a salesman that are additional to those given in the warranty.
10. The reputations of the manufacturer and the dealer for "standing-behind" the articles they sell. Purchase of an appliance with an advertised brand name from a reputable local dealer is likely to have built-in insurance value for the consumer.

Check List After Purchase

The list given below is adapted from a National Safety Council publication:[5]

Check with dealer, retailer, service agency, or manufacturer for special installation instructions including proper fire clearances from combustible surfaces.

Refer to instruction booklets in regard to proper installation, use, care, cleaning, and maintenance. File instruction booklets and warranties so they are readily available when needed.

Mail registration card to manufacturer when required. Record date of purchase.

Special wiring, gas venting, or special fire protective measures should be handled by a qualified person.

Major appliances should be grounded or connected to a three-prong grounded outlet.

Plan for periodic checks of equipment—oiling as recommended by manufacturer, replacement of filters, etc.

[5] *Safe Home Appliances*, Home Department, National Safety Council, Chicago, Ill.

CHAPTER 9
SINKS AND DISHWASHERS

The kitchen sink is one appliance always associated with kitchens. Electric dishwashers, portable or built in, still are a long way from saturation in American homes, but they are becoming increasingly common.

SINKS

The chapter on kitchen design discussed the importance of the sink center as the most often used center and the research finding of different investigators that of the several centers in the kitchen the most trips are made to and from the sink center. Storage and counter requirements in a well-planned sink center and the advisability of having this center close to the cooktop limit the width of sink that should be selected for some kitchen spaces, especially if a built-in dishwasher also is provided.

TYPES AND DIMENSIONS

Counter sinks are those installed in a counter. They are sold separately or as part of sink tops or counters. They may consist of single, double, or triple bowls or wells with one or two drainboards or without a drainboard. When provided, drainboards are made in one piece with the wells. For sinks not part of a sink top, a rim must be purchased to cover the joints between sink and counter, unless the edges of the sink slope slightly downward to the counter. In modern

Figure 9-1. Three-well stainless steel sink; pop-up drains for outer wells are operated by remote control from panel in ledge below faucet. An acoustical undercoating gives rough look to bottom. (Elkay Manufacturing Company)

installations, the space under a counter sink is covered with a recessed or nonrecessed sink front that has two doors (or occasionally one) and ventilation slits.

Cabinet sinks are wells, with or without drainboards, that are mounted in specially made cabinets.

A combination sink and tray is one well and one laundry tray, and sometimes a sliding top part that may be moved to cover either the sink or the tray. Tray is a term used to indicate a laundry sink. In most models the sink is 7 1/2 inches deep and the tray is 10 inches deep. Counter and cabinet types are available. The laundry tray may be at the right or left of the sink. If dishwashing is done by hand, and if soiled dishes are placed at the right of the sink and the clean dishes at the left, the laundry tray should be at the left.

Except for sinks with laundry trays, most of the currently sold sinks have wells 7 or 7 1/2 inches deep. However, sinks are available with wells 4, 4 3/4, 5 1/2, and 6 3/4 inches deep, to quote from only two manufacturers. The more shallow depths are more characteristic of wide, single-well sinks or multiple-well sinks.

Single-well counter sinks are available with wells from 12 to 25 inches[1] or more wide and from 10 to 20 inches front to back. Usable interior dimensions are less than quoted sizes. A 24-inch-wide sink, about 16 inches from front to back, is often a good choice in a kitchen with a small sink wall that also has a built-in dishwasher. (Additional criteria on dimensions are discussed below.)

Double-well counter sinks are available from 32 to 42 inches wide and from 16 to 20 inches front to back. A triple-well sink might be 43 inches wide or much wider (Fig. 9-1). With a two-well sink, dishes can be washed in one well and rinsed in the other. Also, if a dish drainer is used in the rinsing well, this well is available for emptying "forgotten" cups. A limitation is that only a rather small drainer will fit, unless the double-well sink is a wide model. A three-well sink provides the features of a two-well model plus free access to the food waste disposer if one is provided.

Monroe summarizes the matter in this statement: "You can hunt for a sink which allows you to continue with your present methods of washing and rinsing dishes, or you can change your methods to fit the facilities of the sink."[2] A point that might be added is that you may need to have in mind the sizes of dishpans and drainers on the market when you hunt for a sink.

Counter sinks of rather small size, for example, 15 by 15 inches, are used both as extra sinks and in kitchenettes.

The width of cabinet sinks is quoted in width of cabinet; 42- to 72-inch widths are available. The dimensions of the wells vary.

Steidl and Bratton discuss four sink design parameters that affect the posture of the worker:[3] the thickness of the front barrier, the height of the floor or bottom of the well, the height of the top of the sink rim, and the depth of the well. The front

[1] To be consistent with the discussions elsewhere in this book, particularly in the chapter on Kitchen Design, the use of the word width here denotes side to side. Sink manufacturers are likely to call this dimension length and the front-to-back dimension width.

[2] Merna M. Monroe, *Ideas to Consider When You Buy a Kitchen Sink,* Maine Agricultural Experiment Station Bulletin 494, October 1951.

[3] Steidl and Bratton, *Work in the Home.* New York: Wiley, 1968. pp. 298–299.

of the inside of the well should be as close as possible to the front of the counter, 3 inches or less. The height of the bottom of the well from the floor should permit the worker to grasp items from the bottom of the sink with little or no bending. The top of the sink rim must not be so high that the homemaker has to work with raised shoulders. The depth must relate to the type of activity as well as the comfort of the user. A 4-inch depth is likely to be sufficient for food preparation jobs in which little splashing occurs. A 6- or 8-inch depth is adequate for washing most cooking utensils, as well as seldom-used larger vessels.

MATERIALS AND CARE

The materials used for sink bowls are white or colored enamel over cast iron or steel, chrome stainless steel, nickel stainless steel, and Monel metal. Monel metal is very satisfactory, but expensive. Chrome stainless steel has no nickel in it and usually is used in lower-priced sinks. It requires more care than nickel bearing stainless steel. Type 302 nickel stainless steel is used in better quality stainless steel sinks. The U.S. Department of Commerce has set a minimum standard of 20-gauge stainless steel for use in stainless sinks for the home.

The finish material used on enameled sink bowls is not the same for all sinks. Some manufacturers fuse two or three coats of porcelain to cast iron or steel and others apply a "vitrified enamel." When porcelain enamel is used, the final coat should be acid resistant. Vitreous ware has an acid-resistant glaze finish.

Correct care of a sink depends upon the materials used. Porcelain enamel and vitrified enamel will acquire scratches, chipping, and discoloration in time, if care is not taken. Probably the most practical method for maintaining the appearance of an enamel sink for many years is to use plastic mats on the floors of the wells and over the divider(s) of a multibowl sink. This minimizes scratching of the surface.

Because the finish is acid resistant and not acidproof, acidic foods should not stand in the sink for several hours. Furthermore, the sink should be washed with nonabrasive cleaner whenever acidic foods have been in the sink.

Chlorine bleaches are effective in making enameled sinks look brighter, but they should not be used too frequently.

FITTINGS

Fittings for sinks are sold separately from the sink (Figs. 9-2 and 9-3). The fittings in the sink are the combination unit with swing arm or separate faucets with spouts, rinser spray, and sink strainer. The combination unit may have separate faucets or a single control for regulating temperature of water and rate of flow. The latter may be separate from or integral with the spout. The combination control usually is installed in the rear ledge of the sink. It is made of chrome-plated brass. Some

Figure 9-2. Single-well stainless steel sink *without* fittings. (Carrollton Manufacturing Company)

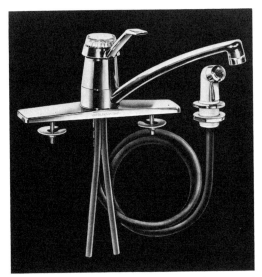

Figure 9-3. Dial-control combination faucet and spray. (American Standard, Inc.)

spouts have nonsplash metal aerators. Rubber aerators for spouts not so equipped may be purchased in hardware stores. Both types of aerators need to be replaced occasionally.

Combination faucets and separate faucets may be the washerless type or standard washer type. The washerless type for a combination or single faucet may have interior valve components assembled in a sealed cartridge (Fig. 9-4) or may use a ball mechanism in place of a washer. A sealed-cartridge type might be guaranteed against drips or leaks for ten years. Both washerless types make stem packing, as well as washers, unnecessary.

A pull-out spray for cleaning fruits and vegetables and for rinsing dishes may be installed to the left or right of the combination faucet. The body of the rinser spray is usually a plastic material. Care should be taken to keep the spray in its storage position when it is not in use. If it were immersed in a sink full of water a "cross-connection" could develop which might pollute the potable water supply. A direct connection between safe water and waste water plus a reversal in pressure can cause a cross-connection. A reversal in pressure may occur whenever there is a sudden heavy demand on the water supply.

Sink strainers fit over the sink drain pipe opening and are basket type with stopper (also called the crumb-cup type), or flat without stopper. They often are made of stainless steel.

Additional fittings needed with but not in sinks are traps and valves in the pipe lines.

SINKS WITH SPECIAL FEATURES

Special features in kitchen sinks are shown in Figures 9-5 through 9-8 and described in the captions.

Figure 9-4. Ceramic cartridge valve faucet. The valve action of this "drip-proof" and wear-resistant "Aquarian" model is described as a shearing type. The polished ceramic surfaces are described as unaffected by sand, silt, or metal shavings. (American Standard, Inc.)

Figure 9-5. Porcelain and enamel dual-level sink, 32 by 21 inches. One bowl is 15 1/4 by 18 1/4 inches and 8 inches deep. Smaller and shallower bowl is useful for food preparation and for installation of a food waste disposer. Elevated water control deck has a soap or lotion dispenser, single-lever faucet, remote pop-up drain control, a second soap or lotion dispenser, hose, and spray. (American Standard, Inc.)

Figure 9-6. Stainless steel, 18-gauge, self-rimmed double-well sink with both faucets in rear ledge at right. Drain is near right rear of left well and is centered in right well. (Elkay Manufacturing Company)

Figure 9-8. Single-well enameled cast iron sink 30 by 21 inches by 8 inches deep, installed with metal rim. The antidrip combination unit is controlled by separate faucets. A similar sink, self-rimmed, is available in a 24-by-21-inch model. (Kohler Company)

Figure 9-7. Two-well, self-rimming porcelain enamel sink with wood cutting board. (Kohler Company)

DISHWASHERS

Nationwide sales of electric dishwashers had increased by 1970 to the point that 26.5 percent of the wired homes in the United States had one.[4] Homemakers give many reasons for satisfaction with electric dishwashers though they also may report limitations of the appliance, according to a small survey by one of the authors (FE) in 1967. One special source of satisfaction was orderliness in the kitchen. Dissatisfactions were associated with poor location in the kitchen, age of the appliance, and failure to consider some relevant factors in purchase or use.

TYPES AND DIMENSIONS

The Association of Home Appliance Manufacturers classifies home dishwashers as built in, front-loading portable, and top-loading portable. The *Handbook of Household Equipment Terminology* uses the following definitions.[5] A built-in dishwasher is one designed to be placed under or in a counter, between or enclosed in cabinetry. A free-standing dishwasher is self-contained, stands independently, and is permanently installed. Top-loading models are opened by lifting a lid. The top racks then are moved out of the way or are raised automatically to allow access to the lower rack for loading. In front-loading models, a hinged door is dropped and racks can be pulled out for loading. Most front-loading models can be converted to a permanent installation, after the casters are removed.

Class laboratory work in an experimental kitchen of the University of Minnesota Household Equipment Laboratories suggests that a built-in or free-standing dishwasher located at the right or left of the sink—whichever side was more conve-

[4] "1971 Statistical and Marketing Report," *Merchandising Week,* February 22, 1971, vol. 103, No. 8, pp. 26–27.

[5] *Handbook of Household Equipment Terminology,* 3rd ed., American Home Economics Association, p. 8. Prepared by a committee of the AHEA Home Economists in Business section.

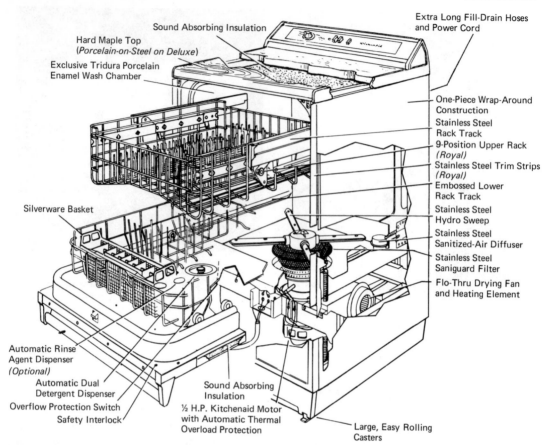

Sound Absorbing Insulation

Hard Maple Top
(*Porcelain-on-Steel on Deluxe*)

Exclusive Tridura Porcelain
Enamel Wash Chamber

Extra Long Fill-Drain Hoses
and Power Cord

One-Piece Wrap-Around
Construction

Stainless Steel
Rack Track

9-Position Upper Rack
(*Royal*)

Stainless Steel Trim Strips
(*Royal*)

Embossed Lower
Rack Track

Stainless Steel
Hydro Sweep

Stainless Steel
Sanitized-Air Diffuser

Stainless Steel
Saniguard Filter

Flo-Thru Drying Fan
and Heating Element

Silverware Basket

Automatic Rinse
Agent Dispenser
(*Optional*)

Automatic Dual
Detergent Dispenser

Overflow Protection Switch

Safety Interlock

Sound Absorbing
Insulation

½ H.P. Kitchenaid Motor
with Automatic Thermal
Overload Protection

Large, Easy Rolling
Casters

Figure 9-9. Cross section of free-standing, front-loading dishwasher. Special features include porcelain wash chamber, stainless steel four-arm wash mechanism, drying fan located outside wash chamber, 1,000-watt heating element in sump below wash arm, and acoustical insulation at top and in door. An overload protection switch insures that excess water does not enter wash chamber during fills; safety interlock insures that water does not spill out when door is open during operation. During drying, fan moves air over heating element and saniguard filter diffuses heated air. Venting is through the bottom front of door. (KitchenAid Division, Hobart Manufacturing Company)

nient to the dining area *and* had adequate counter to receive soiled dishes from the dining table—had considerable convenience.

Built-in models are 24 inches wide, 24 to 25 1/2 inches deep, and about 34 1/2 inches high to fit under a standard height counter. The dimensions do not include counter or side panel thicknesses.

Dimensions of free-standing and portable models are quoted with counter top and with end panels. The latter contribute a total of 1/2 inch or so to width. The height of free-standing models is 36 inches or 36 inches plus the diameter of the cast-

ers. The height of other (portable) front-loading models may be the same or less than the 34 1/2 inches quoted for built ins.

Top-loading portables are more variable in dimensions—they might be approximately 22 to 24 inches wide, 27 to 30 inches deep, and 32 1/2 to 35 inches high.

COMPONENT PARTS AND WASHING MECHANISMS

The basic components of dishwashers are wash chamber plus parts in and on it, control, wash system, filter and drain, and heating element (Fig. 9-9).

Wash Chamber

The wash chamber or tank is the main structural unit of the dishwasher.[6] It is a welded unit of several steel panels which forms the framework for mounting other components. The steel assembly is treated to prevent corrosion. Interior surfaces have a finish coat of polyvinyl chloride (plastisol) or porcelain enamel. Front-loading models have a door at the front of the tank and top-loading models have a lid at the top.

The interior of the wash tank holds the racks into which soiled articles are loaded. Detergent dispenser(s) if provided and the rinse additive compartment if provided are also in the tank for top-loading models and in the inside of the door for front-loading models.

Built-in models usually have acoustical insulating material on the (exterior) top of the tank and may have acoustical material on other exterior parts of the tank. Acoustical insulation *and* thermal insulation, however, are likely to be more important for models that are not built in since these do not have cabinetry to absorb the sound and serve as a thermal barrier.

The motor and pump(s) are assembled to the underside of the sump of the wash tank. The fill valve to which the fill hose is assembled through an air gap also is mounted on the exterior of the tank. (An air gap in the fill line is required by plumbing codes to prevent reverse flow of water from dishwasher to water supply.)

Control

The control which the user operates is at the front. This might be as simple as a start button or a dial that is turned to start, a set of push buttons and a dial, or a set of push

(a)

(b)

(c)

Figure 9-10. Dishwasher controls for different models. (a) On-off control. After user turns it to "on," control advances automatically. (General Electric "Power Shower" model) (b) Five-cycle control plus variable wash and dry. User can select wash time between normal and three times normal; user can also select dry time up to "super dry" or "no dry" when dishes are stored in appliance overnight. The control on right advances automatically to indicate part of the cycle at any time. (General Electric "Versatronic" model) (c) Push-button controls and cancel button for six-cycle dishwasher. Cancel button is used when user changes her mind about cycle. Pushing this button causes timer to move to off. (Whirlpool "Mark 100" model)

---

[6] "P" Line Dishwashers, *Frigidaire Product Information,* Tech-talk 1968, vol. 69, no. 1., reprinted April 1970.

buttons only (Fig. 9-10). After the user operates the control, the dishwasher automatically is programmed through a sequence of operations ending with dry, unless the user opens the door or, on models so equipped, presses a cancel button. Most models have a short interval between the time the dial is set at start and the actual start (fill).

The automatic sequence of operations includes fills (the tank "fills" with about 2 1/2 gallons of hot water for each fill), rinses (washes without detergent), washes, drains, pauses, dries, stops. Deluxe models have several sequences of operation with varying times for different parts of a sequence. The specific sequence programmed into the dishwasher depends on the button pushed. During "dry" the drained dishwasher operates with a heater element on and with or without a fan for moving the hot air.

Essentially the main control depends on a timer cam which as it rotates opens and closes electrical contacts. The timer may be a creep type which rotates with a constant motion or a jump type which rotates in a series of jumps.[7] The fill may be a pressure switch controlled type or a time-fill type. In the latter a definite time interval is allowed for fill and the fill valve has a flow control device to prevent excess fill.

Wash Mechanisms

Wash mechanisms used by two manufacturers are shown in Figures 9-9 and 9-11. A third manufacturer uses the following wash parts on its top-of-the line models: lower wash arm, shower (mounted near the top of the tank), tower with a center spray head in the center of the tank, and silver shower near the bottom of the tank (under the silver basket). Still other manu-

[7] Personal communication, R. R. Johnson, Whirlpool Corporation, St. Paul, Minn.

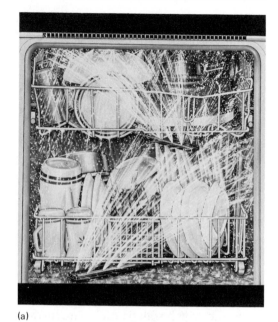

(a)

(b)

Figure 9-11. (a) Wash action with one reversing jet spray located beneath upper basket and another below lower basket. (b) The "Reversa-Jet Spray Arm." (Tappan Division of the Tappan Company)

facturers supply other groups of wash parts. These are not always the same for all models of one manufacturer. In the commonly used multilevel wash systems such as those just considered, a pump recirculates the water (or water with dissolved detergent). A basic (low end-of-the-line) model might use a propellerlike device assembled to the motor shaft rather than a pump to circulate water.

Filter and Drain

The recirculated wash and rinse waters move through a filter (Fig. 9-12) or sometimes two filters located in the sump of the tank into a drain pump. In many models the single filter is designed to be removed and washed under running water by the user; in others the filter is not so designed or a special device is provided. One manufacturer supplies a "soft food waste disposer" which liquifies soft food and

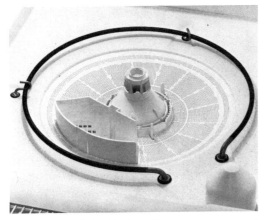

Figure 9-12. Self-cleaning large filter, pump guard, and heating element in sump of dishwasher. Food particles washed from soiled dishes fall onto large filter. After each wash and rinse sequence pump reverses and draws or sucks the water and small food particles into dishwasher drain line. The removable pump guard (part that seems to stand up in lower front section) catches and holds "large" or "heavy" pieces of soil. (Whirlpool Corporation)

flushes it away. Another manufacturer calls the device it supplies a "soft scrap disposer." The dishwasher pump chops residual food scraps from the soiled articles and flushes them down the drain.

Heating Element

A 700- to 1,000-watt sheath or enclosed (Calrod) unit is usual in the wash tank under the bottom of the wash unit. Its functions are to maintain temperature of the water in the tank during wash and rinse, to increase temperature of wash or rinse water on models that have a separate setting with variable time delay to permit this, and to accelerate drying of the washed articles. During the dry part of the cycle the unit heats the air in the wash tank and the washed articles are dried by convection or in models with a fan by forced hot air.

INSTALLATION AND USE

Convenient locations of the built-in model were discussed earlier in this chapter. Criteria for location of a portable may be different. One consideration should be ease of moving the appliance to the sink.

The recommendation for electrical sup-

ply is the same for portable or built-in model—a three-wire, 120-volt, 15-ampere individual equipment circuit. As indicated before, a built-in dishwasher and a food waste disposer might be on one 120-volt, 20 ampere circuit.

Recommended or required characteristics of the water supply relate to temperature, pressure, hardness, and possibly, amount of hot water. The dishwasher manufacturers usually suggest supply water temperatures at the dishwasher of 140° to 150° F. Such high temperatures *at* the dishwasher present difficulties in the home. There is need for experimental work in this area. Some unpublished experimental work of one of the authors indicates that under some conditions a water temperature of 130° F *at* the dishwasher is as effective as 150° F.

The recommended water supply pressure is usually 15 or 20 to 120 pounds per square inch. Concerning hardness in grains per gallon (gpg), the research and development division of one manufacturer of dishwasher detergents states:[8] a water hardness up to 8 gpg does not cause much trouble. A hardness more than this is a problem. Some detergents have been formulated to reduce the problems in hard water, but in some instances a water softener may be needed.

The amount of hot water used and the time over which it is used varies for different models and cycles—4 gallons might be used in a rinse-hold cycle of 6 minutes and 16 or more gallons in a complete (through dry) heavy soil cycle of 60 minutes.

The best counsel on use of a dishwasher is the same as that for other appliances—follow the manufacturer's instruction in the user's booklet.

---

[8] Economics Laboratory, Inc., *Technical Aspects of Automatic Dishwashing*, St. Paul, Minn., 1969.

Some overenthusiastic users wash articles in the dishwasher that were better washed by hand. Cookware with wooden handles should be washed in a dishwasher only if the handles have been treated to resist high temperature.[9] Aluminum cookware should be positioned so that it will not be sprinkled with undissolved detergent when the detergent container(s) open. Anodized aluminum generally should not be washed in a dishwasher. Iron vessels that are not coated with porcelain enamel should not be washed in the dishwasher. Only immersible electric appliances without rubber parts should be washed in the dishwasher.

Although easy to wash by hand, Teflon-coated ware can be dishwasher washed.

Generally flatware or holloware made of sterling silver or silver-plated items should be washed in a dishwasher only after thorough rinsing to avoid tarnish due to contact with such foods as eggs or foods that contain acid or salt. Dirilyte (the trade name for gold-colored metal plating) articles should be washed by hand. Stainless steel of reasonably good quality washes satisfactorily in a dishwasher though many homemakers occasionally hand polish this material.

Underglaze patterns on dinnerware do better than overglaze patterns after repeated washings in a dishwasher. Gold, silver, and platinum on dinnerware will wear after repeated washings in a dishwasher.

Dinnerware made of thermosetting plastic (Melamine is perhaps the best-known trade name) is generally described as dishwasher safe for a specific number

of years. Articles, such as some freezer-ware, made of thermoplastic material, of course should not be washed in a dishwasher.

Glassware carefully placed in the wash tank to minimize chance of breakage washes satisfactorily with some exceptions. Milk glass may yellow. Soft glass etches due to removal of metal ions by the alkaline wash solution.

Some general recommendations that also apply in use of dishwashers are as follows:

1. Load the dishwasher so that the soiled sides of articles to be washed get the best exposure to the washing mechanism.
2. Hard, burned-on, and excess food usually should be removed from pots and pans before they are placed in the dishwasher.
3. Empty liquid from glasses, cups, and miscellaneous containers. Scrape or rinse dishes free of waste such as bones, large scraps of meat, starchy foods, and cigarette or cigar ashes. (If one thinks a moment it seems unlikely that recirculation of cigarette ashes in the wash water could contribute to the cleanliness of the load.)
4. Use the dishwashing compounds recommended by the manufacturer. Several compounds usually are recommended, and since water varies in different areas, it is worthwhile trying several to determine which is best for the local water supply. If the water is hard, some dried articles will show water spotting. A rinse additive device that automatically adds a wetting agent to the last rinse as an aid for avoiding water spotting is likely to be worth the added cost.
5. Wait several minutes after the end of the cycle before trying to remove dishes from the appliance. They are hot! If you are in a hurry , open the door at the end

---

[9] The material on articles not to be washed generally follows that given in *Appliance Sales Training—Dishwashers and Disposers*. Distributive Education Manual of the Distributive Education Instructional Materials Laboratory, Division of Extension, University of Texas at Austin, 1970.

of the cycle and the load will cool faster.

6. Some users store the washed dishes in the dishwasher between uses of the appliance. Small families, on the other hand, may find it convenient and more economical to scrape and store soiled dishes in the dishwasher as they accumulate and operate the dishwasher only once a day. If this procedure is followed, prerinsing of soiled dishes is likely to be helpful. Often prerinsing can be done in the dishwasher by proper selection of the cycle.

7. Clean the removable filter with a brush and sudsy water as necessary. Also clean the container(s) for the dishwashing compound as necessary. Occasionally check bottom of tank and remove insoluble foods, other items, and any film that may have collected. Maintenance of the exterior surfaces depends on the finish provided. Follow instructions in the manufacturer's manual.

Figure 9-13. Undersink dishwasher. The purchaser needs manufacturer's data sheet on suitable sinks and installation requirement. (General Electric Company)

FEATURES

Different features of some models are given in the captions of Figures 9-13 through 9-16. One manufacturer offers a built-in undersink model. Wash mechanisms differ, as do motor and pump characteristics, wattage of heating unit, and flexibility and convenience of interior racking for articles to be washed—large mixing bowls, utensils, tall stemware, and others. Interior volume is approximately the same in all models but number and types of articles that can be racked in different models without blocking wash action are not the same.

As indicated earlier, water demand varies. One model uses 13.5 gallons in its longest cycle, while another uses 16; this does not necessarily relate to usable capacity. Types, number, and times of cycles vary. One model has a full soil cycle and a heavy soil cycle which differ in total times, 52 versus 58 minutes, and in total water demand, 13 1/2 versus 16 gallons. The cycles also differ in that the heavy soil cycle has an extra one-half fill for prewash or

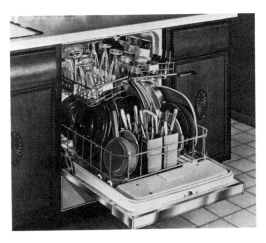

Figure 9-14. Adjustable upper-level racks. This model has multilevel wash system consisting of lower wash arm, top shower, center tower, and silver shower at bottom. (General Electric Company)

(a)

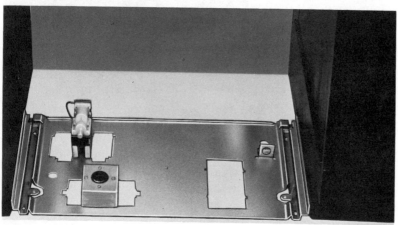

(b)

Figure 9-15. (a) Front-loading portable with reversible wood top and food warmer. Warmer section can be heated to a setting for keeping food warm. It will not operate while dishwasher is operating. (b) Base plate assembly which is connected to electrical, drain, and hot water lines to facilitate dishwasher installation. (Whirlpool Corporation)

rinse and an extra one-half fill for "after rinse." This same model also has other cycles.

A fine china cycle used to differ from a regular cycle in number of prewashes. Now at least two models are available in which the cycle also differs in the force with which the water stream enters the appliance. For the gentle wash, air is admitted with the water. At least one manufacturer supplies an extended wash time up to three times that for regular wash and

a variable dry time from zero to regular.

Some dishwashers have special sanitizing features. The wash cycle may not start until enough time has elapsed for the heating unit to raise the wash water temperature to a predetermined value or the final rinse time may be lengthened to permit the water to reach a predetermined temperature.

Some manufacturers may supply a base plate assembly for built-in models which makes installation easier.

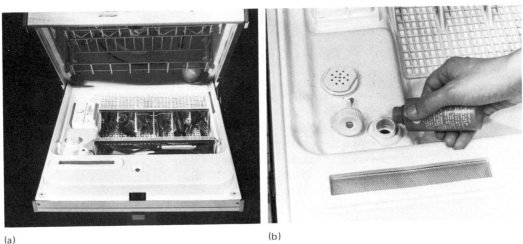

(a)   (b)

Figure 9-16. (a) Removable cutlery and silver baskets installed in door. (b) Rinse conditioner dispenser with "fill indicator" cap. (Whirlpool Corporation)

Provision for stabilizing the front door to insure against tipping of the appliance is especially important for portable models. Means to insure against excess water entering the dishwasher is important. This may be accomplished by the design of the fill valve or by a separate water level safety switch.

Other special features that are available include a rinse additive device which is factory preset to operate in the final cycles, though the user still has to add wetting agent every one to three months depending on frequency of use of the appliance. Forced-air drying, when it is provided, is a more positive approach to the thorough drying of the recesses in items in the time allowed than convection drying.

Some manufacturers provide different combinations of door fronts for various kitchens.

BUYING GUIDE

Kitchen Sinks

Know the kitchen plan for which the sink is to be selected. Research has shown that more time is spent near the sink than any other appliance in the kitchen. A corollary of this is that storage space is needed near the sink. Some space under sink bowls is usable for storage, but this is not as good as separate base cabinets. Also, if a wall cabinet is hung over the sink, it must be hung higher than cabinets over a counter and is therefore likely to be less convenient.

Know whether you will immediately have a built-in dishwasher. If not immediately, do you plan to have space so that one can be installed in the future?

Chat with friends who have fairly new sinks. Try to get brand names. Ask what they like and do not like about their sinks.

If you know the plumber or contractor who will install the sink, find out whether he will install any make you select.

If you live in a fairly large city consult the telephone directory for retailers of plumbing supplies. You may be able to visit a showroom of a large manufacturer. In some communities the only easy places to *visit* will be plumbing shops. Wherever you live, you are likely to have access to catalogues of mail-order houses.

If possible see the sinks that are sold by major manufacturers, that is, those who advertise and sell either nationally or over a large region of the country.

Get information on lengths and depths. If you are considering stainless steel models get specific information on nickel content and type of mechanical finish. (Type 302 nickel steel is desirable.) What gauge metal is used? (No. 20 is more desirable than No. 22.) What finish is used? A difficult-to-maintain finish such as a satin finish is less desirable than one that is easier to maintain.

If porcelain is selected, is iron or steel used for the base? How is the acid resistance of the porcelain described? Is titanium incorporated into the porcelain? What means are used to aid the fusion of the porcelain to the base metal?

Does the underside of the sink have acoustical material?

Is the sink oval or rectangular in shape? If a counter model, how is the sink secured to the adjacent counter? If a rim is used, is the rim flat or raised?

What faucet(s) would be appropriate? Does the faucet have a washer or is it washerless? What is the claimed trouble-free life for the faucet? Is the swing arm such that a reasonably large utensil or a coffee maker could be placed under it?

Where are the strainer holes, or hole of a single-well model, located?

Get information on sink fronts; specifically you want to know what fronts will be available that match the cabinets in your kitchen.

After you have come to a tentative decision check on installation costs.

Before committing yourself to purchase, visualize the sink in the kitchen and consider whether it will meet the food preparation and cleanup work habits of the family members who will use it.

Review in your mind reasons for your tentative choice.

Dishwashers

Review the general buymanship guides (Chapter 8). Make a summary sheet or sheets on the guide points given below for models in which you are interested. If you get all the data, you have facts on which you can base a wise selection. Use your summary and the Check List Before Purchase (in Chapter 8).

1. What is the manufacturer's name, the model number, the catalogue number or name and number if appropriate?
2. Examine the electrical data.
3. What is the recommended pressure range and temperature of water entering the appliance?

4. Is there a pressure or other positive-type water overfill control to shut off water-fill valve when the water reaches a predetermined level?

5. How much water is used per fill and per cycle for each type of cycle?

6. Flexibility provisions: Are the interior racks, silverware basket, and special items basket appropriate for loads to be washed? What is the number of detergent compartments? Are there special construction features for flexibility in use?

7. Materials and dimensions: What is the material of the lining, front, racks, and top (for portables and convertibles)?

   What are the dimensions? For free-standing and portable models depth in inches with door or lid open is likely to be important. Also depth with door open may be important for some built-in installations.

8. What is the function of each wash mechanism?

9. Type of filtering action: Is it easy to clean the filter for type that needs to be cleaned?

10. What are the type and locations of controls for starting and stopping dishwasher?

11. What are the number and type of wash cycles?

12. If regular and gentle *actions* are provided, what is involved?

13. What is the time for each wash cycle and total time for wash and dry?

14. Are wash and/or rinse water heated or is water temperature maintained? (Heating requires that cycle be lengthened by varying amounts of time.)

15. What are the characteristics of "sani cycle" if one is provided? This might be lengthened wash or dry times or high temperature final rinse. (Purpose of this cycle as name implies is increased bacteria and germ kill.)

16. Rinse additive device: If supplied, does it turn on automatically when dishwasher is started or is it necessary to operate a lever or other control?

17. Forced-air drying versus convection drying: If forced-air drying is provided, is the fan outside of the wash chamber?

18. Are there special sound reduction features such as double-wall construction, resilient pump mounting, and others?

19. Does it have instant start?

20. For free-standing and portable models, is there a hot water by-pass connector?

21. For built-in models, is any information available on escape of steam?

22. Are there other significant special features? For example, are different fronts available for a built in appropriate for the kitchen in which it will be installed?

23. What is the content of the warranty?

CHAPTER 10
GAS RANGES

CONSTRUCTION, USE, AND CARE

TYPES

The gas range is available in many different types, each of which may be varied within its group by special features. All should carry the seal of approval of the American Gas Association. The standard type may vary in width from 36 to 42 inches; it has an oven and broiler, usually heated by the same burner, and four surface burners; it may or may not have some of the special features found on the more deluxe ranges.

The apartment-size range varies from 19 to 26 inches in width. It also has an oven and broiler heated by the same burner, and four surface burners. Its chief advantage is that it requires less space and is less expensive than standard-size ranges; in fact, if it is purchased without any special features it may be quite inexpensive. This type of range often is found in furnished apartments but may not be readily available to the average consumer.

The 30-inch range may not be exactly 30 inches in width. The oven width is not 30 inches, but this range usually has an extra-large oven. Some models of this size are equipped with many special features while others are stripped down, known as bottom-of-the-line models. Some are designed to be installed as pictured (Fig. 10-1); others are free standing.

Some double oven gas ranges occupy only 30 inches of space; others are larger. One oven is located above the burners at eye level, the other below in the usual position. Some have a low broiler (Fig. 10-2) while others do not. The eye-level oven should be located at a height that not only is convenient for the user but also does not interfere with seeing and using utensils on the rear surface burners.

Figure 10-1. Built-in range, divided top burner arrangement, low broiler. (Tappan Company)

Figure 10-2. Double oven range, six top burners, self-cleaning broiler/oven, pull-out vertical utensil compartment. (Caloric Corporation)

Domestic Gas Ranges

The American Gas Association defines a domestic gas range as a self-contained, gas-burning appliance designed for domestic cooking purposes and having a top section and an oven section. It may have a broiling section.[1] Free-standing ranges with ovens at eye level provide some of the conveniences of built-in ovens and also are more easily moved. Some gas ranges designed for cooking also have a section used for space heating or for use with another fuel. This chapter considers only ranges designed solely for cooking. Because many ranges do have a broiling section this will be considered, as well as ovens and surface burners.

Built-In Domestic Cooking Units

Built-in surface burners and ovens provide different possibilities in kitchen arrange-

[1] USA Standard for Domestic Gas Ranges, American Gas Association, Inc., 1967, vol. I, p. 58.

ments, although often they are found adjacent to one another in the kitchen. Built-in gas cooking appliances for the home are known as units. The top or surface unit may include burners, griddle, and deep-well cooker, or any combination of the three. The oven unit is designed for installation in a wall cabinet, a partition, on a counter or in the wall. It may be a separate oven, it may have a broiler below, or it may have a broiler in the oven itself. The broiler unit may be a separate broiler (Fig. 10-3), or it may be combined with a rotisserie, or it may be a rotisserie unit. Any combination of oven, broiler, or rotisserie is defined as a unit.[2]

Slide-in units with self-contained oven,

broiler, and surface burners fit into a counter-top opening. They look like built-in models but are not permanently installed. Side panels can be added to make the slide-in range free standing. The drop-in unit is similar to the slide-in range but the top burners, oven, and broiler are supported by base cabinets or special construction.

CONSTRUCTION

Desirable features of construction and standards of performance for gas ranges presented in this chapter have been adapted in part from material of the American Gas Association's *USA Standard for Domestic Gas Ranges.*[3]

[2] *USA Standard for Domestic Gas Ranges*, American Gas Association, Inc., 1967, vol. II, p. 1.

[3] Ibid., vols. I, II, and addendas.

Figure 10-3. Built-in broiler, porcelain-coated cast iron grates, tilt-grill drains fat into removable trough; vent hood required for indoor installation. (Waste King Universal)

Frame and Exterior

The frame of the range is made of steel or iron, welded, riveted, or held together by screws. While the frame is very important to the life of the range, it is difficult for the homemaker to check it when buying a range. Here she must depend upon a reliable manufacturer and dealer and upon the requirements set up by the American Gas Association. The presence of the American Gas Association seal is assurance for the homemaker that the range will give safe and efficient performance and that it is of substantial and durable construction (Fig. 10-4).

Panels of sheet steel or sheet iron attached to the frame give the range the shape commonly attributed to a conventional range. If these panels have been Bonderized or otherwise given a rust-resistant treatment before the application of enamel coatings, they will be rust resistant. If synthetic enamel is used to coat the panels, they will be resistant to chipping and somewhat lighter in weight than if porcelain enamel is used. However, synthetic enamel will scratch more easily, is less stain resistant, and is undesirable for oven linings and the top surface of the range. By studying specification sheets or other information leaflets the prospective buyer can usually determine the kind of finish used on the range.

Range exteriors come in many colors and finishes. Enamel is the most popular finish and can be found in almost any color, although each manufacturer generally limits choice to a few colors. Metallic finishes do not tarnish but sometimes require a little more care than porcelain enamel. Some metallic finishes may look a little water-spotted when wiped with a damp cloth and will need to be wiped with a dry cloth to polish. The latter step is not so necessary with enamel finishes.

Figure 10-4. American Gas Association seal of approval. (American Gas Association, Inc.)

Some thought should be given to color choice, since a kitchen could become a confusing array of colors that neither blend, match, nor complement each other.

For comfort and economy a conventional range should be insulated. According to American Gas Association standards, the insulation must be enclosed and protected from objectionable exposure to air and flue gases and applied so as to produce and maintain uniformity of insulation. The material used for insulation is often Fiberglas. A silicone rubber seal is used on some ranges around the oven door. It is held in place by a stainless steel retainer strip.

Main Functional Parts

Surface burners may be arranged in a number of ways. Some ranges have a built-in griddle and/or a fifth burner. In the cluster arrangement four burners are grouped at one end of the range top, leaving work space on the opposite end, or are located in the center, leaving a smaller work space at each end. The divided arrangement has two burners on either end and work space in the center. Burners may also be placed along the rear of the work surface, leaving counter space at the front or with one burner at the back and three at the front. Another arrangement is a staggered one with two burners at the back of the work surface and two at the front. Choice of burner arrangement is determined largely by individual preference. If several large utensils are used at one time, a divided or staggered arrangement might be preferred, as it allows room for utensils to extend beyond the burners without touching one another. The cluster arrangement gives a larger work space. It is somewhat easier for a tall woman to work at a range with burners at the back than for a short woman who cannot easily reach as far.

On the apartment-size range it may be somewhat difficult to make sure that handles of utensils do not extend beyond the work surface and thereby constitute a safety hazard. It is also a little more difficult to use extra-large utensils on the apartment-size range because the burners are sometimes quite close together.

The gas input for the top, oven, and broiler burners is specified by the manufacturer. The gas input ratings for burners and pilots are checked at normal test pressure. Burners commonly have three "click" settings—high, medium, and low—with an infinite number of "nonclick" settings in between. The American Gas Association minimum requirement for the top burner simmer section is 1,200 Btu's per hour.

Burners usually are equipped with a burner bowl or aeration plate surrounding the burner head. The bowl is made of porcelain-enameled steel, aluminum, or stainless steel. It serves to control the amount of secondary air around the burner, reflect heat, and catch spillovers. If it is very shallow, it is not efficient for the latter purpose. It is highly desirable that a burner tray also be a part of the standard equipment on a range. If the range does not have aeration plates or burner bowls, it must have a burner tray, and this should be large enough to catch a major spillover. Burner trays should be rust resistant and have no edges that are hard to clean or sharp since they might cut the user.

Burners should be so constructed that they can be easily removed for cleaning purposes. At the same time it must be impossible to put them back in the wrong place, and they should fit into the right place easily. If a burner seems to need force to replace it, it probably is not being put in correctly.

When the gas valve is turned on, gas from the manifold flows into the mixer tube. At the same time air enters the mixing tube through the primary air inlet. Thus gas and air are mixed first in the mixing tube. When the mixture reaches the

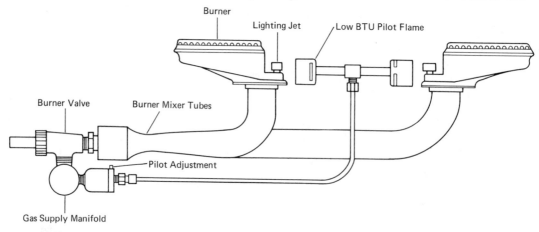

Figure 10-5. Burners with individual pilots. (Lincoln Brass Works, Inc.)

ports in the head of the burner, the gas is ignited, either manually or by a pilot light (Fig. 10-5). Air surrounding the burner head also mixes with the gas and, when the total air-gas mixture is correct, a pretty blue flame with a deeper blue inner cone results. If there is too much air in the air-gas mixture, the flame will tend to dance and not "sit" on the burner. If there is too little air, the flame will be yellow and carbon will be deposited on utensils used on the burner. The air adjustment on most ranges is relatively easy to make but requires patience. It is necessary to put the burner bowl and grate into position in order to check the burner flame properly. However, if the amount of gas coming into the burner needs adjustment a service man should be called.

Grates for gas range burners are commonly made from cast iron. They usually are covered with porcelain enamel, which makes them more attractive, easier to keep clean, and rust resistant. Cast iron will break if dropped at just the right angle, but otherwise will last indefinitely. American Gas Association standards require that grates must be so designed that they cannot be firmly placed in any but the proper position, or if placed improperly that combustion shall not be impaired. The grate arms must support a utensil as small as 2 1/4 inches in diameter placed centrally over the burner. Grates must also be designed so that they do not rock or shift laterally more than 1/8 inch.

Ovens vary in size. A standard size averages approximately 20 inches wide, 16 inches deep, and 16 inches high. Ovens in 30-inch ranges are generally larger. Ovens usually are finished in porcelain enamel. The porcelain enamel may be white, although it is commonly gray or blue or gray or blue speckled with white. Darker enamels show soil less, but all ovens need to be cleaned quite often if they are given a reasonable amount of use. Vapors and greases from food deposit on the oven linings and they are easier to clean if cleaned often. Some oven linings are coated with a fluorocarbon which makes cleaning easier. Other ovens are self-cleaning which eliminates practically all human labor in the cleaning process (Fig. 10-6). To meet American Gas Association requirements, oven linings must be of rust resisting metal or have a rust-resistant finish.

The oven bottom should be easily re-

Figure 10-6. Removable panels in upper oven; place in lower oven for automatic cleaning. (Caloric Corporation)

moved and also easily replaced. The sides with shelf supports are sometimes removable. The shelves in ovens should be non-tipping when pulled part-way out and should be so designed that they cannot be pulled all the way out accidentally. Oven shelves or racks are designed to be pulled part of the way out and then catch with a stop-lock. They should remain level at this point. The shelf should also have a back rail that indicates to the user that the back of the shelf has been reached and prevents foods from being pushed off the back. Oven racks should be made of rust-resistant material. On some models the oven door is removable. It should be reasonably easy not only to remove but also to replace.

Shelf supports are easier to clean if they are somewhat rounded and not too close together. Several rack positions are convenient. There should be one rack position for each full 3 1/2 inches of oven height. However, it is highly unlikely that a user will adjust racks in too many different positions, and too many rack supports make the oven difficult to keep clean.

A glass window in the oven door is a convenience, especially for people who just have to peek. Black glass makes it possible to see the interior of the oven only when the oven light is on. Standard oven temperatures and baking times can be thrown distinctly awry by opening the oven door too much during baking. Some oven windows have two panes of glass sealed in a steel frame at the factory. Oven windows should be especially well sealed so that vapors cannot collect between the layers of glass. The inner pane should be made of glass especially resistant to high heat, sudden changes of temperature, and physical shock.

The vent or flue outlet is the opening provided for the escape of the products of combustion, excess air, and vapors from cooking foods. The vent has an opening in the oven and on the backsplash. Ranges are usually vented on the front of the backsplash. This allows the range to be placed flush against the wall without the wall becoming discolored from oven vapors and heat. It is important that the vent openings in conventional ranges never be closed if the oven burner is to operate as it should and foods are to bake properly.

Broiling is cooking by direct heat, with food placed beneath the flame. The temperature is adjusted by the height of the flame and the distance food is placed from it.

Broilers are described as low or waist high. Low broilers are placed beneath the oven and are heated by the same burner that heats the oven. They must have at least a 3 1/2-inch height for adjusting the broiler pan from the bottom of the flame and three shelf positions.

The waist-high broiler may be located adjacent to and at the same height as the oven or within the oven. It requires less bending or stooping. It has a much greater distance for possible adjustments of broiler pan from the flame. Waist-high broilers located in the oven must be so designed that it is impossible to use the oven burner and the broiler burner at the same time. Waist-high broilers do make possible more flexible pan arrangements than can be obtained with the low broiler and are somewhat more convenient to use. Open-top broilers are located at counter level. When used indoors these must be properly vented with a vent hood.

The broiler pan is usually made of enameled steel or sheet aluminum. Preferably it is 1 1/2 to 3 inches deep and fitted with a grid. The deeper pans allow fat and juices from the food to drip away from the intense heat of the flame. These pans may also serve as roasting pans.

The grid should have enough slits or

openings in it to allow the fats and juices to drip through easily but not so much open space that too much heat can reach the drippings in the broiler pan. A grid that is similar in construction to oven racks is more difficult to clean than those that have wider strips of metal and openings placed farther apart. Broiler grids are usually of cast aluminum, stainless steel, chromium-plated steel, or enameled steel.

Some broiler pans are grooved and the grooves are slanted to a well at one end in which the drippings collect. The well should be so located that it is not directly under the flame.

To change the temperature of broiling foods during cooking it is usually necessary to remove the broiling pan from the oven or broiler. Some ranges have a "high" and a "low" broiler setting. At these settings the flame does not cut down or off as it would at a temperature setting below the regular "broil" position.

The rotisserie is a rotating device which enables foods to be broiled on all sides without the user having to change the position of the foods. It is popular for broiling fowl, kabobs, barbecued ribs, and certain other foods.

Controls

Gas, air, and/or electricity supplied to a gas range are regulated by manual, semi-automatic, or automatic controls. The burners are controlled by turning valve handles, by mechanical timers, or by special thermostatic devices. The most common device used for controlling surface burners is the valve handle, which turns to shut the burner off or turn it on. These valve handles may have different settings indicated either by a click sound, by feel, or both. Such valve handles must be clearly marked so that the open, closed, or intermediate positions can be determined from a distance of 10 feet. All valve handles

which control burners (not pilots) should rotate clockwise to close. Each main burner gas valve must be clearly identified with the burner it serves. Valve handles of the lock type are such that the handle must be pushed in before it will rotate to allow the gas to enter the burner. Oven controls and thermostatically controlled burners have valve handles of this type.

A pilot is a device to light a burner automatically, eliminating the regular need for matches. When it is used all burners must be provided with means for automatic ignition. Pilots supply a very small amount of heat. All constantly burning pilots, except those that are a direct part of an automatic pilot, should have a gas input rating of not more than 175 Btu's per hour. The pilot flame is blue, sometimes with a yellow tip. It has no air-mixing throat as does a regular burner. The adjustment of a pilot may be by means of a small screw near the manifold that can be turned to regulate the rate at which gas enters the pilot tip and consequently the height of the flame.

A small flash tube connects the pilot with the burner head. The flash tube is contiguous to the burner at a point where there is a special port called a lighter port. When the gas-air mixture flows through the mixing tube and into the burner head, it reaches the lighter port first and goes through the flash tube to the pilot. The mixture is lighted immediately by the pilot, and because of the large proportion of air in the flash tube, the flame flashes back to the burner head where the gas-air mixture in the ports is ignited. According to American Gas Association standards this should take place within four seconds after the gas is available at the burner ports. Flash tubes usually are made of a lightweight metal and often are separate from both the burner head and the pilot. Flash tubes that are permanently connected to the burner head do not get out of alignment as easily as those that are separate. Burners with individual pilots do not have flash tubes.

A range may have a separate pilot for the oven burner, broiler burner, and each surface burner; for each two surface burners (Fig. 10-5); or one pilot for all four surface burners.

Electric ignition is provided on some models. When a burner is turned on, an electric connection is made that heats a small coil red hot. This opens a gas valve to a pilot, the pilot flame lights, and then the gas in the burner head is ignited.

When the oven burner is provided with a constantly burning pilot, the burner must be under the control of an automatic pilot. One make of automatic pilot is electromagnetically controlled. When the constantly burning pilot is extinguished for any reason, the electromagnet is deenergized and the gas valve is closed. Gas to the burner, and in some cases to the constantly burning pilot, is shut off. Manufacturers' instructions vary slightly about the procedure to use in getting the automatic pilot functioning again. It is necessary to reheat the thermocouple junction located adjacent to the constantly burning pilot. One method is as follows: (1) Relight the constantly burning pilot and (2) after 30 to 60 seconds depress a valve button—usually a red button. This operates a keeper disk which will then allow gas to flow, and thus the automatic pilot will operate. As long as the pilot flame is lighted, the temperature will be high enough to keep the magnet energized, and the valve will remain open. Always read the specific instructions that come with a range for relighting a constantly burning pilot.

The oven thermostat is usually a hydraulic type and may be described as having four main parts: the bulb, the capillary tube, the diaphragm or bellows, and the dial. The bulb is located in the oven and

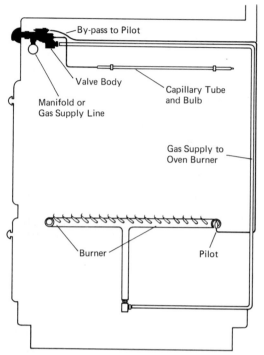

Figure 10-7. Oven burner, pilot, and thermostat. (Robertshaw Controls Company)

the capillary tube leads from it to the diaphragm. The dial is outside the range and its setting determines the tension of the diaphragm. As the oven is heated, the liquid in the bulb expands and pressure is exerted against the diaphragm. This in turn gradually closes a valve which controls the flow of gas to the burner (Fig. 10-7). The amount of pressure that must be exerted to close this valve depends upon the setting of the dial. If the dial is set for 400° F more pressure must be exerted than if it is set for 300° F. As the flame in the burner is cut down, the heat in the oven is reduced. As the oven cools, the pressure against the diaphragm is reduced and the gas valve is allowed to open again. In certain ranges some gas always flows to the burner after it is lighted, even when the main valve is closed by the thermostat. This gas, flowing through a small bypass valve, gives a flame on the burner which is called a bead flame. As a rule this bead flame is about the size of a match head.

Many ranges have an oven thermostat system slightly different than that just described. One type is known as the single thermostat system. The thermostat is of the same type but the oven burner cycles completely off and on instead of maintaining a by-pass flame. A constant pilot is supplied gas directly from the manifold. Normally it is lighted all the time unless there is interference with the gas supply. A heater pilot is ignited from this constant pilot when the burner is turned on or when the thermostat bulb cools and signals for heat. The heater pilot heats a mercury-filled bulb. The mercury vaporizes and opens the cycling valve, allowing gas-air mixture to enter the oven burner. When the oven is heated, the thermostat cuts off the gas supply to the heater pilot and the process is reversed. This system may employ a single-tube pilot unit instead of a heater pilot and a constant pilot.

Occasionally a thermostat may not be calibrated exactly. However, the oven should be checked several times before assuming that the calibration is not exact. (Suggestions for foods to use in checking oven temperatures are given at the end of Chapter 11. If the thermostat does need recalibration it should be done by a qualified service representative.

Oven burners have long had thermostatic controls; the device is used also with some surface burners to control the temperature of the saucepan or skillet in which food is cooked (Fig. 10-8). After the temperature on the heat control is set, the flame goes on full, is lowered, or is turned off—as the sensing element reacts to the temperature of the pan.[4] The sensing element is a spring-backed unit.

Thermostatic burner controls vary. In one type, after the desired temperature has been reached the gas is shut off completely and is reignited by a pilot light when the pan cools. In the other there is a bypass flame so that there is always a burner flame once the dial is turned on.

A flame set is desirable on the thermostatic-controlled surface burners as the size of the flame as well as the temperature can then be regulated. Certain foods, such as puddings and gravies, may stick before reaching the desired temperature if too high a flame is used. On those burners without a flame set the flame is on "high" until the set temperature has been reached. Flame set may be fixed at about half the regular size of the burner flame (Fig. 10-8) or the user may be able to adjust the size of the flame. If so, the dial has several markings in the flame set area.

As in the oven an overshoot in temperature may occur with the thermostatically controlled surface burner. This overshoot seems to be greatest on the first cycle when all components of the thermostat are cold, at high flame settings, high temperature settings, with utensils of poor heat conductivity, and with small loads of food. On the thermostatically controlled burner the flame should be about the same diameter as the bottom of the pan—if

Figure 10-8. Top burner control with fixed flame set; equipped to operate with a standby pilot or a bypass flame. (Robertshaw Controls Company)

---

[4] *Encyclopedia of Top-Burner Cooking with Controlled Heat*, Robertshaw Controls Company, Home Economics Department, 1955, p. 2.

the pan is aluminum; for other kinds of pans, a smaller flame seems to give better results.

Successful use of thermostatically controlled burners requires a clean sensing unit and appropriate saucepans. They should be of material that is a good heat conductor, be flat on the bottom, be of a size to almost cover the cooking area, should not tip, and should have a tight-fitting cover. As in using the oven thermostat, the dial must be set at the correct temperature for the food to be cooked. Large utensils with large quantities of food take longer to cook than smaller utensils with smaller quantities of food, but both can be done successfully. With only a few exceptions the utensil should not be preheated.

The automatic timer is a clock device that will turn the flame on to heat the oven at a predetermined time and turn it off when the cooking time has elapsed. Models differ in details of operation, but all are operated in conjunction with an electric clock. The first step in using an automatic timer is to make sure the electric clock is running and is at the correct time of day. Then the time to start or stop cooking and hours to cook are set. The oven control is set at the desired temperature. On some ranges there is a switch to set for automatic operation. When the cooking period is completed, this switch may turn back to manual control automatically, or it may have to be turned to manual before the oven can be used manually. This type of operation may be known as delay-cook.

Oven programming is possible with the low temperature thermostatic controls. With the single thermostat system the food is placed in the oven, and the heat control and the total cooking time are set. The oven heats to the temperature indicated and cooking begins. When the food

is cooked the control turns itself back to "keep warm" heat. This is usually about 170° F but may be as low as 140° F. This heat will be maintained four to five hours until the oven is turned off. This type of programming often is referred to as cook and hold.

Roasts lend themselves to long programmed cooking; vegetables are better for shorter programmed cooking. Broiled foods and foods that require exact baking time and foods that should be served immediately should not be programmed.

USE AND CARE

Regardless of how well chosen a range may be it will not give its best service unless it is used wisely. One of the first steps in good use is to *read the instruction book carefully and completely.* Almost all range manufacturers issue an instruction book with each new range. Some are much better written and more informative than others, but each is worth reading.

If the range is correctly installed, it will be level. If it is not level, it is impossible to bake nice even cakes.

Burners, grates, drip trays, and burner bowls should be kept clean by frequent washing. If spilled food is allowed to burn on these surfaces, it is much more difficult to remove. Any food spilled on enamel surfaces should be wiped with a dry cloth if the enamel is hot, provided the enamel is not too hot, and thoroughly cleaned with a damp cloth when the surface has cooled. Burners, except those with aluminum heads, may be cleaned by boiling in soda water, rinsing thoroughly, and drying in the oven.

The correct way to use a gas burner is to put the utensil on the burner, *then* turn the burner on. When the cooking period has ended, turn off the burner, *then* remove the utensil. As the gas flame is hot instantly there is no need to allow a burner flame to stay lighted while the pan is off the range—it is more economical and safer to turn the burner off when no utensil is on it.

Do not waste fuel. Water, unless under pressure, will not heat above 212° F. Use the burner on full to bring water to boiling temperature quickly, but turn it down to simmer to hold it there. If this practice is followed water evaporates less quickly, foods are less likely to stick, and there is less vapor in the air to deposit on kitchen surfaces and add to house-cleaning problems.

For economical use and for speed, do not heat more water than is needed. If a pint is needed to make a gelatin salad there is no reason to heat a quart.

Oven doors should fit properly so that heat does not escape. One way of checking this is to observe the casing around the oven door after baking a few products. If this is slightly discolored by vapors from foods baking in the oven, the door does not fit as tightly as it should. Ordinarily this can be easily adjusted by a service man. Doors will not fit properly if the range is not level. Although oven doors are made to hold quite a lot of weight, they are not made for chairs or for use as a step stool.

Cake pans should be filled about two-thirds full. If there is too much batter it will run over the edges of the pan during baking. If there is too little the cake will be thin, poor in texture, and not brown on top if baked a standard length of time at a recommended temperature.

Pans of food should be so placed in the oven that there is good heat circulation around each. This in effect means that pans should not be placed directly above one another, touching one another, or touching the sides of the oven. This caution is more important in baking pies, cakes, or cookies than in oven meals, but all baked foods will be better if this rule is followed.

Do not peek when baking foods. If standard recipes are baked in recommended pans placed in thermostatically controlled ovens for the required length of time, it is not necessary to keep looking at the food. Every time the oven door is opened cool air is allowed to enter and this in itself will change the amount of time required for baking. Excessive peeking means that the fuel bill will be higher than necessary.

The automatic timer can be set to go on and off during any 12-hour period. However, the choice of foods to be placed in the oven helps to determine how far ahead the timer should be set. Some foods reach spoiling temperature relatively quickly. Depending upon the temperature in the house, and the kind of pilot in the oven, the user will need to decide which foods to use for this type of meal. As a rule one should be cautious about foods containing milk, fish, or fowl, and other foods known to spoil rather easily.

A large sheet of aluminum foil should not be used in the oven, as it will interfere with the circulation of heat. If used directly on the oven bottom it may become overheated and adhere to the oven, making cleaning almost impossible. If it "must" be used when baking a juicy pie, cut a sheet just large enough to catch the drip from the pie. It seems a bit impractical for manufacturers to spend great amounts of time, money, and effort in designing an oven in which the heat circulation will always produce evenly browned cakes and then for the homemaker to ruin all this by a few cents' worth of aluminum foil improperly placed in an oven.

It is not necessary to preheat the broiler grid and pan when broiling foods, as the basic idea of broiling is to cook foods by direct heat from the flame rather than on a hot grid. Furthermore, the grid may be easier to clean if foods are placed on a cool grid. When broiling a few hamburgers, if you turn them so that they will be in the same position on the grid as before it will be much easier to wash the grid. After foods are broiled remove the pan and grid from the broiler compartment, let them cool slightly, sprinkle with detergent, and cover with a dampened cloth or paper towel. When the meal is over the grid can be cleaned easily.

Do not store the broiler pan and grid in a broiler compartment that is directly below the oven. Heat from the oven flame is hard on the material of an empty broiler grid and pan.

SPECIAL FEATURES

There are many special features on gas ranges, some in the gadget class and others widely accepted. It is up to the consumer to evaluate these special features according to her own needs and desires. Pull-out towel racks, hooks for hanging stirring spoons and pans, salt and pepper shakers, condiment sets, platform light, clocks, minute minders, and convenience outlets are a few of the many special features available.

Minute minders or time reminders will measure time up to several hours. When the time set has elapsed, a bell or buzzer will sound. Some minute minders ring indefinitely until turned off. They have the advantage of making sure that the user will hear them sooner or later. Others only ring once. Another type will measure two lengths of time intervals, one up to one hour and the other up to 10 or 15 minutes. In appearance the minute minder that measures time for several hours looks much like an automatic timer but should not be confused with it. Minute minders

only signal when a specified amount of time has elapsed; they do not control the flow of gas.

The automatic remote-control roast meter is a device that measures the inside temperature of the roast; it may turn the oven off or the oven may automatically be programmed to a keep-warm setting. A metal probe is inserted into the center of the roast in the same fashion as a meat thermometer is placed in a roast. The metal probe is connected by wire to a plug that fits into an opening in the oven wall and is connected to the automatic roast control on the front of the range. The roast-control dial is set for the degree of cooking desired to make the meat rare, medium, or well done.

Electric convenience outlets are provided on many gas ranges. The wiring within the range is for a 15-ampere circuit, and care should be taken not to overload it. One outlet may be controlled by the automatic timer. Wiring to the range depends upon the house circuit to which the range is connected. The overload protection is not located in the range, but at the fuse panel in the house.

Platform lights are found on the backsplash of the range. They should be located so that light shines into pans or utensils used on the work surface but not in the user's eyes. If the light is to shine effectively upon the work surface, it needs to be several inches above that surface. A number of platform lights have as their chief contribution attractiveness and glamor rather than usefulness. If the bulb is covered by a glass or plastic diffusing cover it is easier to keep the bulb clean. Almost any platform light makes a good nightlight if one is needed in the kitchen.

A gas range should hardly be chosen for the type of electric clock on it. However, if it is to serve as a kitchen clock it should have numbers large enough to be seen a few feet away. The face of the clock should be covered so that it is easy to clean. Neither the clock nor any other special features should be located so that heat from the oven vent will cause them to soil and darken.

LIQUEFIED PETROLEUM GASES

British thermal unit ratings at which various performance tests should be conducted vary with the type of gas. The American Gas Association lists a heating value of 3,175 Btu's per cubic foot for butane gas, 2,500 Btu's for propane gas, 525 and 1,400 Btu's per cubic foot for butane-air mixtures, 1,075 Btu's for natural gas, 535 Btu's for manufactured gas, and 800 Btu's for mixed gas.

Because the British thermal unit content of liquefied petroleum gases is different from that of natural or manufactured gases, some adjustments in burners need to be made when changing from one type of gas to the other. Although these adjustments may be relatively simple it is highly recommended by authorities in the field that they be made by service representatives familiar with the appliance.[5]

The American Gas Association states that main burners which are to be used with liquefied petroleum gases shall be provided with fixed orifices. If double coaxial orifices are to be acceptable the orifice in the needle must be sized for liquefied petroleum gases.

Each range must be permanently marked as to the type or types of gas for which it is to be used.

The National Fire Protection Association issues a pamphlet entitled *Recommended Good Practice Rules for Liquefied Petroleum Gas Piping and Appliance Installations in Buildings,* and these standards are followed by most states.

AMERICAN GAS ASSOCIATION STANDARDS

The standards set up by the American Gas Association are now subject to the approval of the American National Standards Institute. They are reviewed and revised every two or three years. These standards represent the basic requirements for safe operation, substantial and durable construction, and acceptable performance. Since many of the construction and performance requirements are too involved for the average person to evaluate, the blue star seal of approval is placed on each gas range that meets the minimum requirements of the association (Fig. 10-4). It is an important seal and no gas range should be purchased that does not carry this seal of approval.

The standards for domestic gas ranges are written in two volumes. Volume II is a completely separate set of standards for built-in domestic units, though they are much the same as those for other domestic gas ranges. Separate performance requirements are given in both volumes for ranges using liquefied petroleum and liquefied petroleum gas-air mixtures. Each volume has standards for outdoor broilers.

[5] Personal communication, Arthur C. Kreutzer, Liquefied Petroleum Gas Association, Inc., May 1956.

EXPERIMENTS

Experiments 1-3 require no special instruments or equipment and could well be done by beginning classes in household equipment. Experiments 4 and 5, for ranges using natural gas, are designed for more advanced students and some special equipment is required. For these students, the American Gas Association's *U.S.A. Standard for Domestic Gas Ranges*, Volumes I and II, would be a valuable detailed reference.

Experiments have been adapted in part from tests used by the American Gas Association, with modifications to fit college conditions of time, laboratory space, techniques of students, and ranges and testing equipment available.

Experiment 1. Uniformity of Oven Heat Distribution by Evenness of Browning of Plain White Cakes

Either a good plain white cake mix or a standard white cake recipe may be used.

Adjust rack to approximate center of oven. Turn on oven to temperature recommended by recipe or cake mix.

Prepare pans for baking. Use aluminum pans 8 or 9 inches in diameter. All pans should be alike.

Prepare cake mix or cake recipe, following directions carefully so that all cakes will be made by the same procedure.

Divide batter equally into two pans (for a box cake mix).

Place pans on the center rack. Pans should not touch each other or the sides of the oven. If using more than one oven, all pans should be placed in the same relative position in each range. Place at least two pans in each oven.

Bake for the required length of time. Be accurate in measuring the time.

Judge for evenness of browning, texture, and amount and type of crust.

This experiment may be varied to study effects of materials of pans, or diameter and depth of pans. For the former use pans as nearly the same size as possible but made of different materials. For the latter use pans of the same material but of different diameters and depths. For both experiments use ranges that have given comparable baking results in the first part of this experiment.

Experiment 2. Uniformity of Broiler Heat Distribution by Brownness of Toast

Follow directions for Experiment 6, Chapter 11, on electric ranges, with the following changes.

Place broiler pan between 2 and 5 inches from burner ports, close broiler door, and permit bread to toast for eight minutes (or less)—not more than ten minutes.

Does the uniform broiling area meet the AGA requirement of 80 percent of the grille area?

Experiment 3. Temperatures Maintained by Surface Burner Thermostat

(American Gas Association Standards state at the lowest thermostat setting the reading should be within ±5° F of the temperature set; at the highest temperature setting within ±20° F. The American Gas Association tests are done with a larger vessel; both water and oil are used; a thermocouple is used instead of a thermometer. This suggested experiment has been adjusted not only to classroom conditions but also to some gas company recommendations for checking a surface thermostat in a home.)

Place 1 1/2 quarts of tap water in a 4-quart covered saucepan. Insert a thermometer through a cork in a 1-inch opening near the center of cover. Set the control dial for the lowest setting. Turn on the flame. Read the thermometer at five-minute intervals for six consecutive readings.

How much does the average of the last three readings vary from the thermostat setting? The average of the first three readings? Repeat with water; set the thermostat at 200° F. If desirable to run tests at the highest thermostat settings it will be necessary to use a vegetable oil.

The bulb of the thermometer should be completely immersed about midway in the water.

Experiment 4. Thermal Efficiency of a Surface Burner

Determine gas pressure with gauge or manometer. If necessary, adjust pressure regulator to obtain 7 inches' pressure.

Record gas meter reading when utensil is put over heat. Turn burner on full. Heat water to 200° F. Turn off burner and record gas meter reading.

Compute thermal efficiency using the following formula.

$$\text{thermal efficiency} = \frac{(W \times W_v) \times (O_2 - O_1)}{\text{Btu/cu ft} \times Q \times CF}$$

where

W = weight of water in pounds
W_v = water equivalent of utensil (weight of utensil in pounds × 0.22 (0.22 is specific heat for aluminum)
O_1 = initial temperature of water in degrees Fahrenheit
O_2 = final temperature of water in degrees Fahrenheit
Q = gas consumption as shown by meter in cubic feet
CF = correction factor to reduce observed gas volume to 30 inches of mercury pressure at 60° F

Experiment 5. Broiler Performance

Adjust gas pressure to 7 inches water column.

Turn broiler flame on full. Close both oven and broiler doors. Let broiler heat for 15 minutes.

Read gas meter at the end of that time.

Make six lean ground beef patties that weigh about 1/3 pound each and are approximately 4 inches in diameter and 3/8-inch thick.

Place the meat patties on the broiler grid. Place one meat patty under the pilot or as close to it as possible.

Place broiler pan in broiler compartment so that the top of the meat patties will be 2 inches from the burner ports.

Broil patties for five minutes on each side.

Turn off flame and read gas meter.

Check to see if patties are done. Check gas consumption, and effectiveness of burner and pilot adjustments. Neither the burner nor pilot flames should go out nor should the automatic pilot shut off the main gas supply while patties are broiling. Check meter hands for speed of rotation to see if the burner flame is cut down during the cooking process.

BUYING GUIDE

Here are some questions to consider and some information to seek that should help you decide which range best suits your needs.

Type

What type of range do you want—free standing, built in or slide in?

Size

1. How much space do you have for the range? (Measure it, to be sure!)
2. What is the overall height of the range to top of back splash? The counter height? How wide is it? How deep is it?
3. Will it fit flush to the wall?
4. How far into the room do the oven doors extend when opened?
5. Does the range have toe space?

Materials and Construction

1. Does it have a welded steel frame?
2. Are parts treated to be rust resistant? What parts are finished with porcelain enamel, synthetic enamel, paint, japan, or other finishes? What care will each require?
3. How much will soil affect appearance and efficiency of the oven? (Washing the oven lining *often* is a lot easier said than done.)

4. Are there many places for dirt to collect? (Raised letters, crevices, and extra decoration and trim can be most difficult to keep clean.)

Seals and Service

1. Does the range carry the AGA seal? (If not, do not buy it.)
2. Does the range have other seals? (Do you know what they stand for?)
3. What does the guarantee actually guarantee? (Did you read it?)
4. Who pays for the installation of the range? The installation of the gas service line?
5. Where can you get service? Will a factory-trained service man be available at your dealer's or do you have to locate a service man?
6. What do you know about the reliability of the manufacturer and the dealer? (Time spent in investigating these is well spent.)

Surface Burners

1. How many burners does the range have? What Btu size are they?
2. How many settings are marked on the control? Are they well marked? Are they removable for cleaning underneath?
3. Are the control valves located where they will not interfere with use of the cooking surface or where they will not get excessively warm or easily soiled?
4. What burner arrangement is best for you? (Rear burners are not as useful for the short person or the person with certain physical handicaps.)
5. Are the burners the lock type? (Preferable if there are small children in the home.)
6. How many burners are thermostatically controlled? Are you willing to take time to learn to use the thermostatically controlled burner so that it will be truly worth the money? How are the burner and the sensing units cleaned?
7. Is there a burner tray? Is it well made? Can it be easily removed from the range? Does it have sharp corners or edges to make cleaning a problem?
8. Are the burners easy to remove and replace? Try it! (Recessed and lift-up tops make cleaning easier.)
9. Are the aeration bowls designed to catch a reasonable amount of spillover? Would the size of the aeration bowls be convenient for you to handle when washing? Are the bowls sturdy, not easily dented or bent out of shape? Will the finish be easily cleaned or will it require special care?

Oven

1. How large is the oven—depth, width, and height? Will it hold the Thanksgiving turkey? Large cooky sheets?
2. Do you need two ovens? One large oven? Would a small oven be sufficient?

3. Does the oven door fit tightly? At how many positions will the door stay open? Will it stay just ajar? (Makes it easy to cool the oven.)
4. Is the glass window in the oven and/or broiler door guaranteed for quality and for vapor-proof installation?
5. How many rack positions are in the oven? Would the rack supports be easy to clean?
6. Are the racks rustproof? Nontipping? Do they have an adequate back rail? Do the racks slide out smoothly and easily? Could you place an unbaked pumpkin pie on the rack and slide it into the oven without spilling the filling?
7. Does the oven burner light automatically? Is it easy to light manually or must you use the constantly burning pilot? Can you see the burner flame easily through an observation hole or by means of the door to the oven burner compartment?
8. Does the flow of gas to the pilot stop if the pilot is extinguished accidentally? (In order to meet AGA specifications, flow of gas to the burner must stop if pilot goes out.)
9. Does the oven have a light? How is it controlled? How do you change the lamp bulb? Is the oven light protected from accidental breakage when pans are placed in the oven? Is the thermostatic bulb likewise protected?
10. Can *you* remove the oven bottom and replace it? Is the door removable? (When you clean the oven, the fact that the salesman could do this will not mean much. You need to be able to do it easily.)

Broiler

1. Do you broil lots of food? How many broilers are there?
2. Can you stoop easily or do you need a waist-high broiler? Is the broiler conveniently located?
3. How many positions are available for the broiler pan? How is the height of the broiler pan changed?
4. Would the broiler grid be easy to clean? (Too many openings in the grid can be as unsatisfactory as too few.)

Other Considerations

1. What special features are on the range? Will you use them? (Some involve time to learn how to use or may require some adjustment in present cooking techniques to get their full benefit.)
2. Are the directions for use and care in the users' booklet clear? Are you sure which model of range is being discussed in the booklet? *Will You Read It Carefully Before Using Your New Range?*

CHAPTER 11
CONVENTIONAL
ELECTRIC RANGES

Cooking equipment need not be a separate range or built-in oven and cooktop. Three or four portable electric appliances might be the cooking equipment for a young couple, for elderly people, and for some other families.

For the large number of families who prefer a complete range, numerous features are available. For the vacation home, or for the principal home, one homemaker might choose a used range. Others will want a spacesaver-type that is approximately 21 inches wide, has three surface units, a thermostatically controlled oven, and some storage space. Homemakers in still other families consider that the cooking equipment appropriate to their family's life style is an eye-level or three-cooking-level range with one oven combination electronic and conventional type, a built-in barbecue grill, and a number of portable electric cooking appliances. A fourth group of homemakers wants a conventional one-oven range and a portable-type electronic oven. A fifth group wants a 30-inch wide, divided top, single-oven range.

Especially in homes where the homemaker is handicapped, a built-in cooktop and oven each located at a convenient height for the handicapped person might be considered.

The general guide is to buy according to needs, budget, what contributes to the well-being and happiness of the family. The determination of which needs are real and which fictitious is not easy. Two possible evaluation measures are results (kinds of cooked products) expected and convenience in use associated with different means of realizing results, for example, portable electric equipment

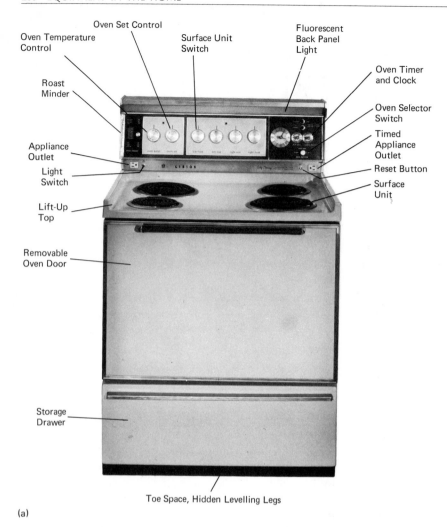

Oven Set Control

Oven Temperature Control

Surface Unit Switch

Fluorescent Back Panel Light

Roast Minder

Oven Timer and Clock

Oven Selector Switch

Appliance Outlet

Timed Appliance Outlet

Light Switch

Reset Button

Lift-Up Top

Surface Unit

Removable Oven Door

Storage Drawer

Toe Space, Hidden Levelling Legs

(a)

Figure 11-1. Free-standing range with parts labeled. (a) Exterior view. (b) Interior view. (Gibson Refrigeration Sales Corporation)

versus a camping stove versus a range. The homemaker and other members of the family should consider how different types of cooking equipment actually will serve in the home. An advertisement that shows an attractive model dressed in fashionable, casual clothes removing what seems to be a delectable meal from a range does not also show who planned, shopped, and prepared the food.

CONSTRUCTION, USE, AND FEATURES

The essential characteristic of conventional electric ranges is rather simple. Heat generated when electric current flows through surface or oven resistance elements is used to cook foods. Conventional electric ranges are available as free-standing, built-in, slide-in, or set-in models.[1] A free-standing range stands independently and consists of surface units, oven(s), and

[1] American Home Economics Association, *Handbook of Household Equipment Terminology,* 3rd. ed., section on range terminology.

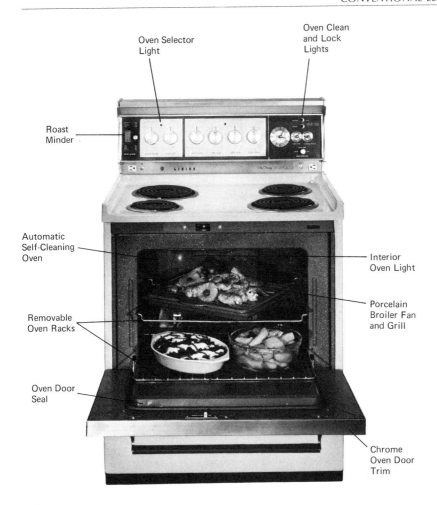

Oven Selector Light

Oven Clean and Lock Lights

Roast Minder

Automatic Self-Cleaning Oven

Removable Oven Racks

Oven Door Seal

Interior Oven Light

Porcelain Broiler Fan and Grill

Chrome Oven Door Trim

(b)

broiler(s). A particular type of free-standing range is the eye-level or high-oven type in which one or more oven(s) and broiler(s) are above the cooking top. Most are three-level type with a second oven and broiler below the surface cooking top; other free-standing ranges have an eye-level oven plus cooking top mounted on a special base cabinet that is used for storage.

Built-in cooktops are designed to be built into a counter top and built-in ovens have a frame that is installed in a wall or special cabinet. Slide-in ranges are complete units that stand independently and fit snugly between base cabinets. They may have unfinished sides. Set-in ranges are complete units that are supported by base cabinets, counter top, or special con-struction. The cooking top may be lower than the standard 36-inch counter height. Sides may be finished or unfinished.

Ranges have the following operating parts: switches and other controls for oven, broiler, and surface cooking; flat electric surface units or units under a smooth (glass ceramic) top; and oven units that cook foods by radiation and conduction (Fig. 11-1).

CONSTRUCTION

Free-standing electric ranges are available in the following widths: approximately 20–21 inches, 30 inches, 36 inches, and 39–42 inches. For several years the 30-inch width has been especially popular. At one time the 36-inch width was "standard" and older kitchen planning guides specify that 36 inches be reserved for the range. Depth including handles is about 26 1/2 to 27 1/2 inches. Height to surface cooking top usually is between 34 1/2 and 36 inches. Overall height is about 41 to 66–72 inches. (The maximum height given is for an eye-level range with built-in vent hood over the eye-level oven.) Dimensions for specific models are given on specification sheets.

The underlying structure or framework usually is a one-piece welded steel unit. The exterior side panels are synthetic enamel or occasionally porcelain enamel on steel. The back exterior panel generally has a synthetic enamel finish. The cooking surface and the oven lining usually are porcelain on steel, though some models have a special synthetic finish for the surface or a Teflon finish for the oven lining. Most commonly the oven lining is made in one piece, sometimes with extra removable panels for ease in cleaning. Also, some manufacturers may still provide extra disposable oven liners made of aluminum.

Self-cleaning ovens have smooth porcelain finishes. Continuous clean ovens, at present much less common for electric than gas ranges, have a dull finish that may be slightly textured (section on features in this chapter).

Sides, back, top, and bottom of the oven(s) have insulating material between lining and frame. The oven door is insulated. The door often also has a silicone seal. The insulation decreases heat loss from oven to room. Bottom or lower electric ovens usually are vented through a surface unit. Eye-level ovens do not have a special vent.

A storage drawer is common under the lower oven. The front leveling feet on ranges that have them sometimes are accessible behind the storage drawer front.

Ovens

Oven size varies for different ranges and oven volume is not easily estimated by eye. An apartment-size range may have interior oven space comparable to that of a 36-inch range. Clear interior width of 30-inch ranges may be 24 inches or less. Extra-large or odd-shaped baking pans do not fit into all ovens.

Two shelves and four pairs of shelf supports are usual. But more than four shelf *positions* may be possible because some ovens have more than four pairs of shelf supports; also, one shelf may be an offset, reversible type which can be at either of two heights in the oven, depending on the way the user places it on the shelf supports.

Shelves should be made of corrosion-resistant metal and should have stops so they cannot be accidentally pulled all the way out of the oven. Also, they should remain level when pulled out to the stop position.

The door for a bottom oven is hinged to the framework so as to stay flat in the fully open position. Usually provision is made for the door to remain ajar about 3 inches for broiling. However, for some models the door may be counterbalanced to remain open at several positions between ajar and fully open; and for some deluxe models with several heat settings for the broil unit the door of the lower oven is not open during broil. The door for an eye-level oven is hinged to the framework on the left side (Fig. 11-2). Door handles are attached by screws, in case replacement of the handle is necessary.

ELECTRICAL CHARACTERISTICS

Electric ranges usually are rated to operate on a three-wire, 118/236 to 120/240-volt supply. The three-wire supply permits two-wire 118–120-volt circuits in the range,

Figure 11-2. Eye-level range has hinged recessed porcelain top, plug-in surface units with infinite temperature-setting controls, lift-off lower oven door, storage drawer, and leveling feet. Teflon or catalytic oven panels are optional. (Admiral Corporation)

two-wire 236–240-volt circuits, and combinations of these—for example, simultaneous use of two-wire 118-volt and two-wire 236-volt circuits. Ranges may be designed to operate on the 120/208-volt supply used in some buildings, at wattages corresponding to these voltages.

The plug of the pigtail or range cord has three prongs and the cord has three wires. Assume voltage supply for the range is 120/240 volts. When the range is connected to the supply, one prong of the plug is at plus 120 volts, another is at 0 volts, and the third is at minus 120 volts. The three wires in the range cord connect the contacts in the house receptacle for the range with terminals on a terminal block inside the range.

The framework of the range may be grounded to the neutral or 0-volt contact on the terminal block.

Surface Units and Controls

The resistance element(s) of surface units that are time controlled for infinite heat settings or thermostatically controlled are likely to be part of the two-wire 240-volt circuit. The resistance elements of surface units not thermostatically or time controlled are usually part of the three-wire,

120/240-volt circuit. One arrangement is two resistance elements with a common terminal connected to the neutral of the supply while the other two terminals are connected to plus 120 and minus 120 volts respectively.

Four surface units are usual: three or two "6-inch" and one or two "8-inch" units. Surface-unit assemblies commonly consist of one or two tubular resistance elements connected to a small terminal block, a support for the element or elements, a metal ring, and a reflector pan which may be shiny or dark and which reflects or radiates heat back to the bottom of the utensil. If one tubular element is used, it may contain two or three wires insulated from each other (or for infinite heat setting and thermostatically controlled units, one wire).

The electrical connections between the surface units and the main terminal block in the range are controlled by the surface-unit selector switch.

Standard or conventional 6-inch units have wattages between 1,200 and 1,600 or so; a rated value of approximately 1,400 watts is common. Eight-inch units have wattages between 2,100 and 2,700 or so. Precise values are given on specification sheets. They are quoted at design voltage for the highest heat setting. Units with five or three heat settings in which the two elements of the unit have the same resistance are likely to develop wattages which decrease by a factor of two for successively lower heat settings. For example, a conventional unit with a rated wattage of 2,000 at the high setting of the selector switch would be likely to develop about 1,000 watts at the next-to-high setting, 500 watts at the next lower (middle) setting, 250 watts at the next to the lowest setting, and approximately 125 watts at the lowest setting.

Wattages of thermostatically controlled and infinite heat setting (single element) surface units usually are the same at all heat settings. Some models have a single surface "unit" that consists of three tubular elements and is designed for 4-inch, 6-inch, or 8-inch thermostatic control. The wattages of the different elements are different, but only one wattage is used for one size.

At least one manufacturer provides a "speed-heat" unit which does not have an extra-high *rated* wattage. The speed-heat 6-inch unit is rated at 1,250 watts but initial wattage is about 5,000. The length of time this high wattage is developed is short—15 to 30 seconds or so.

Surface units under smooth tops may be ribbonlike elements and both the 6-inch and 8-inch ones may be rated for 120 volts rather than 120/240 volts (p. 215).

Selector switches with infinite heat setting positions may have certain settings marked but they can be rotated to any position between the marked ones (Fig. 11-3). The position of the switch controls on-off cycling of the surface unit *unless* the unit is thermostatically controlled. At the highest setting of the infinite heat switch the unit is on all the time; at lower settings electric current is supplied intermittently, and relative lengths of on and off times vary according to the setting.

Thermostatically controlled surface units, on the other hand, do not have factory preset on-off cycling. Instead, a hydraulic-type mechanism usually controls the cycling. The movable and shielded "sensing element" is mounted in the center of the controlled unit and makes direct contact with the bottom of the utensil, except in the case of smooth top ranges. Spring-type mounting of the sensing unit provides for positive contact between pan and sensing element. A capillary tubing filled with a heat-sensitive liquid connects the surface-type sensing element

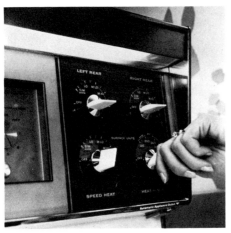

(a)

(b)

Figure 11-3. Surface unit controls with infinite heat settings. (a) Range with speed-heat surface unit. Right rear unit is a nominal 6 inches and left rear unit a nominal 8 inches. Infinite heat 6-inch left front unit is a speed-heat unit and infinite heat 8-inch right front unit is thermostatically controlled. (Frigidaire Division, General Motors Corporation) (b) Controls for glass-porcelain top range. Both 8-inch and 6-inch cooking areas are thermostatically controlled. (Corning Glass Works)

with a diaphragm below the cooking surface. The liquid expands when heated and contracts when cooled, thereby causing the diaphragm to move so as to open or close the electric circuit of the surface unit in accordance with changes in temperature of the bottom of the pan.

Details of construction and operation vary for different models. The selector switch usually is marked in numbers or temperatures but may be marked for cooking "zones," such as fry, boil, and so on. Recommended settings for different foods are given in the user's guide for the range.

Different settings may be given for uncovered versus covered pans. Recommended settings are established by measuring the interior temperature while cooking different foods. Aluminum utensils are usual and the range guide may recommend that aluminum or aluminum-clad utensils be used. Well-balanced pans with flat bottom surfaces are recommended for good contact with the sensing element. The user places the "filled" utensil on the surface unit and turns the selector switch to a particular temperature, number, or cooking zone—for example,

"lo-boil." Pickett found that temperature was controlled most effectively when the selector switch was set at the start and not changed during the cooking process.[2]

Philson found that good use of a manual control was more economical of fuel and produced less humidity and heat in the kitchen than use of a thermostatic unit. However, use of a thermostatically controlled unit was better than poor manual control.[3] Thus the way a homemaker uses the surface units on an electric range influences comfort conditions in the kitchen as well as the amount of electric energy used for cooking.

Oven Units and Controls

The oven units are the top unit, called the broil unit, and the bottom unit, called the bake unit (Fig. 11-4). These are parts of two-wire 240-volt circuits; that is, the units are connected across the plus 120-volt and

[2] Mary S. Pickett, "Controlled Heat Burners," *Journal of Home Economics*, Jan. 1962, vol. 54, pp. 38–42.

[3] Kathryn Philson, *Temperature and Humidity Studies*, Agr. Exp. Station, Home Economics Research Dept., Series No. 2, Auburn University, June 1964.

Figure 11-4. Lower (bake) unit and upper (broil) unit with reflector pan. (Corning Glass Works)

(a)

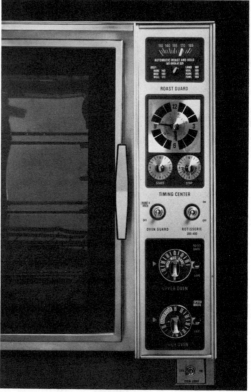

(b)

Figure 11-6. Oven controls. (a) Two-unit oven control. Note oven vent in top of back splash. (Corning Glass Works) (b) Oven controls for eye-level range next to upper oven. (Westinghouse Electric Corporation)

the minus 120-volt terminals of the terminal block. The bake unit is rated at some value between 2,000 and 3,200 or so watts; the maximum wattage of the broil unit(s) may be between 2,000 and 4,000 or so. During preheat and bake the bottom unit and all, part, or none of the top unit may be on; during broil only the top unit is on except for a few models that broil both sides at once (Fig. 11-5). Also the broil unit is used with a rotisserie when one is provided.

Most commonly the oven has a single dial control which is turned to the desired bake temperature or to broil. When the oven control is at some temperature other than broil, a signal light usually indi-

Figure 11-5. "No-turn speed-broil" unit. (Westinghouse Electric Corporation)

cates the end of preheat. Some models have more than one selector control (Fig. 11-6). In the United States, the oven temperature most commonly is controlled by an hydraulic thermostat. The hydraulic component is a small-diameter liquid-filled metal tube closed at one end by a diaphragm located in the switch component and at the other end by a small (pinched) bulb located at the center of one side or the center of the back of the oven.

As the oven heats, the liquid in the tube expands and exerts pressure on the diaphragm. When the temperature of the small or pinched bulb at the other end of the tube reaches the dial value set by the user, the diaphragm causes a switch to open and interrupt the oven circuit. As the temperature in the oven falls, the liquid in the tube contracts, the pressure on the diaphragm diminishes, and the electric circuit is closed.

During broil, on the other hand, power usually is supplied to the broil unit continuously, except that the unit will cycle on and off as a safety measure if the range becomes too hot.

Automatic timer controls are electric clocks which turn oven units, appliance outlets, and other components on and off at preselected times. For automatic oven control, the user generally follows three steps: checks that the electric clock indicates the correct time of day; adjusts two dials, one for the time the bake unit will go off and one for the number of hours the food is to bake; and sets the oven thermostat dial for the appropriate baking temperature. The bake unit goes on at the correct time (which may be immediately) to permit the food to cook for the preselected cooking time and goes off at the preselected off time. An oven set for timed control cannot be operated manually until the automatic timer is cleared.

When the control is used for an appliance connected into the appliance outlet, oven temperature is not set; instead, an on–off switch if provided on the appliance must, of course, be on. One control cannot be used for an appliance outlet and the oven at the same time.

Some automatic timers permit oven temperature to decrease to a factory-preset second temperature when the food is nearly cooked. The purpose is to maintain food at serving temperature. This usually

Figure 11-7. Eye-level range with automatic roast and hold, "Roast Guard," shown in Fig. 11-6(b). Note two-level, "Terrace-Top," cooking surface. (Westinghouse Electric Corporation)

is described as "programmed cooking." Also, a meat probe or thermometer may control temperature. As the temperature of the meat rises, the temperature of the oven decreases to reach and maintain a preselected meat doneness (Figs. 11-7 and 11-6B). More commonly, however, a meat probe *indicates* temperature or degree that the meat is done *without* controlling oven temperature.

Recommendations on use of automatic oven timers are given in users' booklets. Some suggest types of food suitable for automatically controlled oven cooking. For safety, thought must be given to types of food placed in an oven several hours before the food is to start cooking. Food that should not stand at room temperature for the time it would under the conditions selected for automatic control should not be used. Freshly prepared warm casseroles should not be allowed to stand at room temperature for several hours. Chilled or frozen casseroles may be used with safety.[4]

Electrical Accessories and Total Wattage Rating

Electrical accessories such as the range oven or back splash light, and electrical components of a built-in hood, rotisserie, or electric timer are parts of two-wire, 115–120-volt circuits; that is, the electrical accessory is connected across the grounded terminal and the plus or minus 115–120-volt terminal. Each appliance outlet also is part of a two-wire circuit and is protected by an accessible plug-type fuse or circuit breaker of 15- or 20-amperes capacity according to whether the two-wire circuit is 115 volts or 120 volts.

The kilowatt rating for a range (total wattage rating) given on the nameplate is the sum of the maximum wattages of surface units, bake and broil units, and electrical accessories. With one exception pertaining to an unbalanced load, Underwriters' Laboratories specifies that the nameplate rating shall include 1,400 watts for each 15-ampere attachment-plug receptacle circuit and 1,920 watts for each 20-ampere attachment-plug receptacle circuit.[5]

One manufacturer gives a total wattage of 19.79 kw at 230 volts for a self-cleaning

eye-level range. Wattages of the separate parts are as follows: three 6-inch surface units, 1,425 each; one 8-inch surface unit, 2,320; upper oven—bake unit, 2,220 and broil unit, 3,020; lower oven—bake unit, 2,660 and broil unit, 3,800; appliance outlet allowance and lights are not specified separately.

Another manufacturer specifies 15.09 kw for a single-oven 30-inch range corresponding to surface units of 2,700, 2,600, 1,250, and 1,500; bake, 2,600 and broil, 3,000; appliance outlet, 1,440.

USE

Top-of-range cooking includes simmering, boiling, frying, and grilling. Oven cooking includes baking, roasting, broiling, and "rotissing."

Top-of-Range Cooking

Good use of an electric range requires that pans fit the units. Pans with a base diameter of 5 to 7 inches "fit" 6-inch units, and those with a base diameter greater than 7 inches fit 8-inch units. Pans with flat bottoms make good contact with the unit and food is likely to cook more uniformly in them than in pans that are warped. When glass utensils are used, it is desirable to place a wire grid between the utensil and the unit if the rating of the unit is 1,500 watts or more. (Wire grids should be available at stores that sell glass utensils.)

Use of a thermostatically controlled unit was considered in the earlier section on electrical characteristics. For standard units with five heat settings, general recommendations are as follows:[6] high heat—quick boil, start cooking; medium high—brown meat, deep fat frying; medium—frying,

---

[4] C. Helgeson Huppler and M. McDivitt, "Bacteriological Implications of Holding Casseroles in Automatic Ovens," *Journal of Home Economics,* Dec. 1964, vol. 56, pp. 748–751.

[5] Underwriters' Laboratories, Inc., *Standards for Safety—Household Electric Ranges,* UL 858, July 1969, paragraph 239.

[6] The recommendations in this section generally follow those in *The Electric Range Book* prepared by the Edison Electric Institute—EE1 70 R-314.

heating milk, cooking eggs and cheese; low or simmer—finish cooking, slow boil, braise; warm—melt chocolate, keep foods warm.

The general procedure for pressure cooking is to place the cooker on the unit at high heat until steam escapes, then switch to medium or low to maintain pressure.

Grills and griddles are available on some models. As applied to ranges in the United States, a grill usually is a solid metal component permanently installed over direct heat such as a surface unit. A griddle is a nonpermanently installed accessory designed to be placed over one or two surface units. A griddle used with a thermostatically controlled unit provides controlled-heat cooking for pancakes, hamburgers, French toast, cheese sandwiches, etc.

Oven Cooking

Foods that are to be baked usually are placed in an oven that has been preheated to the baking temperature. This also is the usual case for roasting, though refrigerated roasts may be started in a cold oven. Actually most foods probably will cook satisfactorily from a cold start and foods are so cooked in ovens operating under automatic or programmed control. But modern recipes usually specify times and temperatures for a preheated oven.

The results obtained in oven cooking do not depend solely on the range; preparation techniques and quality of the uncooked food are also important. On the other hand, proper adjustment and use of the oven are important factors in obtaining acceptable products.

For both oven *and* surface cookery the range should be level. A homemaker can check levelness with a bubble-type level or, less accurately, by placing a pan of water on an oven rack. If the water level is

uneven the range or rack needs leveling. Adjusting screws are provided at the bottom corners of some ranges. If no screws are provided, thin wood shims (sticks) can be used.

The calibration of oven thermostats is sometimes incorrect over part of the temperature dial. A homemaker can check correctness of thermostat calibration by following standard recipes. If foods in specified utensils consistently bake in a shorter or longer time than the recipes indicate, she might logically question the correctness of the thermostat calibration.

The following are some general recommendations for baking: Arrange racks before heating the oven. When using only one rack, position it in the center of the oven. Place a single pan at the center of this rack. Position two or more pans so they do not touch each other or the sides of the oven. When using two racks, divide the oven in thirds. Avoid placing pans directly over each other.

Use tested recipes from reliable sources, measure ingredients accurately, and use recommended utensils. For glass utensils, reduce oven temperature by 25 degrees; for ceramic utensils make temperature adjustments suggested by the utensil manufacturer. (This might be a reduction of 25 to 75 degrees.) Set a timer to signal when the food is to be removed from the oven.

Roast suitable cuts of beef, pork, lamb, veal, and poultry, fat side up, in an uncovered, shallow roasting pan. Use a rack under boneless cuts. The broiler pan and its grid are useful for large boneless cuts. Use a regular meat thermometer or the oven meat thermometer, when provided, to tell accurately when the meat is done. The oven meat thermometer cable has a skewerlike probe at one end and a plug at the other. The plug is inserted into the special receptacle inside the oven and the probe is inserted into the center of the thickest muscle of the meat cut.

The *Handbook of Household Equipment Terminology* defines a rotisserie as a motor-driven device for rotating food during cooking; it defines rotissing as cooking food on a rotisserie. A spit is run through the center of the meat, poultry, fruits, and vegetables to be rotissed. Most manufacturers use the broil and bake units as the sources of heat, but some use only the broil unit. The foods cook as the spit rotates in the oven. The broiler pan is used to catch drippings.

According to *The Electric Range Book,* rotissing meats is about one-third faster than roasting. Small items should be placed in the center of the spit. Meat is anchored by a pair of prongs and thumb screws supplied with the unit. Wings and drumsticks of poultry are tied with string to prevent "flapping." Some manufacturers have the rotissing temperature—about 300 to 400 degrees—marked on the oven control. An ordinary roast meat thermometer —not the automatic one that might be supplied with the range—is useful for determining when the food is done. The users' booklets give temperature settings and minutes per pound for turkeys up to about 12 pounds.

To broil is to cook by direct heat. For broiling in the oven usually only the upper (broil) unit is on. (An exception is illustrated in Fig. 11-5.) If the oven is designed for broiling with the door open, high-temperature or fast broiling is obtained by placing the broiler pan so the top of the food is close to the upper unit; for low-temperature or slow broiling the pan should be placed so the food is 3, 4, or more inches from the upper unit. At moderate temperatures, meat cooks more uniformly and shrinks less than at high temperatures.

Some ranges provide a control for different broiling rates. To test whether meat cuts that are at least 1 1/2 inches thick are done, an automatic or nonautomatic meat thermometer can be inserted through the side of the cut.

A rack or grid is used in the broiler pan to allow meat drippings to fall into the pan where they are protected from the direct heat of the upper unit by the meat itself. Aluminum foil should not be used to cover the entire grid because the drippings will remain on it and may get hot enough to smoke and possibly burst into flame.

MAINTENANCE

A standard comment in the literature on cleaning electric ranges is that care is almost entirely a matter of maintaining appearance. The comment is correct in that no regular servicing is required; however, any implication that the appearance of the oven is always easy to maintain is not correct.

The appearance of an oven that is not the self-cleaning type (see the next section on features) is most easily maintained by cleaning after each use (and after the oven has cooled). The oven liner should be wiped with warm sudsy water, rinsed, and dried, thus removing grease and spillover before the food is burned on in the next use. Sometimes wiping is not enough. If any special oven cleaner is needed, directions on the package should be followed very carefully. Oven racks too may need cleaning after each use and the broiler grid and pan always do.

Some oven doors are removable and this is a help (Fig. 11-8). Some ovens have removable plain or Teflon-coated components (Fig. 11-9). The bake unit or the broil unit may be removable to make cleaning easier. If removed, the unit should be pushed firmly back into place after the

Figure 11-8. Door lifted from hinges for easy access to oven. (Magic Chef, Inc.)

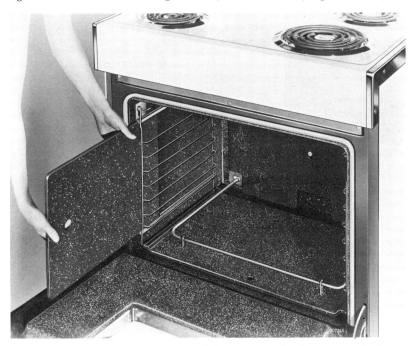

Figure 11-9. Removing oven side panel for ease in cleaning. (Monarch Range Company)

oven is cleaned. One home service department (General Electric) suggests that the proper materials for cleaning a rotisserie, spit, thumb screws, and so on, are soap and water, a stiff-bristled brush, and/or a scouring pad. A general suggestion is to soak parts in hot, sudsy water before scouring.

The principal other parts that require cleaning are reflector pans and metal rings around surface units, drip trays under these units, if provided, exterior surfaces, and accessories such as range hoods. Except for hood filters, these parts are likely to be metal or have a porcelain or synthetic enamel finish. Enamel finishes are cleaned with a detergent-dampened cloth,

Figure 11-10. Cooking surface tilted for cleaning surface with formed reflector pans. (Magic Chef, Inc.)

rinsed, and dried. Abrasives are not recommended for the enamel finishes; mild abrasives sometimes are needed, however, for metal parts.

Heating elements, such as surface units, are self-cleaning since food on them burns off. The tilt-top cooking surface provided on some ranges makes cleaning built-in drip trays easier and this is especially helpful when the reflector pans are formed in the surface (Fig. 11-10). Alternately, plug-in surface units that can be removed completely make access to the surface under the units easier.

Avoid applying cold water to hot or warm porcelain. When a spill happens in surface cooking, transfer the utensil to another unit if possible in order to avoid having food burn on the reflector pan. Wipe spilled foods off the surface as soon as practicable. Acids in milk, lemons, tomatoes, and other foods can etch porcelain. Do not slide utensils over porcelain or enamel surfaces.

If the range has an accessory hood, its filter(s) should be cleaned or replaced as directed by the manufacturer. Also, grease-covered surfaces in the hood should be cleaned not only for aesthetic reasons but as a precaution against fire. If the blower or fan motor needs periodic oiling, the manufacturer's guide should so indicate.

FEATURES

Different features appeal to and meet the needs of different users. Built-in models are important to some; eye-level ranges with attached hoods are desired by others; "self-cleaning" ovens are wanted by many; glass ceramic tops are liked by others; and so forth.

"Automatically Cleaning" Ovens

Ovens that clean automatically are continuous clean or self-cleaning type (Fig.

Figure 11-11. A budget ("stripped") model 30-inch self-cleaning range. Maximum connected load is 10.0 kw. Oven door is lift-off type and cooking surface can be tilted. (Norge, by Fedders Corporation)

11-11). Self-cleaning was introduced for an electric range oven in about 1964 and, as noted earlier, is still very much more common for electric ovens than continuous clean. Information on which type, if either, is used on a range, is available at the point of purchase. Also for self-cleaning models, a careful observer sees controls on the exterior of the range which must be set by the user to initiate the cleaning cycle.

Basically, the oven temperature is raised to about 900 degrees for a period of 30 to 90 minutes, depending on amount of soil to be removed.[7] Allowing for heat-up and cool-down time, the total cleaning cycle is about 2 to 4 hours. The soil in the oven is incinerated. Thus, the principle is simple. Design considerations involved in ac-

complishing the cleaning include the following: The oven must be brought up to temperature at a reasonable rate. The amount of air in the oven must be controlled for proper oxidation. Usually this is accomplished by openings at the bottom of the door. A temperature control must be used that will withstand the high temperature (the organic fluid now used in the hydraulic type will not). A smoke eliminator is needed. Typically this is a precious metal or a hot wire catalyst in the oven vent. The catalyst should raise the exhaust temperature sufficiently to cause complete combustion of the smoke. A special oven insulation is required. Exterior surfaces of the range must not become excessively hot. Special provision must be made in the range to protect an oven door window, if such a window is provided.

Users' guides give instructions on how to initiate the cleaning cycle. Self-cleaning ovens are claimed to get ovens completely clean, except for some ash which is picked up with a bit of paper, if the cleaning cycle is long enough and if the oven has been reasonably treated with respect to soil build up.

Parts of the range—such as reflector pans, oven racks, removable oven panels of a second oven in the range, broil-unit reflector pan (metal component over broil unit), and others—may be cleaned in some self-cleaning ovens. However, racks may discolor. Also, the oven facings and the part of the door around the seal do not clean automatically. In particular, the seal itself must be cleaned carefully by hand.

Continuous clean ovens may be so labeled or may carry a brand name that implies that they *stay* reasonably clean.[8] The

[7] A. W. Vonderhaar, *Oven Cleaning—A Developing Art,* 1970 Technical Conference on Household Equipment. Proceedings published by Association of Home Appliance Manufacturers.

[8] W. A. Hubbard, "Self-cleaning and Continuously-cleaning Ovens." Proceedings, College Equipment Educators' Conference, Iowa State University, 1970.

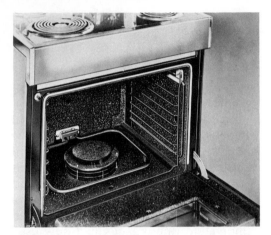

Figure 11-2. A continuous clean 30-inch electric range with lift-off oven door and removable side panels. This range oven also has a fan to diffuse heat in oven during bake and roast operations. (Monarch Range Company)

walls of the continuous clean oven have a mat finish characteristic of the poly-crystalline structure (Fig. 11–12). The surface has many microscopic openings and on that account offers a very extensive area on which soil can deposit. The extended surface is responsible for continuous clean action. Mechanisms that might be involved are evaporation of the soil, "cracking" (breaking down of large molecules into small ones), and oxidation. It has been suggested that, because the soil is spread over a large surface area, oxygen can get to it to form carbon dioxide and water.

The continuous clean oven works best for soils which are fluid at baking temperatures. This type oven is claimed to do a good job against spatter from broiled or roasted meat and poultry. It does not do well against thick deposits such as spill-overs of melted cheese or cherry pie filling.

Exhaust System

An exhaust system (hood accessory) that is an integral component is available with some eye-level ranges. One type has exhaust openings with filters in the range over the cooking top (below the eye-level oven) and an exhaust opening with filter in the hood over the upper oven. (See Fig. 12-3.) Inside the hood are a motor and

blower; the system functions only when the blower operates. When the blower is operating and the hood over the eye-level oven is open, the blower exhausts cooking vapors through openings in the surface units and an opening in the oven simultaneously; in the closed position of the hood, the blower exhausts over surface units only.

The exhaust system should be operated whenever odors, vapors, or smoke are expected during cooking. Further it usually is recommended that the blower be turned on at the *start* of the cooking operation. A caution noted by at least one manufacturer is to be sure to be at hand when the hood is operating to stop it in case vapors from a spillover catch fire.

The effectiveness of the system depends partly on clean filters. The users' guides tell how to remove them for washing or replacement. If the motor is not permanently lubricated, instructions may be given on this point.

Range with Glass Porcelain Top

Several manufacturers have reported plans to make a range with a glass porcelain cooking top. "Second-generation models" of the manufacturer who pioneered the concept are shown in Figures 11-13, 11-3(b), and 11-6(a). The "Counterange" electric range is available in several colors in free-standing and slip-in models. All have (pyrolytic) self-cleaning ovens. The range plus starter set of cooking utensils supplied with it is described as a "total cooking system." Utensils made especially for use with the range have ground and polished bottom surfaces. Other utensils, metal ones made by various manufacturers and Corning Ware utensils made by the range manufacturer, will work, but contact with the cooking surface is not likely to be as good as that for the utensils made for this range.

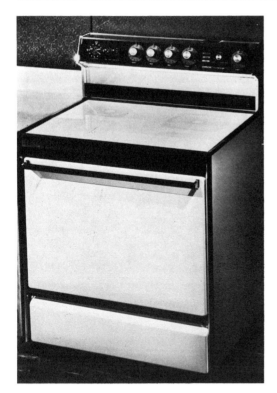

Figure 11-13. Self-cleaning range with one-piece glass porcelain cooking top. (Corning Glass Works)

Electrical characteristics are: power supply—120/240 volts, 3 wire, single phase, 50 amperes; total connected load at 240 volts, 11.7 kilowatts; upper oven unit at 240 volts—bake, 782 watts, and broil, 3130 watts; lower oven unit at 240 volts—bake, 2,760 watts. Surface units are rated at 120 volts and wattages are: two 8-inch cooking areas, 2,000 watts each; two 6-inch cooking areas, 1,200 watts each. Dimensions of the free-standing range are: width 30 inches; height to top of cooking surface adjustable by four leveling screws from 1/16 inch less than 36 inches to 36 5/8 inches. Height to top of back splash is 46 1/2 inches.

The outstanding special feature is that the entire top is sealed and the heating units are mounted underneath the top (Fig. 11-14). Sunburst designs show the 8-inch and 6-inch cooking areas. The metal trim around the glass-ceramic surface and on the front edge of the range top is brushed stainless steel. All the surface

cooking areas are thermostatically controlled. Indicator lights illuminate the setting selected. The clock space has a minute timer and an automatic oven control. As is usual with such an oven control, a timed appliance outlet is provided. An unusual feature is the oven indicator light that remains on as long as the oven is on, in addition to the usual cycling light that goes on and off during oven use.

Other signal lights indicate when the oven is at bake or broil, or when the oven

(a)

(b)

Figure 11-14. Components under cooking surface. (a) Insulation that forms thermal barrier below the heating elements. (b) Cutaway showing one of the mica cards with flat resistance wire grid and thermostatic control. (Corning Glass Works)

(a)

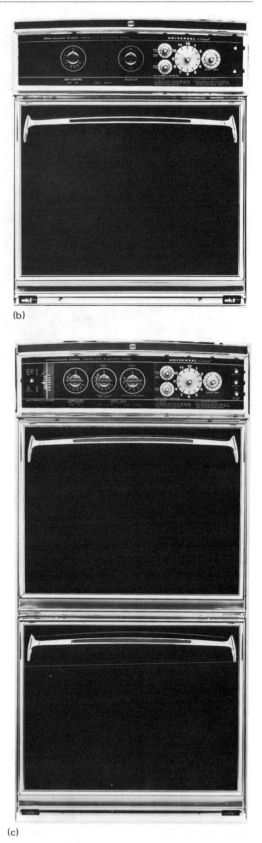

(b)

Figure 11-15. (a) Built-in double oven. Upper oven is (pyrolytic) self-clean and has electric meat thermometer. (General Electric Company) (b) Built-in single oven, continuous clean type. This oven has low-temperature broil control for closed door broiling at temperatures below 500° F. (Waste King Universal) (c) Built-in double oven. Control panel states continuous clean and meat minder. (Waste King Universal)

door is locked (throughout self-cleaning) and when the oven temperature is above 550° F. The locking control used in the cleaning cycle is a knob on the control panel on the back splash. Air flow through the oven to the room during the self-cleaning process is designed so the outside door temperature does not exceed 175° F.

The storage drawer in the range has kick space below it.

(c)

Figure 11-16. Built-in cooktop with staggered surface units and separate controls (not shown). (General Electric Company)

Built-In and Set-In Models

Built-in electric range units include single and double ovens and cooktops. Both 30-inch and 24-inch built-in ovens are available. Those illustrated in Figure 11-15 are 30-inch models. The cooktop shown in Figure 11-16 would have the controls for the units mounted elsewhere—for ex-ample, on the front of the hood. Controls for most cooktops are part of the cooktop.

Set-in ranges usually are 30-inch models, as shown in Figure 11-17.

Accessories and Additional Features

The range hood may be an accessory or it may be an integral part of the range as is true for the hood described in the earlier section on exhaust system and shown in Figure 12-3.

Other accessories are automatic meat thermometer, rotisserie, and double broiler

(a)

(b)

Figure 11-17. (a) Set-in 30-inch range with surface unit controls and plug-in surface units. (Whirlpool Corporation) (b) Self-clean set-in 30-inch range with push button surface unit controls on top of back splash. (General Electric Company)

Figure 11-18. Deluxe eye-level range. Total connected load is 20.7 kw. Bottom (master) oven is self-cleaning and broil unit in this oven can be set at high, medium, or low. Other features include two appliance outlets and plug-in surface units. (Whirlpool Corporation)

pans which give "cool" broiling when water is added to the lower pan.

Additional features include removable oven door, tilt-up cooking surface, reversible oven rack, light on back splash for cooking surface, oven light(s), storage drawer, oven fan, adjustable broiler control for high, medium, or low heat settings, two appliance outlets (one that can be automatically timed), plug-in surface units, and many others. The eye-level range shown in Figure 11-18 has many of these special features.

A different type of "special feature" is the kind of guarantee. The guarantee for the range shown in Figure 11-1 states that for one year the manufacturer will pay all costs for repairing or replacing any parts it finds defective. For the second through fifth year from the date of purchase the manufacturer will provide a replacement surface heating element, oven heating element, or surface element switch for any part found defective.

SAFETY STANDARD

The current safety standard is the Underwriters' Laboratories Standard for Safety—Household Electric Ranges, UL 858—1969, with subsequent addenda.

The UL standard notes that its requirements "section" is also approved as a USA standard. The requirements section covers scope, construction, performance, rating, and marking. The scope of the requirements covers household cooking equipment that is free standing, wall mounted, counter mounted, and rated 600 volts or less. Also covered are ventilating hoods that are provided as an integral part, or that are separately supported but arranged for factory-provided electrical connection to the cooking equipment with which they

are intended to be used. The requirements do not cover electronic or self-cleaning ovens.

The construction section covers mechanical and electrical parameters. The following are examples of electrical requirements (including several new ones):

1. A grounding lead of a flexible cord shall have a green color, with or without one or more yellow stripes required by the National Electrical Code for such a lead, and no other lead shall be so identified.
2. Except for the leads to a blower motor, the wiring of a range shall be so located that it will not be exposed to the vapors from the vented oven.
3. A heating element shall be so con-

structed that any motion that might occur during normal cleaning will not concentrate any stress on electrical connections.

4. An appliance employing a motor rated 1 horsepower or less shall incorporate adequate thermal or overcurrent protection as part of the appliance.

5. If the maximum current of the complete appliance when calculated in accordance with the National Electrical Code demand load is more than 60 amperes, it shall have two or more circuits provided as part of the appliance, each of which shall have overcurrent protection rated at not more than 50 amperes.

An example of a mechanical requirement is: With one special exception, each filter employed in a ventilating hood shall be readily removable for cleaning or replacement.

Performance requirements are of a different type than a homemaker might consider (See p. 227 in the Buying Guide for the latter type.) The Underwriters' Laboratories performance requirements relate to power input, temperature increases at different locations within and on the surfaces of the range, fire or shock hazard when a broiling element is operated continuously under specified conditions, and others.

The performance section also describes installation forms. Form O essentially describes flush-to-wall and cabinet installa-

tion. Specifically, "the range is installed close against a vertical wall at the back. The end (or ends) of the range in which surface units are spaced 6 inches or more from the end is also installed close against a vertical wall. The end (or ends) of the range in which surface units are spaced less than 6 inches from the end is installed close against a base cabinet. . . ."

Form E covers eye-level ranges. "The range (with either one or two ovens) is installed close against a vertical wall at the back. . . ." The remainder of the form pertains to cabinets above or close to the sides in relation to door opening and is described in paragraph 203 of the standard.

The rating section specifies that the appliance shall be rated in amperes or watts, and also in volts. Further, the spread of a voltage rating shall be not more than 20 volts. The ratings of appliance outlets were noted earlier (p. 208).

The marking section includes the material relative to catalogue number, etc., as given later in the Buying Guide at the end of this chapter. A requirement effective January 1970 is that a ventilating hood if provided shall have a permanent marking so located as to be readily visible after the hood has been installed and shall call attention (1) to the need for frequent cleaning of all grease from the fan itself and from all other grease-laden surfaces, and (2) to the need for frequent removal and cleaning or replacement of any filter unit provided.

EXPERIMENTS

Experiment 1. General Laboratory Tests—Surface Temperature, Electric Leakage, Accuracy of Automatic Oven Control, Oven Cool-Down Time

1. "Ready" five thermocouples for use in a later part of these tests. Install the insulated thermocouples in a central plane in the oven as follows: one at the center, one midway between the center and

right corner, one midway between the center and left rear corner, one midway between the center and right front corner, and one midway between the center and left front corner. The thermocouples can be secured to the rack with freezer bag fasteners or paper clips, but the junctions should not contact the rack. In step 2, the five positive free ends of the thermocouples will be connected to the positive terminal of the temperature-indicating instrument and the five negative free ends will be connected to the negative terminal of the indicating instrument.

2. Check how closely automatic oven controls can be set as follows. Set oven thermostat for 350° F. Adjust automatic oven control to turn on oven "immediately" and to stop in "one" hour. Within how many minutes can the controls be set? After approximately 40 minutes connect free ends of thermocouples to temperature-measuring instrument. As soon as the control turns the oven off or after one hour by the laboratory clock, proceed with step 4.

   Note: Some range ovens will be at a "steady state" temperature in one hour. For other ranges, the oven temperature may be decreasing to a keep-warm or serving temperature.

3. While the oven is heating, turn on all surface units *except* a fast-heating surface unit to "high" and let heat for seven minutes. With units on, check: (a) surface temperatures (use Alnor or other surface-temperature measuring instrument) and (b) electric leakage with neon tester and ground wire or current instrument, resistor, and ground wire. Do surface temperatures seem excessive? At what locations, if any, did you find electrical leakage? Turn off units.

   Caution: "Watch" time while doing the temperature and leakage tests so that you do not "miss" the time the automatic oven control turns off the oven units. Also take care not to touch the hot units.

4. Observe the time for the oven temperature to fall 100° F below the temperature at the end of the "one-hour" cooking period by taking readings at one-minute intervals.

   Note: The cool-down time is one measure of the effectiveness of oven insulation. Extremely slow cooling of the oven means that an adjustment in standard cooking times may be necessary when foods are cooked with the oven set for automatic control, to compensate for continued cooking during the cool-down period.

Experiment 2. Accuracy of Thermostat Calibration with Temco Pyrometer, Leeds Northrup or Other Temperature Indicator

1. Adjust instrument according to directions in instruction booklet or on indicator case.
2. Install five insulated thermocouples in a central plane in the oven

as follows: one at the center, one midway between the center and the right rear corner of the oven, one midway between the center and the left rear corner, one midway between the center and the right front corner, and one midway between the center and the left front corner. (The thermocouples can be fastened to a rack with paper clips; however, the junctions of the thermocouples should not contact the rack.)

3. Connect the free ends of the thermocouples to the temperature-indicating instrument through a selector switch.

4. Set thermostat dial at 400° F. (The oven should be at room temperature when the test is started.)

5. Observe temperatures at one-minute intervals through preheat and through the first three on-off cycles. Record preheat time. Record durations of successive on-off cycles.

6. After the thermostat has opened the circuit four times, record oven temperatures at half-minute intervals for three cycles. That is, record temperatures for the last three of six on-off cycles after preheat temperature has been reached.

7. Average the maximum and minimum temperatures during the last three cycles for which temperatures were recorded. If the average thus found differs from 400° F by more than plus or minus 15° F, change the thermostat calibration if possible. Let the oven cool; then recheck the 400° F setting.

8. Set thermostat dial at 300° F. The oven should again be at room temperature when the test is started. Check accuracy of thermostat calibration at this setting by the procedure used for the 400° F setting. Calibration is assumed correct at 300° F if the average value found at that setting differs by not less than 90 degrees or more than 110 degrees from the average value found for the 400° F setting. For example, if the average temperature corresponding to a 400° F dial setting is 405° F, the average temperature corresponding to a 300° F dial setting should be between 315° F and 295° F. If not, no change should be made in thermostat calibration at 300° F, but the user should compensate for this inaccuracy of the calibration for recipes that call for 300° F.

Experiment 3. Thermal Efficiency of Surface Units

The thermal efficiency of a surface unit is determined by heating a specified amount of water in a utensil of known weight and specific heat. The change in temperature of the water and the electric energy used are recorded. The thermal efficiency expressed as a percent is

$$\frac{[\text{wt of water} + (\text{wt of utensil} \times \text{specific heat of utensil})] \times \text{temp change} \times 100}{0.239 \text{ watts} \times \text{seconds}}$$

Weights of water and utensil with cover are in grams. The specific heat of the utensil depends on the material of the utensil. The specific heat of aluminum is 0.22. The specific heat of stainless steel is

0.12. The temperature change is in degrees centigrade. Specific procedure is outlined below.

1. Set up an electric circuit to measure watts supplied to range.
2. Weigh the utensil and lid. (Weight of thermometer is assumed negligible.) For a 6-inch surface unit, use a 2-quart utensil. The lid of the utensil should have a hole in it for a mercury-in-glass thermometer.
3. Add 900 grams of cold tap water to the utensil.
4. Place the thermometer in a cored rubber cork and place cork with thermometer in the hole in the lid. Adjust the thermometer so that the bottom of its bulb is at least 1/8 inch above the bottom of the utensil.
5. Place utensil with thermometer on the surface unit whose thermal efficiency is to be measured.
6. Record temperature of water.
7. Turn control on surface unit to "high" and start a stop watch.
8. Record wattmeter reading.
9. Turn surface unit off when temperature of water reaches 90° C. Record number of seconds required.
10. Compute thermal efficiency.

If the thermal efficiency of an 8-inch unit is measured, a 3- or 4-quart utensil and 1,800 grams of water should be used. If the material of the utensil is not aluminum or stainless steel, get the specific heat of the material from a table in a handbook.

Experiment 4. Effectiveness of Pyrolytic Self-Clean Ovens

1. Use the following materials for applying soils to ovens: canned cheese soup, barbecue sauce, gravy mix, paste of flour and water.
2. Remove racks from oven. Apply soils in strips about 2 inches wide to bottom, sides, and back of oven with wax paper and/or toweling as convenient. Use about 2 teaspoons per strip. Make a chart to identify locations of different soils.
3. Heat oven to 350° F from a cold start and permit oven to operate at this setting for 20 minutes.
4. Turn off oven and let cool to room temperature.
5. Set self-clean control to operate for two hours. While the oven is in the self-clean mode, observe *exterior* surface temperatures with a suitable instrument.
6. At the beginning of the next laboratory period check cleanliness of ovens. If an oven is not clean repeat the cleaning cycle.

Experiment 5. Uniformity of Oven Heat Distribution by Brownness of Cookies

The uniformity of temperatures at one height in an oven may be checked by baking sugar cookies. Use a commercial refrigerated bake and slice dough or prepare the following recipe.

Sugar Cookies (about 30)

1/2 C. shortening	1/2 t. salt
1 C. sugar	2 1/2 C. cake flour
2 eggs	(sifted before measuring)
1 t. flavoring	2 t. baking powder
1 T. cream	

Cream shortening and sugar. Blend in eggs and beat with an electric mixer at high speed for about two minutes. Add vanilla, cream, and sifted dry ingredients. Beat at low speed just long enough to work in dry ingredients (two to three minutes).

Chill batter in frozen food compartment of a refrigerator or in a freezer until it can be shaped easily into a roll about 2 inches in diameter. After the batter has been molded into a roll, chill again.

Preheat oven to 375° F and let the oven operate at this setting for at least 15 minutes. Place uniform slices of cooky batter, about 3/16-inch thick, on identical baking sheets. Place cookies in oven at the end of fourth "on" cycle.

If a roll of prepared bake and slice dough is used, cut slices by placing thread around the roll and pulling the two ends of the thread together.

Bake for approximately eight minutes on a rack placed about 7 inches from upper unit.

Remove baking sheets from oven. Place cookies on racks in same relative positions as the cookies had in the oven.

Observe brownness of tops and bottoms of cookies baked in different locations in the oven. For which locations in the oven, if any, is brownness nonuniform?

Experiment 6. Uniformity of Broiler Heat Distribution by Brownness of Toast

Measure area of broiler grid. Cover entire grid with slices of white bread. Place broiler pan with grid on a rack about 5 inches from upper unit. Toast bread on one side only for about seven minutes (less time if you smell smoke or more if toast is not brown). Record distance from top unit and time it took for bread to toast.

Note: The oven door should be ajar for most ranges. However, check user's instructions for the particular range you use.

Remove broiler pan with toast. Measure area of uniformly toasted bread.

Was the area of uniform broiling central with respect to the entire broiling area? What percent of the total broiling area gave uniform broiling?

Experiment 7. Uniform Broiling of Beef Patties

Determine how far meat must be placed from the upper unit to get uniformly done beef patties, as follows.

Shape ground beef into patties about 4 inches in diameter and 1 1/2 inches thick in a mold, if possible.

Broil different groups of patties in successive trials at several distances from the upper unit.

Cut through patties vertically to observe whether they are cooked uniformly.

Broil patties with oven door fully open unless user's booklet specifies otherwise. Observe uniformity of cooking and shrinkage for patties placed at different distances from the upper unit.

Experiment 8. Oven Cookery

The following procedures will give an indication of the performance of the oven and the correctness of the thermostat calibration at two temperatures.

Check medium-temperature performance with plain cake. Use a cake mix, a favorite recipe, or the following recipe.

Plain Cake (two 9-inch layers)

3 C. cake flour, sifted	1 C. sugar
2 1/2 t. baking powder	1/2 C. shortening
1 1/2 t. salt	2 eggs
1 t. vanilla	1 C. milk

Sift flour, baking powder, and salt. Add vanilla and sugar to shortening and cream until mixture is light and fluffy (four to six minutes with electric mixer). Add eggs and beat two minutes at high speed. Add sifted ingredients to shortening mixture in about three parts, alternating with 1/2-cup milk after first and second additions of shortening mixture. Mix at low speed for about one-half minute after each addition.

Bake in two lightly greased, paper-lined or Teflon-lined aluminum cake pans for 25 to 30 minutes in an oven that has been preheated to 375° F.

Test to see if cake is done by inserting a wire cake tester or a toothpick. If the tester comes out clean, the cake is done. It should be lightly browned and should have shrunk slightly from the sides of the pan. The surface should spring back when pressed lightly with a finger.

If the cakes for the recipe given above are not done in 25 to 30 minutes, the actual oven temperature is probably less than 375° F. Other indications of too-low temperature may be coarse texture or sogginess on the bottom of the cakes. If the cakes are overbaked in 25 to 30 minutes, the actual oven temperature is probably higher than 375° F. Other indications of too-high temperature may be humping in the center, cracking on top, and dryness.

Check high-temperature performance with baking powder biscuits. Use commercial refrigerated biscuits or the recipe given below.

Baking Powder Biscuits

2 C. all-purpose flour, sifted	4 T. shortening
2 1/2 t. double-acting baking powder	3/4 C. milk
1 t. salt	

Sift baking powder, flour, and salt. Cut shortening into dry ingredients with a pastry blender until mixture has the consistency of coarse corn meal. Form a cavity in the center of flour mixture, pour in milk, and stir with a fork. Turn out on lightly floured board and knead *gently* about one minute. Pat or roll dough 1/2-inch thick and cut with biscuit cutter. Place on a 10-by-14 inch cooky sheet.

Bake in a 450° F oven for 12 to 14 minutes.

The appearance and eating quality of the biscuits will depend to a large extent on technique in preparation. However, if the actual oven temperature is very different from 450° F, an acceptable product is unlikely with any conventional preparation method.

BUYING GUIDE

General

1. Does the range have the Underwriters' Laboratories seal of approval? Note that this means varying requirements are met according to whether the range was manufactured before or after July 1, 1969 (for the 1969 standard). Also, additional requirements are added between revisions.
2. Is the following marking information available at the front of the range as required by UL: Name or trademark of the manufacturer, the catalogue number or the equivalent, the installation-form designation, and the electrical rating (total connected load in kilowatts at rated voltage for range or for cooktop and oven if these are to be connected separately)?
3. What are the exterior dimensions in inches: width, depth or depths with oven doors(s) closed and open? Height to cooking surface? Overall height with and without vent hood, if the hood is to be installed? Height to top of hood for an integral hood?
4. If an eye-level type, what is the clearance in inches between the surface cooking top and bottom of the eye-level oven? Is this satisfactory for utensils used on cooking surface?
5. What are the finish materials on front and sides? Is the hue of front and sides the same? (If not, will it matter for the proposed installation?)
6. Is there a drawer or other storage compartment? Is it likely to be useful?
7. Where does the oven vent to the outside of the appliance?
8. Are there leveling feet?

Cooking Surface and Surface Unit Controls

1. What materials are used for the cooking surface, reflector pans, and separate drip trays if provided?
2. Is the cooking surface recessed, that is, is it lower than 36 inches but with 36-inch high edges?
3. How are the units arranged: in a cluster in the center, left, right: divided with two at each side; in a row; staggered; or in another way?
4. Six-inch units: Where are they located—left front, right rear, and so on? Are these locations convenient for the size of utensils used frequently on the units.
5. What is the wattage at design voltage of each unit?
6. Thermostatically controlled unit: Is the location convenient? Nominal size—8-inch only? Or 8-inches, 6-inches, and 4 inches?
7. Is there an extra-fast unit? What is the nominal size? Is the location convenient?
8. Surface-unit controls: Is the location convenient? What are the number of heat settings—five plus off, infinite, or other? If an infinite number, how many positions are marked on the control?
9. Do the units have one, two, or three coils? If there is more than one coil, which ones are on at different heat settings? Does each coil "cover" almost the entire unit or is one coil in the center part of the unit only and another coil in the outer part only?
10. Does a signal light glow when a surface unit is on?
11. Is a lamp provided for illuminating the cooking surface? Does it seem adequate for its purpose?

Oven(s) and Controls

1. Are there one or two ovens?
2. What are the wattages of top and bottom units of each oven?
3. Is there an *interior* oven vent in each oven and what is the location?
4. Where is the heat-sensitive part of the oven thermostat for each oven?
5. What are the interior usable dimensions in inches: width, height, depth? Is one oven large enough for the Thanksgiving turkey and is this important?
6. How many inches is the center plane of each oven from the floor of the room? Do these heights correspond to convenient reaches for the person who will use the range most?
7. What materials are used for the oven liner, interior of door, oven insulation, rack supports, racks, broiler grid, and broiler pans?
8. Oven racks: Do they have effective stop positions when pulled out? Is one designed to be positioned at more than one height?
9. Is there an oven light? From where is it controlled?
10. In how many positions will each oven door stay open?

11. Is a special seal used for each oven door? What material(s) are used?

12. Special provisions to aid in oven cleaning: If the range has more than one oven and if one is self-cleaning or continuous clean, what is provided to make cleaning the other easier? For example, are the side liners removable?

13. Controls: Does the control for each oven seem straightforward to operate? What is the lowest marked temperature? How many broil settings are provided? What temperature range is marked for rotissing? What kind of automatic programming is provided? Is the minute minder electrical or mechanical? What is the maximum number of minutes that can be set? Is any special procedure necessary when the minder is set for less than ten minutes? What signal lights are provided?

Special Selection Points

1. What group of features is desired? For example, a consumer might want in one range two ovens either side by side or one above the other, a particular size for each oven (relative to use); a rotisserie that comes with the range or an outlet to provide power to an oven rotisserie that might be purchased as an optional feature; an automatic meat probe; a broiler at a convenient height; a self-cleaning or continuous clean model; a thermostatically controlled surface unit; a particular type of programmed oven cooking; or other features. All or none of the features listed may be desired. If all those listed including the "other" are wanted, the consumer will likely learn that no range has all.

2. What models are available at local reputable dealers?

3. Who will service and install the range?

4. What kind of warranty is given?

5. Does the prospective purchaser really want something quite special—for example, a range that has a glass porcelain cooking top? At a particular time a highly specialized requirement may limit the choice of all available ranges to one model.

6. What performance information of the following type is available from unbiased sources: uniformity of oven heat distribution and of broiler heat distribution; effectiveness of the thermostatically controlled unit, that is, does it overshoot or undershoot? Does the self-cleaning or continuous clean perform as implied in the sales literature?

CHAPTER 12
MICROWAVE OVENS
AND RANGES

The microwave range or oven, also called the electronic oven,
uses radiation in the microwave portion of the electromagnetic
spectrum. This portion has frequencies of megacycles or mega-
Hertz. Microwave ovens use either 2,450 or 915 megaHertz
(mHz). The Federal Communications Commission assigns the
megaHertz because they are close to those used in communications
systems.

Two types are sold—fixed, complete cooking appliances and
microwave ovens. The complete cooking appliance includes conventional surface units and one or two ovens. If the complete appliance has one oven only it is a combination oven that may be
used either as an electronic or conventional unit. If the fixed appliance is an eye-level range with two ovens, one is conventional and
the other is electronic only or electronic and conventional. The
complete appliance operates on 120/240 or 236 volts.

The microwave oven, on the other hand, generally would be an
extra cooking appliance, that is, additional to a conventional range.
It is designed for use on 110, 115, or 120 volts.

PHYSICAL PRINCIPLES AND THE COOKING PROCESS

The microwave oven depends chiefly on one of the three familiar
methods of heat transfer, namely, radiation. The special operating
characteristic is that electromagnetic waves of microwave length
penetrate below the surface to depths comparable to the wave
length and are absorbed by materials such as food which have a
high water content. Further, the microwaves penetrate the food
from all angles because they are reflected within the metallic oven
cavity. As the microwaves are absorbed, polar food molecules tend
to align themselves with the electromagnetic fields of the microwaves. The fields change each half cycle and the polar food molecules oscillate. The rapid oscillation of the food molecules raises
the temperature of the food, thereby cooking it.

Microwave Ovens—Physical Components and Their Functions

The microwave oven has eight major components:[1] (1) The auxiliary circuit adapts the line power to the generator (magnetron) requirement. (2) The magnetron is a special vacuum tube that converts the power supplied to it into microwave energy by acting as an oscillator. (3) The transmission system—a small antenna and a waveguide (pipe-like tube)—propagates energy into the oven cavity. (4) Coupling devices permit the transfer of energy to the load, that is, the food. (5) The stirrer distributes the microwaves in all directions for even energy distribution within the cavity. The stirrer rotates and reflects a portion of the generated power back into the magnetron, which causes the operating frequency to change. The rotating stirrer also causes the standing wave patterns to move somewhat within the oven. Thus the stirrer, or the rotating food platform used in the combination oven, (p. 231), make for more uniform heating of the load than would otherwise be the case—that is, than would be the case without the action of the stirrer or the rotating food platform. (6) The cavity or oven itself acts as a resonant structure. The energy entering the cavity is reflected off the metal walls until it reaches the food. (7) Energy-sealing or trapping devices are provided to prevent stray radiation. (8) Operating controls permit the user to select cooking conditions and safety devices protect the user.

Operation of the Microwave Oven

The microwave oven uses a time setting instead of a temperature dial (Figs. 12-1 and 12-2). Two or three controls are usual.

[1] Sol M. Michaelson, "Biological Effects of Radiation," College Equipment Educators' Conference, Iowa State University, 1970.

Figure 12-1. "Minute Master" electronic oven. Electric supply—110 to 120 volts AC, two wire with ground, 15 amperes, 60 cycles. Power consumption idle: 250 watts; operating 1,750 watts. Power output: 600 watts at 2,450 megacycles per second. Dimensions: width 24 inches; height 14 3/4 inches; depth 16 1/2 inches (allow 1 inch above unit if sides are restricted). Cavity: width 15 inches; height 8 5/8 inches; usable depth 12 inches. Net weight 80 pounds. (Litton Industries)

If two are used, one is an on-off switch which in the on position activates the auxiliary circuit and "readies" the magnetron for oscillation. The other is a cook-time dial which (1) closes contacts to allow the magnetron to broadcast microwave energy and (2) closes a circuit to the timer motor which begins to return to the zero position of the timer. If three controls are used, one may be an on-off switch, another a "start-cooking" control, and the third the cook-time control.

Interlocking mechanisms are so de-

Figure 12-2. "Radarange microwave oven" model RR-3: 115 volts, 60 cycles; 1,450 watts; 14.5 amperes. Exterior dimensions: height (including feet) 15 inches; width 22 3/4 inches; overall depth 15 1/8 inches; door open 29 1/4 inches. Usable oven dimensions: height 9 1/8 inches; width 14 5/8 inches; depth 13 3/8 inches. Net weight 87 pounds. (Amana Refrigeration, Inc.)

signed that the magnetron will oscillate only when the door is in the fully closed position.

USE

The cooking process proceeds differently in a conventional oven than in a microwave oven. When the oven units are on, the air in a conventional oven is hot and the exterior surfaces of the food are heated by conduction of heat from oven air and by absorption of radiant energy at the surfaces. Heat is conducted from the exterior to the interior of the food. Conduction is a relatively slow process and exterior food surfaces have an opportunity to brown. In a microwave oven, however, the microwave energy penetrates a significant distance into the food and heat is produced almost instantaneously. Heat is transferred to exterior surfaces and, in the case of large and/or nonuniform food masses, to interior surfaces by conduction. Because conduction of heat from the site of absorption is not instantaneous and because the air in the microwave oven is nearly the same temperature as the room air when only the microwave unit is on, foods such as white cakes do not brown as they do in conventional ovens. Also foods such as bacon that are prepared in thin sheets have a different texture than the same foods cooked by conventional methods.

Suitable utensils for microwave cooking are those that absorb little radiant energy (in the microwave part of the electromagnetic spectrum) and do not reflect microwave energy. Otherwise stated, suitable utensils are somewhat "transparent" to microwave energy. These materials include plastic-coated paper dishes; glass, pottery, and glass ceramic that does not contain lead; and china without metallic trim. As is true with other equipment, the prudent user follows instructions of the manufacturer. (See also p. 232). Use of an electronic oven is not the same as use of a conventional oven. Cooking times are very much shorter than in conventional ovens and more critical. The manufacturer supplies a booklet with recipes and times. The time given for a particular recipe sometimes has to be increased, but this should be done only as experience indicates need. General procedure is to prepare the food, turn the oven on, place the food in or on a suitable container in the oven, set the cooking time given for the recipe, and close the door. At the end of the set time the microwave circuit turns off automatically. The user turns the oven off (an extra safety measure) and removes the food. If the door is opened during the cooking cycle, the microwave unit will go off automatically and the timer will stop.

Recipes in the users' booklets cover the usual categories of cooked foods but not all foods are best when cooked in an electronic oven and recipes are not given for all foods. Categories include appetizers and hors d'oeuvres, soups, meats, poultry, fish, seafood, sauces, vegetables, beverages, desserts, and cereals. Examples of average times for the range shown in Figure 12-2 are stated to be: individual hamburger, one minute; six strips of bacon, four minutes; one baked potato, four minutes; frozen vegetables, five minutes. (A *very* complete book on microwave cooking for the electronic oven shown in Figure 12-1 is available from the manufacturer.)

Menus may be suggested with a recommendation on sequence for cooking the different foods if only one food is cooked at a time.

Although the electronic oven, like other appliances, has limitations, it is a versatile unit that by virtue of the speed with which it cooks permits a user extra flexibility in planning time for cooking. A fun aspect of the oven is the speed with which appetizers may be cooked or reheated for a large group.

Complete Cooking Appliance with One Oven, Electronic and Conventional

The lower electronic and conventional oven for the range illustrated in Figure 12-3 has a round shelf that revolves during electronic cooking for even distribution of energy in the food. The caption of the illustration indicates that the electronic (microwave) frequency used is 915 mega-Hertz. This lower frequency (longer wavelength) penetrates farther into food than the 2,450-megacycle frequency used in the ovens shown in Figures 12-1 and 12-2.

Instructions for cooking meat such as rolled beef rib roast specify for different weights: "oven-set" at broil, oven thermostat at broil (cold start) or at 450° F (preheated oven), high electronic power, cooking time on one side and a spread or range of cooking times on the other, and standing time out of the oven before carving. Times inside and outside of the oven (before carving) are different for meat from rare to well done. Further, it is recommended that a meat thermometer be inserted into the roast after the suggested minimum time on the second side and, if necessary, additional time be allowed to get the meat done to the degree desired. The user is most definitely cautioned to remove the metal meat thermometer *before* cooking in the oven is continued. Numerous hints are given for this and other foods to get the most satisfactory products. For example, a recipe is given for making gravy with drippings of electronically cooked roasts.

Instructions for roasting a frozen unstuffed turkey specify 1 hour 30 minutes to 1 hour 45 minutes for a 14-pound bird. Again after minimum cooking time, the

Figure 12-3. "Versatronic" range with eye-level conventional oven and combination electronic-conventional lower oven. Total connected load at 236/208 volts is 18.4/17.7 kw. Electronic system frequency is 915 megaHertz. A high-low electronic power selector and a separate timer are provided for microwave cooking. Lower oven is self-cleaning. Two-level exhaust system is rated 190 cfm (cubic feet per minute) for top exhaust and 220 cfm for rear exhaust. (General Electric Company)

user is told to check internal temperature with a meat thermometer. The oven-set is at bake, the oven temperature at 350° F and the electronic power at high.

Recipes for other main dishes include noodles Romanoff, lasagna, scalloped oysters, and lobster tails. The latter uses electronic power only. Two frozen rock lobster tails (about one pound each) cook in water in the oven in approximately 19–22 minutes.

A few recipes are given for cooking two or three foods at a time, such as chicken, baked potatoes, and peas; chicken halves and rice; roast meat and baked potatoes.

An investigation was carried out in a meal management class to evaluate the use of microwave ovens.[2] The menu of chicken, baked potatoes, and peas was prepared by four cooking methods: I. a combination of microwave (915 mHz) and conventional cooking; II. microwave cooking at 2,450 mHz; III. conventional cooking in and on the combination electric range used for method I; IV. conventional cooking in and on a gas range. The sensory evaluation by 35 students indicated, in general, that the different methods all produced meals of comparable acceptability. Method I required the shortest cooking time and method II the least amount of energy.

Safety Aspects

Since microwaves generate heat when absorbed by tissue, it is reasonable to be concerned about possible injuries to human beings. According to Michaelson,[3] existing standards in the western world permit a maximum safe level of 100 milliwatts per square centimeter for continous whole-body exposure for periods in excess of six minutes. The industry standard for several years was one-tenth as much power, namely, 10 milliwatts per square centimeter at 2 inches.[4]

In a release dated January 1970, the Bureau of Radiological Health of the U.S. Department of Health, Education, and Welfare made the following suggestions: Before using the microwave oven, one should (1) read the instruction manual to acquaint himself with operation of the oven and the manufacturer's recommendations for safe use and (2) examine the oven for evidence of shipping damage. It is suggested also that these recommended practices be followed: (1) never tamper with or inactivate oven safety interlocks; (2) never poke anything, such as a fork prong or a wire, through the door grille (on older ovens that have an unprotected grille) or around the door seal; (3) try to stay at least a full arm's length away from the front of an operating oven; (4) do not allow children to watch the cooking of food near the grille [for ovens that have a grille]; and (5) switch the oven off before opening the door.

A March 1971 compendium of regulations under the Radiation and Control for Health and Safety Act of 1968 has the following requirement for microwave ovens manufactured after October 6, 1971:

The power density of the microwave radiation emitted by a microwave oven shall not exceed one (1) milliwatt per square centimeter at any point 5 centimeters or more from the external surface of the oven, measured prior to acquisition by a purchaser, and thereafter, 5 milliwatts per square centimeter at any point 5 centimeters or more from the external surface of the oven.[5]

OUTLOOK FOR MICROWAVE OVENS AND RANGES

Production models of household units became available for sale through distributors in 1956. One company manufactured units that used the 2,450 mHz waves

[2] Dorothy Davis, Dan E. Pratt, Elwood F. Reber, and R. Gordon Klockow, "Microwave Cooking in Meal Management," *Journal of Home Economics,* Feb. 1971, vol. 63, pp. 97–100.

[3] "Biological Effects of Radiation," op. cit.

[4] American National Standards Institute, C 95.1– 1966 *Electromagnetic Radiation with Respect to Personnel, Safety Level of.*

[5] *Regulations for the Administration and Enforcement of The Radiation Control Health and Safety Act of 1968,* Part 78, March 1971, p. 6 *et seq.* Washington, D.C.: U. S. Department of Health, Education, and Welfare, Public Health Service—Bureau of Radiological Health.

for other companies and also for sale under its own name. Later, after considerable independent research and development, another company started nationwide sale of the fixed, combination microwave-conventional range unit. The actual number of microwave ovens in use in homes, while substantial, is still small compared to the number of conventional ranges. The electronic oven or range fills a need in the life styles of different families. Also, the "portable" type is widely used as an adjunct to machine vending of foods in such public applications as office buildings, college eating centers, and other locations. Thus an increasing number of persons get experience with the appliance. It is reasonable to expect that the number of fixed and portable electronic ranges in household use will continue to grow.

CHAPTER 13
REFRIGERATORS

Refrigerators are used primarily to store fresh foods and foods that stale or develop "off" flavors, to freeze fresh and leftover foods, and to store frozen foods. Fresh foods are refrigerated not only because we enjoy them cold but because refrigeration retards the growth of the microorganisms which cause food spoilage.

Since models with different operating characteristics are available, it is advantageous for a prospective purchaser to consider what his (her) family will store in the refrigerator. A family that has a separate freezer might store very little frozen food in a refrigerator, preferring instead that almost all the refrigerator storage space be available for nonfrozen items. For another family storage of frozen food in the refrigerator might be quite important. Informed decisions on selection are made easier by understanding industry usage of terms before seeking in the marketplace what adequately meets the family's needs among the models available.

Definitions used in the current ANSI standard[1] also are used in the industry. A household refrigerator is a cabinet or any part of a cabinet which is designed for the refrigerated storage of food at temperatures above 32° F (0° C), which has a source of refrigeration, and which is intended for household use. It may include a compartment for freezing and storage of ice and for short-term storage of food at temperatures below 32° F (0° C).

An "all-refrigerator" is a household refrigerator which does not include a compartment designed for storage of food at temperatures below 32° F (0° C). It may include means of freezing and storage of ice.

---

[1] American Society of Heating, Refrigerating and Air-Conditioning Engineers, Inc., *American National Methods of Testing for Household Refrigerators, Combination Refrigerator-Freezers, and Household Freezers, ANSI B38.1—1970.*

A household combination refrigerator-freezer is a cabinet which consists of two or more compartments, with at least one of the compartments designed for the storage of foods at temperatures above 32° F (0° C) and with at least one of the compartments designed for the storage of foods at temperatures 8° F (−13.3° C) average or below, and which has a source of refrigeration and is intended for household use.

A limited freezer combination refrigerator-freezer is a household combination refrigerator-freezer which has at least 90 percent of its total net refrigerated volume always remaining at a temperature above 32° F (0° C).

Different manufacturers use different average temperatures in the freezer section of refrigerators and combination refrigerator-freezers. Experiments in household equipment classes of the type described on page 250 have shown that interior temperatures vary with model, cold-control settings, load, length of time the appliance has been operating at a given setting, and other factors. For some conditions, some combination refrigerator-freezers reach average temperatures below 0° F in the freezer compartment.

CONSTRUCTION, USE, AND FEATURES

CONSTRUCTION

Standard and combination refrigerator-freezers are available in free-standing and built-in models. Sizes are quoted in interior volume to the nearest 0.1 cubic foot (nearest 2,500 cubic centimeters). The size or volume quoted is the sum of the usable volumes of the general refrigerator and freezer compartments, meat compartment, chiller tray where provided, and such convenience features as crispers or hydrators, ice makers, and door shelves.

Contractors or builders who provide refrigerators in apartments or houses may purchase smaller and less deluxe models, perhaps 12.0 cubic feet or less. Families and individuals on the other hand have purchased increasingly larger sizes over a period of several years. One manufacturer quoted a 17.0-cubic-foot size as the most commonly purchased one in the United States in 1969. The same manufacturer stated that eight out of ten refrigerators sold that year were 14.0 cubic feet or larger. Families buy smaller models—4.0 to 10.0 cubic feet—for special purposes such as use in an office, a recreation room, or a vacation home.

Certain parts common to all refrigerators and combination refrigerators are described below. A knowledge of basic construction is helpful in evaluating sales literature and is an aid in understanding why certain use and care procedures are necessary.

Cabinet and Door

The shell of the cabinet is usually a welded steel structure that supports the inner food compartment(s), the door, and the refrigerating mechanism. The steel shell usually is treated chemically, as by Bonderizing, to increase resistance to rust. Exterior panels on the shell are generally baked-on synthetic enamel, but may be of porcelain enamel.

The thickness of the insulation between the shell and the food compartments usually is less in the refrigerator section than in the freezer section. The usual materials are formed-in-place plastic foams or Fiberglas. The insulation is installed so as to permit any moisture vapor that gets into it from the outside air to migrate to the inside of the cabinet.

The insulation in the door is installed between the outer steel panel of the door and an inner plastic panel. A flexible door gasket fits over the outer edges of the plastic panel.

Part of the inner plastic panel and plastic strips on the front of the cabinet may serve as breaker strips between the steel door and the steel cabinet.

The cabinet of a combination refrigerator-freezer usually has two insulated outside doors, one for the freezer and one for the refrigerator compartment. Some combination models have one outside door only; in this case the freezer has an interior insulated door.

Except for side-by-side models, refrigerators usually have door(s) hinged on the right side. For many models left-hand doors (hinged on the left) can be obtained. Also, many models have hinges which permit the door or doors to be mounted on the right or left side, the shift from one to the other being done in the home by a service man.

Storage Compartments

The compartment for short-term storage of foods below 32° F in a household or standard refrigerator is always above the fresh food compartment. The freezer of a combination model is above the fresh food compartment, adjacent to it (on the left side in side-by-side models), or, in a very few models, below the fresh food compartment.

The compartment for short-term storage of foods below 32° F in a standard refrigerator may be a U-shaped compartment on the least expensive models and an approximately full-width compartment on more expensive models. The freezer section of a combination model with freezer above the refrigerator may have partial or full-width wire-type or solid shelves, as well as door storage. The freezer of a side-by-side model always has shelves and door storage.

The refrigerator section usually has two or more rust resistant wire-type shelves plus door front storage on the inside of the door. In addition the top of the crisper(s) may serve as a shelf. All the shelves may be full width or one or more may be half width. A pull-to-you shelf may be provided. Supports may be provided to make shelf and door storage adjustable. Some manufacturers provide cantilevered adjustable shelves supported at the rear of the liner.

The finish usually used on liners of storage compartments is an acid-resistant porcelain enamel. However, synthetic organic finishes may be used.

Various special compartments may be provided within the refrigerator compartment. One is a meat storage section maintained at a temperature close to freezing and at a high relative humidity. Compartments, drawers, or covered pans are provided for leafy vegetables. These also are high-humidity compartments, since in a closed container the cooling surface is the entire surface of the container. A compartment with a heating coil may be provided in a door front to maintain butter at suitable spreading temperatures.

REFRIGERATING MECHANISM

Refrigerators may have a compression refrigeration mechanism with a motor-compressor unit or an absorption refrigeration mechanism with gas or electricity for energy. In the United States the compression mechanism is the usual electric type for household models.

Figure 13-1. "Electrigas" camper refrigerator. Flipping a switch changes from butane on road to electricity in park. (Norcold, Inc.)

Refrigerators for mobile campers are available that utilize an absorption refrigeration mechanism with gas burner while traveling and an electric unit when parked (Fig. 13-1).

The principal functional parts of a compressor-type refrigerating mechanism are a motor-compressor assembly, condenser, restrictor or expansion device, evaporator(s), refrigerant tubing, refrigerant, and control(s).

No-frost models may also have a fan mounted behind the liner to move cold air inside the box and a temperature control for the refrigerator compartment as well as the cold control for the freezer.

The motor-compressor assembly frequently is called the unit. The compressor is operated by the motor, and the two parts (compressor and motor) are assembled in an air-tight housing supported externally by rubber or spring mountings. In current free-standing models, the unit is near the bottom of the refrigerator. The compressor may be a reciprocating type or a rotary type. The motor is an induction start and induction run or a capacitor start and induction run.

The consumer fact sheets currently used by the industry often describe the electric characteristics as 60-cycle AC 115 or 120 volts. The nameplate of the appliance will also give total maximum wattage or current used (p. 249).

The condenser on some free-standing refrigerators is installed on the back of the cabinet. Three types are used. One is finned, steel tubing, brazed or welded to a plate; the second is wire-type tubing spot welded to vertical rods; and the third is a pair of plates formed to provide passage between them for the refrigerant. A condenser mounted on the back of the cabinet, unless it is a wire type, is likely to have a curved metal panel surrounding it. Spacing bolts may be used with the wire type.

Other refrigerators have the condenser at the bottom of the cabinet to permit installation of the cabinet flush to the wall.

Condensers are designed for free (natural) or forced-air circulation. When the condenser is on the back of the cabinet, the curved panel or spacing bolts insure free space for air circulation. When the condenser is mounted in the unit compartment in the bottom of the cabinet shell, louvered plates or grilles are provided for natural air circulation. (A fan may be provided for forced circulation of warm air from the unit compartment.)

The restrictor or expansion device is a

capillary tube installed in the refrigerating circuit between the condenser and evaporator tubing or coil. Evaporator "tubing" may actually consist of tubing fastened to a sheet, tubing alone, or two plates so formed as to provide refrigerant passages between them. Combination refrigerators frequently have two functional evaporator components.

The two refrigerants used are Freon-12 (dichlorodifluoromethane) and Freon-114 (dichlorotetrafluoromethane). Freon-114 is used in refrigerators with rotary compressors only; Freon-12 is used with rotary and reciprocating compressors.

The cold control starts and stops the compressor motor.

REFRIGERATING SYSTEMS

General principles of refrigerating systems are described here. The various auxiliary components found in actual refrigerators are considered in specialized texts on refrigeration. Three types of household refrigerating systems are used: the conventional manual defrost system, the no-frost system used in some combination refrigerator-freezers, and the cycle defrost used in other combination models.

The conventional system has a single evaporator located in the upper part of the cabinet. Figure 13-2 shows schematically the path of the refrigerant in the closed refrigerant "circuit" of such a refrigerator. A cam assembled to a motor shaft drives the rotary compressor. (The motor is not shown in the sketch.) Motion of the cam causes the ring to move in such manner that one point always contacts the housing. The eccentric movement of the ring in the housing compresses the refrigerant vapor in the enclosed space ahead of it so that vapor leaves the compressor at increased pressure and temperature. At the same time, the pressure of the refrigerant

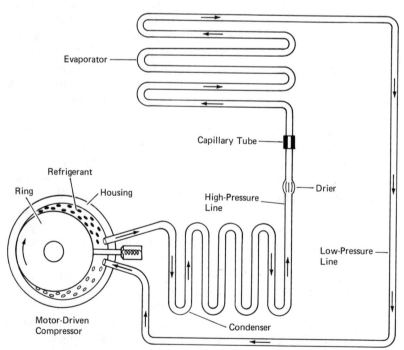

Figure 13-2. Schematic diagram of a conventional compression refrigeration system.

on the inlet side of the compressor is reduced.

The high-pressure refrigerant vapor moves from the compressor to the condenser and part of it condenses to a liquid, still under high pressure. The liquid refrigerant moves through the high-pressure line, through a drier or strainer (to remove moisture), through a capillary tube, and into the evaporator. The capillary tube slows the passage of the refrigerant into the evaporator so that the liquid does not enter the evaporator as rapidly as it is formed in the condenser.

In the evaporator the pressure of the liquid is low enough that it boils at the temperature at which refrigerators are designed to operate. The change of state of the refrigerant (liquid to gas) removes heat from the refrigerator box and the foods in the box. The refrigerant vapor then returns through the low-pressure line to the inlet side of the compressor. (Means are provided to insure that only vapor enters the compression chamber. A steel vessel called an *accumulator* may be used to hold *liquid* refrigerant in the suction line or the return may include a pass in the warm motor-compressor compartment which will vaporize the liquid component.)

The movement of refrigerant—compressor to condenser to evaporator and back to compressor—continues as long as the motor operates. When the temperature of the evaporator compartment that holds foods below 32° F for short periods falls to the value for which the *cold control* is set to open the electric circuit, the motor and compressor stop. Some of the liquid refrigerant continues to move through the capillary into the evaporator. (Sometimes one can hear the refrigerant unloading into the evaporator after the compressor stops.) But the movement of liquid refrigerant into the evaporator does not

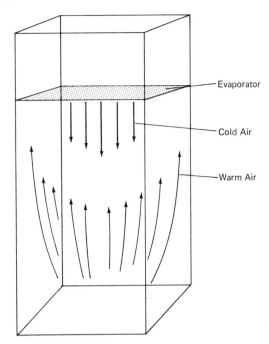

Figure 13-3. Air flow in a conventional refrigerator.

continue because vapor is not being drawn off from the evaporator. The refrigerator therefore starts to get warm. When the temperature rises to the value for which the cold control is set to close the electric circuit, the motor starts and the movement of refrigerant described above is resumed.

Figure 13-3 shows the air flow within the box for a conventional refrigerator. The colder air near the evaporator falls and displaces warmer air below it. The warmer air rises and when it reaches the evaporator coil moisture is deposited and this moisture turns to frost. The frost builds up until it is liquefied when the user manually initiates a defrost cycle or turns the cold control to "off."

The no-frost combination refrigerator-freezer may also have a single evaporator coil. However more extensive evaporator tubing is usual than in the conventional system. The tubing is not visible in the box and the system is designed to prevent "permanent" formation of frost on refrigerated plates or coil. The water from defrosting is disposed of automatically.

The refrigeration system for one no-frost model with freezer at top is shown schematically in Figure 13-4. Hot high-pressure

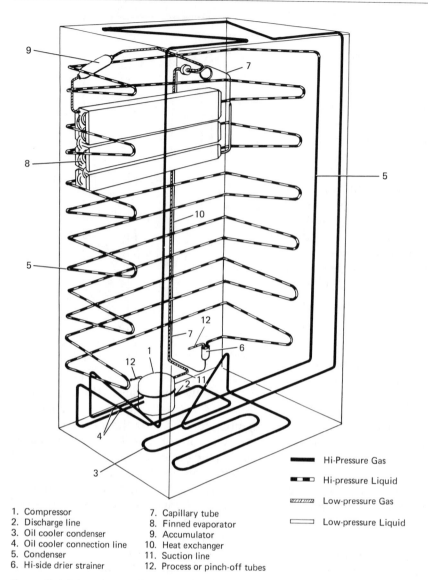

1. Compressor
2. Discharge line
3. Oil cooler condenser
4. Oil cooler connection line
5. Condenser
6. Hi-side drier strainer
7. Capillary tube
8. Finned evaporator
9. Accumulator
10. Heat exchanger
11. Suction line
12. Process or pinch-off tubes

Hi-Pressure Gas
Hi-pressure Liquid
Low-pressure Gas
Low-pressure Liquid

Figure 13-4 Schematic diagram of refrigeration system for a top-freezer no-frost refrigerator. (Franklin Manufacturing Company)

refrigerant gas is discharged from the compressor 1 into the oil cooler "condenser" 3 which also serves as a heater for the drain pan to evaporate defrost water. The partially cooled refrigerant gas then passes through oil cooler connection lines 4 into coils submerged in compressor oil in the compressor. The refrigerant gas then enters the "regular" condenser where additional heat is given up. As the gas cools, it condenses into a liquid and flows through the drier into the capillary tube 7. This tube permits a continuous flow of liquid refrig-

erant into the evaporator, but because the flow is restricted the refrigerant pressure is reduced and the refrigerant boils, thereby absorbing heat from the food compartments.

The cycle continues until the cabinet interior temperature falls to the point where the cold control operates to open the circuit to the motor and stop the compressor.

A fan, mounted above the evaporator on the upper rear wall of the freezer-section liner is operated by a 10-watt permanently lubricated motor that has a switch wired in series with the compressor switch. If either the freezer door or the refrigerator door is opened during the "on" cycle the fan switch located in the hinge-side breaker strips of each compartment will automatically stop the fan motor. This prevents cold air from being discharged into the room.

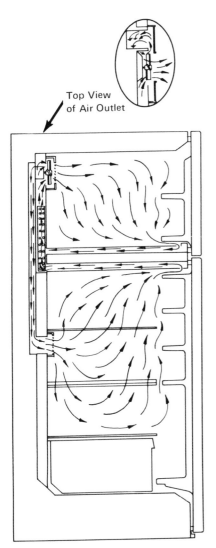

Top View of Air Outlet

Figure 13-5. Air flow for a no-frost refrigerator with top freezer. (Franklin Manufacturing Company)

Figure 13-5 illustrates how air flows within the cabinet. Refrigerator and freezer sections are insulated from one another by a Styrofoam divider formed to provide two return air ducts. Air drawn into the return ducts through openings at the front of both sections moves upward through the finned evaporator. The air cools and gives up moisture which appears as frost on the evaporator. Some of the cold dry air is discharged into the freezer and the remainder into the refrigerator. The amount of cold air that enters the refrigerator is controlled by a refrigerator air-flow control. (This is the *temperature control.*)

Removal of frost and ice on the finned evaporator is accomplished by a 300-watt heater pressed into the fins of the finned evaporator. A defrost cycle is initiated automatically every 12 hours by a defrost timer. Defrost is terminated by a termination thermostat which interrupts the power to the defrost heater when a preselected design temperature is reached. The defrost water moves through a drain tube into a drain pan located at the bottom of the appliance in the motor-compressor compartment.

The air flow in a side-by-side (vertical freezer) combination is shown schematically in Figure 13-6. This type of combination model, like other no-frost models, has a single evaporator.

The cycle-defrost refrigeration system used in some combination models has two evaporators—one in the freezer section and one in the refrigerator section. The latter may be a cold plate as shown in Figure 13-7. It defrosts automatically during each cycle when the compressor is off. The small amount of defrost water drains into a pan in the compressor compartment. The evaporator in the freezer section does not defrost during each cycle; it may require manual defrosting, or it may

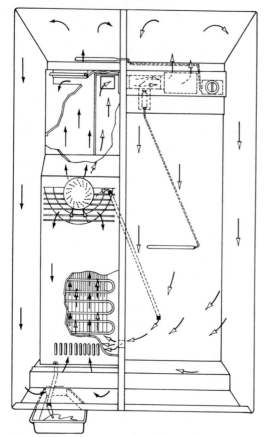

Figure 13-6. Air flow in a side-by-side (vertical freezer) combination. (Frigidaire Division of General Motors Corporation)

have an automatic defrost initiated by a timer and aided by a heater or other means.

INSTALLATION AND USE

User's booklets or instruction folders give instructions on use and care for specific models. Booklets for combination refrigerator-freezers include, in addition, instructions on preparation of foods for freezing. The objective of the general recommendations on installation, controls, placement of foods, and care given below is to provide a general guide for maintaining quality of stored foods, keeping operating cost down, and avoiding unpleasant refrigerator odors.

Clearance Space(s) and Location

Most household refrigerators now have recessed hinging of the door(s) thereby permitting a 90-degree or near 90-degree door opening within the width of the appliance. Nevertheless economical operation *may* require that space be left clear at the door-opening side and above the refrigerator even for some models with forced-air dissipation of heat from the unit compartment. Manufacturer's recommendations on number of inches of clearance should be followed.

The refrigerator center should be located conveniently to the other work centers in the kitchen. It is also desirable that it be conveniently located relative to the eating areas if this is practical.

Refrigerators preferably should not be close to ranges or other sources of heat and should be located so they will be out of the path of the direct rays of the afternoon sun.

Current practice is to have a three-prong grounding-type plug on the refrigerator and to connect it to a three-wire 115–120-volt branch circuit. It seems reasonable to recommend further that a combination refrigerator-freezer be installed in a 115–120-volt individual equipment circuit.

Refrigerators should stand level. This may be especially important for a side-by-side model to avoid having water collect and freeze in the bottom of the box. Many refrigerators have leveling screws at two corners. These should be adjusted when the appliance is installed or moved, to compensate for unevenness in the floor. Refrigerators on rollers or wheels normally would be expected to be self-leveling. (Some have leveling feet at the front, as well as rollers.)

Controls

For normal conditions the cold-control dial and the temperature-control dial when provided should be at the normal setting(s). Use of a colder setting than needed to maintain quality of fresh and frozen foods or to freeze foods satisfactorily is an unnecessary waste of energy. Instructions in the user's booklet on setting of the temperature control in the refrigerator section of a combination model should be read carefully. Occasionally, satisfactory performance of an automatic ice maker may require that the cold-air opening to a meat keeper be closed.

If a heating element used to prevent excessive sweating on the exterior of the box has a manual control, it should be turned to the "off" position when heat is not needed.

Placement of Foods

Placement of certain foods generally will be in the specialized compartments to the extent that these are adequate. Examples are placement of fresh vegetables in the hydrator(s), certain types of cheese in the cheese compartment, and eggs in the egg compartment.

Additional recommendations depend on the type of appliance: no-frost models versus standard and cycle-defrost combination models. In the no-frost type the temperatures within the refrigerator and within the freezer should be approximately uniform due to the circulation of cold air by the fan. Thus foods to be refrigerated might be placed in any convenient location within the refrigerator and frozen foods in any convenient location within the freezer. An exception suggested by the Frigidaire Home Economics Department is not to store ice cream or sherbet in the freezer door since frequent door openings might produce change in texture due to the exposure to room air temperature. The same source suggests that foods should be spread through the compartments so that air can circulate between top and bottom and along sides and rear walls.

For standard refrigerators and cycle-defrost combination models the temperatures within sections are not uniform. The general rule for these types is to store heat-sensitive foods and/or foods that spoil easily in the coldest parts, as far as practical. The coldest parts are those nearest the single evaporator (in standard refrigerators) or evaporators (in cycle-defrost combination models).

Store fresh and cured meat in the coldest part or in the meat keeper, if the appliance has one. Remove market paper from fresh and cured meat that is not prepackaged and rewrap loosely in waxed paper. Fresh meats that are prepackaged in plastic wrap may be stored in the refrigerator in the plastic wrap unless otherwise marked. When in doubt it is reasonable to loosen the plastic wrap on fresh meat prepackaged by the dealer to be sure that enough air can circulate over the meat to prevent surface sliming, without excessive drying out of the surface.

Store cooked meats in the coldest part of the refrigerator but package tightly to prevent drying of surfaces.

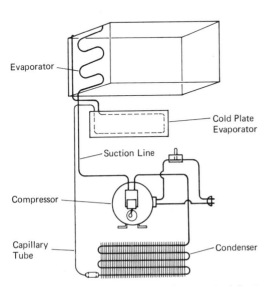

Figure 13-7. Refrigeration system for a cycle-defrost model. (Franklin Manufacturing Company)

Ice cream and sherbet should be stored in the coldest part of the freezing section. Milk and cream are stored near but not in the coldest part of the refrigerator. Good storage for cheese depends on the kind of cheese. Some connoisseurs prefer not to store certain aged cheeses in the refrigerator at all. Cream cheese, on the other hand, usually is stored near the coldest part of the refrigerator. Processed cheeses are stored on any convenient shelf in the food compartment; cut surfaces should be covered closely with aluminum foil or other freezer paper.

Store green or leafy vegetables, carrots, and radishes in the crisper without wrapping. Store fruits on a convenient shelf away from evaporator coils. Spread berries on a plate and cover lightly with waxed paper; they should not be washed or stemmed until they are to be used.

Maintenance

Normal maintenance consists of cleaning for all models and manual defrosting of the evaporator section of standard models and the freezer section of cycle-defrost combination models that do not have automatic defrosting of the freezer section. In manual defrosting ice cubes and other frozen items are removed, the cold control is turned to "off" and the defrost water is collected in a container under the evaporator. When the frost is melted the evaporator is washed and dried and the cold-control dial reset. Defrosting can be hastened by placing utensils of warm water in the freezer. Do not use hot water unless this suggestion is given in the user's booklet.

Frost should be removed from the evaporator section of a standard refrigerator whenever it has built up to 1/4 inch, or the thickness of a lead pencil. Frost should be removed from the freezer of a cycle-defrost combination model that does not defrost automatically when its thickness is about 1 inch, or less if it interferes with the placement of frozen foods. Manual defrosting of a standard refrigerator may be necessary every few weeks in the winter and every few days in a hot, humid summer. Manual defrost of the cycle-defrost model, on the other hand, may be necessary four or so times per year.

The interior and exterior of all models require regular cleaning. Walls, shelves, and other interior parts are cleaned with a solution of baking soda in water (a tablespoon or so per quart of warm water), rinsed, and dried. Abrasives should not be used on either the interior or exterior of a refrigerator. Rubber door gaskets are cleaned with a solution of mild soap and water, rinsed, and dried. The exterior surfaces of the cabinet are wiped with a damp cloth. Also, it may be desirable to wax the exterior surfaces once or twice a year with a product recommended by the refrigerator manufacturer.

Where practical, the condenser of refrigerators should be cleaned once or twice a year with a long-handled brush such as an automobile snow-removal brush or a vacuum cleaner tool while the refrigerator is disconnected.

A refrigerator to be unused for a few weeks or longer should be disconnected, emptied, cleaned, and the door kept open.

For most electric refrigerators, the motor-compressor assembly should be bolted to cross members when the appliance is moved in a van or railroad car. The shipping bolts then are loosened before the refrigerator is placed in service at the new location. The door, of course, is closed during transport.

Preventive maintenance involves avoiding odor buildup through prompt cleanup of spill, regular cleaning, use of foods before they have spoiled, extra-tight covering of strong smelling foods and, especially in no-frost models, effective wrapping of foods to be frozen.

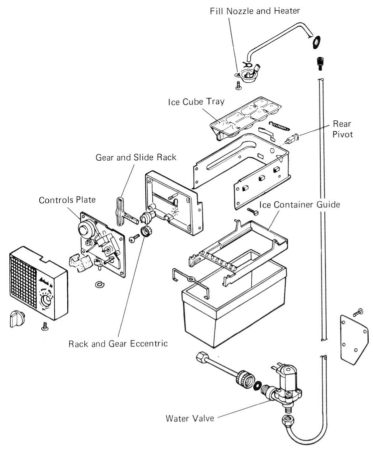

Fill Nozzle and Heater

Ice Cube Tray

Rear Pivot

Gear and Slide Rack

Controls Plate

Ice Container Guide

Rack and Gear Eccentric

Water Valve

Figure 13-8. Automatic ice maker. (Frigidaire Division of General Motors Corporation)

FEATURES

Size is not always designated as a feature; it is however an important characteristic. In addition one should assess interior appointments relative to suitability of storage for items regularly stored by the family. Some of these appointments or features will matter to some families and not to others. Adjustable shelves in the box and on the door(s) may or may not make a model suitable for storage of large quantities of milk in the "bottle" size usually stored by the family. Maximum shelf area may matter more to another family.

Automatic Ice Maker

Automatic ice makers are available on several current deluxe models as an accessory; this permits a user to buy a particular model with or without this feature. Advantages of the ice maker noted by *Con-*

sumer Bulletin are that it eliminates the chore of filling trays with water and allows the family to always have about 5 pounds of ice on hand. Disadvantages noted by the same source are that automatic ice makers make ice at a much slower rate than the ice cube trays included in the price of the box; they also occupy a large amount of space in the freezer section which could otherwise be used for the storage of frozen food.[2]

The component parts of one accessory-type ice maker are shown in Figure 13-8. The control housing encloses the basic electrical and mechanical parts. A U-shaped frame fastened to the side of the control housing encloses and supports an eight-cube polypropylene tray. The ice server or storage bin is suspended from the

[2] *Consumer Bulletin*, June 1970, pp. 9–10. Published by Consumers' Research.

(a)

(b)

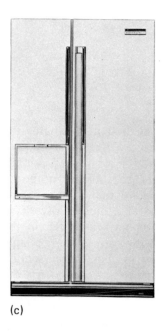

(c)

Figure 13-9. Three models of one manufacturer. (a) Conventional 10-cubic-foot refrigerator with full-width "freezer" and hydrator. (b) No-frost 18-cubic-foot refrigerator. Top freezer has a partial shelf, as well as floor, two door containers, and three
ice cube shelves with storage bucket or optional automatic ice maker. Refrigerator
section has cantilevered shelves. (c) the 24-cubic-foot side-by-side model with
exterior ice service has approximately 9 cubic feet of freezer space and 15 cubic feet
of refrigerator space. (Hotpoint Division of General Electric Company)

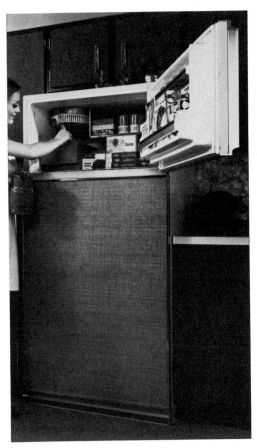

Figure 13-10. Frost-free refrigerator with decorator-type front and optional automatic ice maker in top freezer. (Westinghouse Electric Corporation)

Figure 13-11. Side-by-side model with "add-on" ice maker in place. (Amana Refrigeration, Inc.)

for cleaning purposes. Adjustable shelves in the box and adjustable shelves or containers on the door(s) are likely to be useful, even though they may not be changed frequently. Reversible doors which can be hinged at the top right (right-hand opening) or top left (left-hand opening) are a help to a consumer who has not carefully considered the convenient opening

bottom of the mechanism. In operation a water line and shut-off valve must be connected to the ice maker. The water line tubing is small in diameter, 1/4 inch. A solenoid-operated fill valve regulates flow of water into the ice tray. The capacity of the tray is approximately 0.33 pounds of ice. Total cube production per 24 hours is affected by several conditions—freezer temperature, load, ambient temperature—but 5 pounds or 120 cubes can be produced in a 24-hour period. The user can operate the control to make and store less than the maximum amount.

Other Special Features

Some of the other features available on different models are listed below and/or illustrated in Figures 13-9 through 13-12. Refrigerators on wheels can be rolled out

Figure 13-12. Compact electric refrigerator with net refrigerated volume of 3.4 cubic feet is approximately 19 inches wide, 33 1/2 inches high, and 21 1/2 inches deep. (Frigidaire Division of General Motors Corporation)

for a specific kitchen and may be a help when moving to a different home.

The meat keeper permits storage of fresh meats for periods up to seven days or so. In some models this compartment can be changed easily by the user to an additional vegetable bin (hydrator). A chill compartment is available in the refrigerator of some no-frost models. This compartment has a control which permits it to be used either for quick cooling of meats, vege-

tables, fruits, and beverages or for conventional refrigerated space.

Some refrigerators have door stops to prevent the door(s) from swinging too far. "Automatic" door closures are provided on some models.

Refrigerators are available in a limited number of colors. As is true for other major appliances, however, the stripped or least expensive models are usually only available in white.

STANDARDS

Refrigerator standards are ANSI B38.1–1970 cited on page 234 and the Underwriters' Laboratories, Inc., standard.[3]

AMERICAN NATIONAL STANDARD INSTITUTE STANDARD

The scope of the ANSI standard includes methods of determining volumes and shelf areas of freezing and storage spaces for household refrigerators, combination refrigerator-freezers, and household freezers. The standard has sections on definitions, methods for determining the utility aspects of shelf spacing and volumes occupied by special features, performance test procedures, and durability test procedures.

The utility aspects of volumes occupied by such special features as baskets, chiller trays, meat pans, and other containers not intended for ice or liquid storage are determined by measuring the volumes of the individual features. These are reported to the nearest 0.01 cubic foot (nearest 250 cubic centimeters) separately in freezer

compartment(s) and general refrigerated compartment(s).

Performance test procedures are no-load pull-down test, simulated load test, and ice-making test. The simulated load uses packages filled to a density of 35 ± 5 pounds per cubic foot with hardwood sawdust which has been water-soaked for three days or packaged frozen chopped spinach.

The durability test procedures include handling and storage tests, external surface-condensation test, internal moisture-accumulation test, current-leakage test, and others.

UNDERWRITERS' LABORATORIES, INC., STANDARD

The parts of this standard described here are approved as ANSI B97.1–1971. The construction section specifies that a compressor motor shall be protected by thermal (inherent-overheating) protective devices or other devices giving equivalent protection. Also, a fan motor shall have overcurrent protection or be of the impedance-protected type.

The performance section covers amperes input, overheating during defrost,

[3] Underwriters' Laboratories, Inc., *Standard for Safety—Household Refrigerators and Freezers, UL 250,* Dec. 1968.

defrost-control endurance tests, production tests of refrigerant-containing parts, and others. The measured ampere input shall not exceed the total ampere rating marked on the refrigerator nameplate (exclusive of defrost rating) by more than 10 percent when tested under specified conditions; except that the input to an electric absorption refrigerator shall not exceed its marked rating by more than 5 percent when tested at rated voltage in the normally heated condition. The measured ampere input to the defrost system shall not exceed the defrost ampere rating marked on the refrigerator by more than 10 percent when the refrigerator is subjected to a defrost test.

A normal hot-gas, reverse-cycle, or electric-heater defrost cycle shall not cause temperature rises in insulating materials above specified values.

A control for an electric-defrost heater shall be capable of withstanding an endurance test under the load which it controls for a specified number of cycles of operation.

The manufacturer shall factory-test each refrigerator to determine that it is free from leaks at the following pressures. The test pressure applied to the high and low sides of a compression system employing Refrigerant 12 (Freon-12) is to be not less than 235 pounds per square inch and 140 pounds per square inch, respectively. The corresponding figures for a compression system that uses Refrigerant 22 (Freon-22) are 300 pounds per square inch and 150 pounds per square inch.

Refrigerators shall be plainly marked with the manufacturer's name or trademark, a distinctive type or model designation, the electrical rating, the kind and amount of refrigerant in pounds and/or ounces, and the factory test pressures for the high- and low-pressure sides. The nameplate shall be so constructed and fastened as to form a permanent part of the assembly.

The kind of refrigerant shall be designated. Examples for refrigerant marking are as follows: R 12, Refrigerant 12, or 12 Refrigerant; (trade name) 12, etc.

A refrigerator for free-standing installation only shall have a legible marking to that effect. A refrigerator intended for recessed installation and tested with the enclosure spaced from the top of the refrigerator shall specify the minimum installation clearances to be maintained.

Refrigerators shall be marked with the operating voltage frequency, and with the total or individual loads in amperes. The load of small fan motors may be shown in watts. An electric-defrost heater load may be marked in amperes or watts. However, the marking need not include a separate defrost-heater load if the refrigerator total ampere rating is marked.

If a manufacturer produces refrigerators at more than one location, each refrigerator shall be marked to identify the particular factory.

EXPERIMENTS

Experiment 1. Food Storage Volume and Shelf Area

1. Record manufacturer's name and model or catalogue number given on nameplate. Refer to specification sheets for rated capacity and shelf area.
2. Measure shelf area and volume of refrigerator, freezer or ice cube

compartment, and forced-air cooled fresh meat compartment.

> *Note:* Storage space on door or doors is part of the volume rating.

3. Compare your measured values with manufacturer's ratings.

Experiment 2. Door Closure

1. Tightness of gasket seal: Check tightness of door gasket seal with a 100-watt lamp. Place lamp in a socket that is connected to a plug by a flat extension wire, to permit easy closing of the door on the wire. Locate the lamp on the upper shelf of the refrigerator, as near the front as possible. Avoid contact of lamp with plastic parts. Close the door and connect the plug on the extension wire to an electric outlet. Darken the room and check whether you can see light coming out of the refrigerator. Record the location, if any, at which you see light in the appliance.

   > *Note:* For some door closures this test is good only if the refrigerator is level.

   Alternate methods can be used. One method is to put a flashlight in the refrigerator in such a way that the light is directed toward the gasket. Another is to examine the fit of the gasket from outside the appliance by "sighting" along the length of the gasket.

2. Pounds of force needed to open door: Check whether the appliance is level. If it is not, level it by turning the adjusting screws at the bottom.

   Attach the hook of a spring balance to the handle of the refrigerator and measure the force in pounds needed to open the door from the outside. (The spring balance should have a range at least to 16 pounds.)

Experiment 3. Electrical Characteristics and Interior Temperatures

The tests will be done with "no load" (empty refrigerator).

1. Starting current: Measure starting current with an ammeter that has an adjustable pointer-stop. To operate, move pointer-stop manually to position the pointer at a reading just below the anticipated starting current. (By moving the indicating pointer to this position, the distance the pointer travels due to the starting current is shortened and a reading is got more easily than with an ammeter that does not have a pointer-stop.)

2. Power factor: Install the refrigerator in an electric circuit with ammeter, voltmeter, and wattmeter. Calculate power factor in percent from the equation that defines power factor.

$$\text{Power factor} = \frac{\text{measured watts} \times 100}{\text{measured volts} \times \text{measured amperes}}$$

> *Note:* The wattmeter should be of a type that is reasonably accurate for low power factor.

An alternate method for getting the power factor is to use a watt-var meter.

$$\frac{\text{vars}}{\text{watts}} = \text{tangent (power factor angle)}$$

From a table of trigonometric values find the cosine of the angle whose tangent is vars divided by watts. (This is the power factor, expressed as a fraction.)

3. Cycling, interior temperatures and kilowatt-hours for warm, normal, and cold settings.

Cycling only: The "on" and "off" times of the compressor motor can be measured by listening for the "on" sound of the motor and "clocking" times. If on and off times are observed for several *complete* cycles a reasonably good estimate of percent on time is possible.

Cycling and interior temperature at one location: A Bristol time-temperature recorder will give average temperature at one location in the refrigerator and *on-off* times.

A Honeywell electronic recorder and a recording wattmeter will permit temperature measurements at several locations *and* determination of kilowatt-hours used in 24 hours, as indicated below. Set thermostat dial, install thermocouples in refrigerator, close door, and permit refrigerator to operate for 6 hours or longer to establish thermal equilibrium inside the appliance. Use a recording wattmeter such as a General Electric inkless instrument to obtain a 24-hour record of how many watts the appliance uses and "on" and "off" times. Use the Honeywell electronic recorder to get interior temperatures and room temperature. Calculate electric energy in kilowatt-hours from average watts and total *on* time in 24 hours.

Notes: a. A refinement of this calculation is to calculate energy used for each on cycle in the 24-hour period and to sum these values for the total period. This refinement gives information too on energy used during defrost period(s) of appliances with freezer section that defrosts automatically.

b. If one worker in the group marks unit start and stop times manually on the temperature record, temperature fluctuations can be correlated with on and off times.

Experiment 4. Menu Project

1. Obtain a weekly market order from a homemaker who does planned, once-a-week shopping except for occasional items. Purchase the foods that need refrigeration or improvise packages similar to those in which the foods are usually sold. (Empty cartons and bottles can be used.)

2. Can all the items that will be used in one week and that require refrigeration be stored in their correct locations in the refrigerator? Is the refrigerator or the freezer too closely packed? Can the items be stored so they are readily accessible?

> *Note:* In addition to the market order foods that are refrigerated, store simulated leftovers and staples. (Ask the homemaker what she has in her refrigerator.)

Experiment 5. Effectiveness of Storage

1. Store three similar heads of lettuce in refrigerators as follows: one uncovered on a shelf, one in a moisture-proof wrap on a shelf, and one in the vegetable crisper or other closed container. Observe appearance after two, three, four, and five days. Compare results for a no-frost versus a conventional refrigerator and/or a cycle-defrost combination model.
2. Store several pints of one flavor of a good-quality ice cream or sherbet in the freezer. Use the original cartons without overwrap. Observe flavor and graininess after varying lengths of storage, such as one, two, three, and four weeks. For test purposes remove a fresh package each week.
3. Store ice cream using the original cartons. For one package, cover the exposed surface with aluminum foil, after some of the ice cream has been used. Do not use foil for the other package. Sample the ice cream from the two packages after different intervals of storage. Does the foil help?
4. Store similar patties of frozen ground beef prepared from a single lot of ground meat in the freezer. Broil patties after different lengths of storage such as one, two, three, and four weeks. Note appearance, flavor, texture, and juiciness.

BUYING GUIDE

(Review Chapter 8.)

1. Does the refrigerator have the Underwriters' Laboratories seal of approval?
2. Space requirements:
   a. What is the height?
   b. Width?
   c. Depth?
   d. Width with door or doors open?
   e. Depth to extreme swing of door or doors?
   f. Required installation clearances?
3. Capacity:
   a. What is the AHAM (Association of Home Appliance Manufacturers) rated total capacity in cubic feet?

b. What is the capacity of the refrigerator section?

c. Capacity of the freezer section?

d. Capacity of the ice cube compartment?

e. Capacity of the forced-air-cooled meat compartment, if one is provided?

4. Is the nameplate accessible? Does it give manufacturer's name and address, model, voltage rating, amperes, serial number, other information? If a gas refrigerator, check for the American Gas Association seal of approval and for the British thermal unit rating of the burner on the nameplate.

5. What type(s) of insulation are used?

6. What type of refrigerator is it: a conventional refrigerator or a combination refrigerator-freezer?

7. If it is a combination refrigerator-freezer, is it a no-frost type with one evaporator or a cycle-defrost combination model with two evaporators?

a. If it is the cycle-defrost combination type does the freezer defrost automatically?

b. If it is a no-frost type with one evaporator does it have a cold control and a temperature control?

8. What is the average temperature in the freezer for the normal or average setting of the cold-control dial? If temperature information is not available check *use* information on type of foods recommended for storage and length of storage. If neither temperature or use information is available, a purchaser might find that performance of the appliance in the home is different from what was anticipated.

9. What is the average temperature in the refrigerator for a normal or average setting of the cold-control dial?

10. How many fans or blowers are in the refrigerator?

11. What is the number of heating elements?

12. For a no-frost model, how frequently (times per day or other unit) does the appliance defrost?

13. Is sound insulation used in the machine compartment?

14. Ease of cleaning and maintenance: Are the materials used in the refrigerator easily cleaned? Are the fixed interior appointments such as egg storage section easily cleaned? Do the removable interior parts seem to be easily removable for cleaning? If the refrigerator defrosts automatically, is the defrost water tray easily removed? Are interior vents and baffles easily adjusted for different conditions of use, temperature, and humidity?

15. What is the weight in pounds of the appliance?
Ordinarily this is important to a consumer only if the bearing strength of the floor on which the appliance is to be placed is somewhat questionable or if the family is likely to move several times.

16. How informative is the user's booklet or pamphlet (very, somewhat)?
17. Acceptability of overall design of the refrigerator for your home:
    a. Are the door(s) hinged on the right or left side? Are they reversible?
    b. Do the door(s) extend the full width of cabinet or are they enclosed by trim or flange on the cabinet?
    c. Is the design of handles and other hardware acceptable or not acceptable?
    d. What type of door closure is used (short magnet, full-door opening magnet in gasket)?
    e. If the appliance has a foot pedal, is adequate clearance provided between it and the bottom of the door?
    f. Is an accessory trim kit available for built-in appearance?
    g. If an accessory kit is available, what are the space needs (height, width, depth in inches) with the kit?
    h. What is the exterior finish (porcelain enamel, baked-on white enamel, baked-on colored enamel, or acrylic)?
    i. What is the interior liner of the refrigerator (porcelain enamel, baked-on enamel, or another)?
    j. What is the interior liner of the freezer (porcelain enamel, baked-on enamel, aluminum, steel, or another)?
    k. Are interior appointments of refrigerator satisfactory for the size, quantity, and shapes of foods you store in a fresh food section? How many shelves are there? Are the shelves spaced appropriately to receive articles you refrigerate? Are the shelves pull-out type, roll-to-you type? Are adjustable locations provided for the shelves? How many hydrators are there? Is the door storage useful for your needs?
    l. Are interior appointments of freezer satisfactory for the size, quantity, and shapes of foods you store in a freezer? Is the door storage useful for your needs? Are there shelves, roll-to-you baskets, an interior light, and other convenience features?
18. Special features:
    a. Are the ice cube trays designed for easy removal of ice cubes? Are the ice cubes such that they can be stored in a container without freezing together?
    b. If an ice maker is part of the appliance or available as an accessory does it take only a "reasonable" amount of freezer space? What are its characteristics—that is, the time it takes to make 1/3 pound of ice, the number of pounds that can be made per 24 hours, care cautions, and others?
    c. Is a chill compartment for rapid cooling of desserts and beverages provided in the refrigerator section?

d. Is there a control for the heating element used to prevent sweating?
e. If a fresh meat (meat keeper) compartment is provided can this be converted into an extra hydrator?
f. What other features are provided—for example, an activated charcoal filter in an air duct of the refrigerator section?

CHAPTER 14
FOOD FREEZERS

Food freezers are self-contained appliances used for freezing foods and storing frozen foods. The choice of two appliances—refrigerator and freezer—versus combination refrigerator-freezer may relate to life style and/or need. If the need exists for more freezer space in the home (more than 10 cubic feet or so) the two appliances are indicated if the family has space for both.

If the family's living style calls for a substantial amount of refrigerator and freezer storage and if the kitchen space will not permit a good working arrangement or design with a very large combination appliance two appliances are again indicated provided space is available elsewhere for the freezer. A family's budgeting pattern may call for a nondeluxe freezer and refrigerator. The same total capacity is likely to be obtainable with less money for two appliances than a single one though convenience features will be lost.

CONSTRUCTION, USE, AND CARE

Chest and upright models are available. The chest has a lid and the upright has a door. Upright freezers include small capacity (about 4 cubic feet) undercounter models which, as the name implies, fit under a counter. Uprights of 10 to 23 cubic feet or so may be no-frost type or conventional (manual-defrosting) type. Chest freezers are designed for manual defrosting. Available sizes vary from about 5 to 28 cubic feet.

The overall size of both types is quoted in interior volume to the nearest 0.1 cubic foot (nearest 2,500 cubic centimeters), or in nominal pounds of food that can be stored in the storage and freezing sections. The food-weight rating is equal to the volume rating times 35. Thus a 10.0 cubic-foot freezer has a *nominal* capacity of 350 pounds; but a cubic foot will not store 35 pounds of irregularly shaped and/or low-density foods such as poultry.

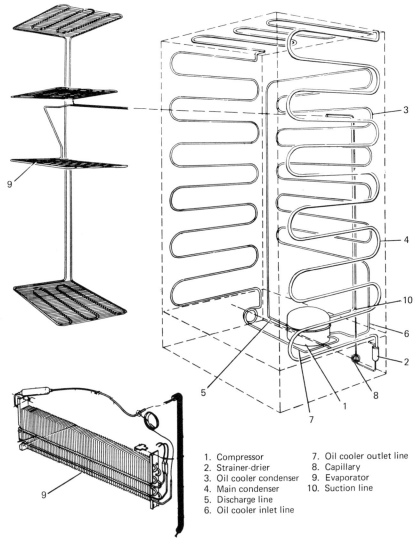

1. Compressor
2. Strainer-drier
3. Oil cooler condenser
4. Main condenser
5. Discharge line
6. Oil cooler inlet line
7. Oil cooler outlet line
8. Capillary
9. Evaporator
10. Suction line

Figure 14-1. Refrigeration pictorial for upright freezer with shell condenser. (Frigidaire Division of General Motors Corporation)

When a special freezing section is provided, its capacity also is stated in cubic feet or in pounds of food. The pound rating in this case indicates the number of pounds of compact packages that can be placed in the freezing section. It is not a direct measure of the actual time required to freeze a given quantity of food.

CONSTRUCTION

Component parts, refrigerating mechanism, and construction features are similar for freezers and refrigerators, except for the differences required by the lower interior temperatures of freezers. The cold control is similar to that on refrigerators. The refrigerating mechanism of freezers is designed so that the difference between maximum and minimum temperatures during a cycle of operation is less than for refrigerators. Freon-12 and Freon-22 are the refrigerants commonly used. Insulation may be heavier than in refrigerators, because the temperature difference between room air and the interior of the appliance is greater for freezers.

Figure 14-1 is a pictorial representation of the refrigeration system for one type of no-frost upright food freezer (right side

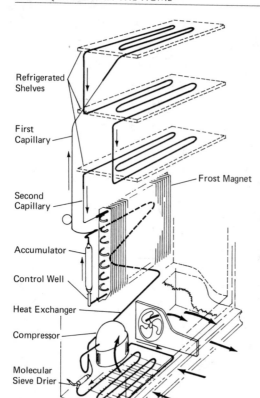

Refrigerated Shelves

First Capillary

Frost Magnet

Second Capillary

Accumulator

Control Well

Heat Exchanger

Compressor

Molecular Sieve Drier

Two-Row Condenser

Figure 14-2. Refrigeration pictorial for upright freezer with compact condenser in unit compartment. (Amana Refrigeration, Inc.)

upright freezers use a compact, fan-cooled condenser which may be in the unit compartment at the bottom of the appliance (Fig. 14-2) or a tube and fin type on the back of the appliance.

The capillary tube through which refrigerant passes from the strainer-drier at the end of the condenser tubing into the evaporator regulates the amount of refrig-

and lower left) and a conventional type (upper left). The *general* refrigeration scheme is of course the same as for a refrigerator (p. 237). The function of the "oil cooler condenser" shown in this sketch is to cool compressor oil by routing partially condensed refrigerant to an oil cooling loop in the compressor.[1] Here the refrigerant still at high pressure absorbs heat and evaporates. The vapor then moves through the balance of the condenser to be condensed again.

The type shown has a *shell* or *hot-wall condenser.* The condenser tubing is attached to the inside surface of the cabinet shell and thus offers an extended surface for heat dissipation to the room air. Other

[1] American Society of Heating, Refrigerating and Air-Conditioning Engineers, Inc., *ASHRAE Guide and Data Book—Systems and Equipment 1967,* pp. 485–486.

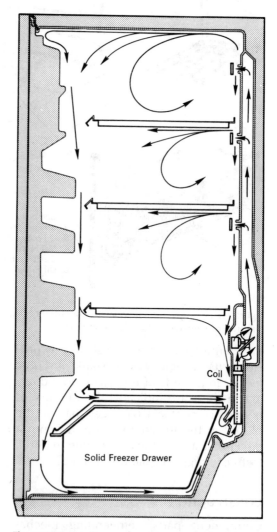

Coil

Solid Freezer Drawer

Figure 14-3. Air-flow system for a no-frost refrigeration system with solid shelves. Air flow reaches door shelves and surrounds lower storage drawer. (Admiral Corporation)

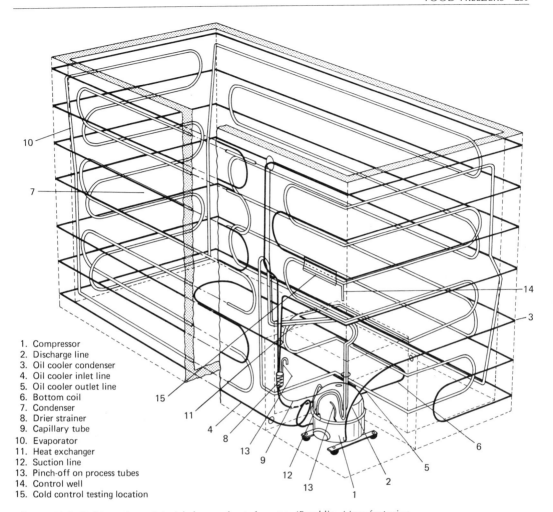

1. Compressor
2. Discharge line
3. Oil cooler condenser
4. Oil cooler inlet line
5. Oil cooler outlet line
6. Bottom coil
7. Condenser
8. Drier strainer
9. Capillary tube
10. Evaporator
11. Heat exchanger
12. Suction line
13. Pinch-off on process tubes
14. Control well
15. Cold control testing location

Figure 14-4. Refrigeration pictorial for a chest freezer. (Franklin Manufacturing Company)

erant entering the evaporator in a given time. The cycle is completed when gaseous refrigerant from the evaporator returns through the suction line to the compressor.

On-off cycling of the compressor motor is regulated by a cold control.

The no-frost type shown in Figure 14-1 has a fan (not shown) to force cold air across the evaporator into the food compartment and a sheathed heater for defrosting that is mounted in the lower part of the evaporator coil. The frost on the evaporator is melted when the heater goes on automatically twice a day. Defrost water moves through a drain trough and tube to a pan located in the machine compartment.

The air-flow system in the storage space for a no-frost refrigeration system with solid shelves is shown in Figure 14-3.

Although the conventional upright freezer does not defrost automatically, it has a small hole with cover in the freezer floor with a tube leading from the hole to a pan in the unit compartment. During manual defrost most of the melted frost from the evaporator tubing is caught in a utensil the homemaker places in the freezer; the remainder flows through the drain tubing into the pan in the machine compartment.

Figure 14-4 shows a pictorial sketch for a chest freezer. The heat exchanger is not a separate part; rather it consists of suction (low-pressure) line and capillary soldered together. As the refrigerant in the suction line returns from evaporator to compressor it absorbs additional heat from the capillary which transforms any remaining liquid refrigerant into a gas. This eliminates the possibility of a liquid reaching a compressor designed to pump gas. Some liquid refrigerant will also be changed to a low-pressure vapor in the accumulator, when one is provided. *Note:* Some refrigerant can remain liquid in the evaporator.

Different models differ in various aspects. Reference has already been made to no-frost versus conventional models of uprights.

Freezers differ also in the spacing of evaporator tubing around the liner and in the number of interior shelves (upright) or dividers (chest) that have evaporator tubing. Some uprights have evaporator tubing in the interior floor and ceiling; others do not. Some models have refrigerated plates with passageways for the refrigerant formed in them rather than tubing. These differences affect the quantity of food that will freeze in a given length of time.

Especially in a chest freezer, a separate freezing section may be useful for easy separation of stored frozen foods and fresh foods that are to be frozen.

A signal light to show that electric power is reaching the freezer is useful. Battery-operated buzzers that sound when interior temperature exceeds a preset value are no longer supplied. Key locks are still available. However, freezers manufactured after December 1, 1971, either have a key lock that can be released from the inside or the key slot is such that the key must be manually held in the lock in any position of the lock. In addition the key is to be permanently marked with a notice equivalent to the following: Caution—to prevent a child from being entrapped, keep key out of reach of children and not in the vicinity of freezer (refrigerator).[2]

[2] Underwriters' Laboratories, Inc., *Household Refrigerators and Freezers Standard, UL 250,* 1968.

Doors or lids that shut securely with minimum effort by the user are a protection against semiopen closures and spoiled food.

USE AND CARE

As for other appliances, the user's booklet or folder gives instructions on use and care. In addition, an extra booklet on freezing foods for storage may be supplied, or suggestions on freezing may be included in the booklet on use and care.

Installation

The manufacturer's recommendations on clearances (below, at the sides, back, and top) should be followed. These recommendations vary with different models, depending in part on the location of the condenser. Minimum or maximum adjustment of leveling screws may be specified to insure proper air circulation across the bottom of the freezer. Manufacturers provide bolts on the back of some models to insure adequate clearance between the freezer and the wall.

The freezer should also be installed level. Some upright models have casters. The door seal should be carefully checked at the time of installation.

Operating costs are likely to be less when a freezer is installed in a cool location, such as an unheated basement. However, a cool location may not be convenient. Some models must not be placed in a location where the ambient temperature falls below freezing, while the operating mechanisms of other models are not affected by below-freezing temperatures unless the temperatures are well below freezing, say 0° F. If possible, the freezer should not be installed close to a source of heat such as a radiator or where it receives the direct rays of the sun.

Because a freezer has a starting wattage that is high relative to operating wattage, it should be connected to an individual-equipment circuit protected by a fuse or circuit breaker with an appropriate current rating. If a freezer is installed in the same circuit as other appliances, the circuit may be overloaded each time the freezer motor starts.

Safety note: New freezers manufactured in the United States have the cabinet grounded to a grounding-type three-prong plug. The plug should not be changed or fitted with an adapter for use in a two-wire circuit unless the receptacle cover screw is grounded (Fig. 14-5).

Temperatures

For normal-use conditions, a correctly calibrated cold-control dial is turned to the normal setting. When large loads of

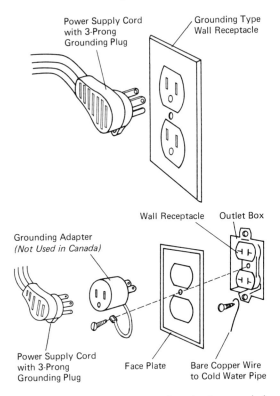

Figure 14-5. Three-prong grounding plug in grounded (three-wire) wall receptacle versus receptacle with bare copper wire connecting face plate cover screw to water pipe. (Amana Refrigeration, Inc.)

food are to be frozen, turning the dial to the coldest setting may accelerate the freezing process. After freezing, return the dial to the normal setting. One manufacturer notes in a service manual that a "common misuse of freezers is expecting them to freeze too large a quantity at once."[3] This same manufacturer suggests approximately 3 to 3 1/2 pounds per cubic foot per day. Other manufacturers may or may not suggest precise poundages. But within limits, the less the quantity frozen at a time, the faster one would expect freezing to take place.

Placement of frozen foods in freezers is a matter of convenience. Although temperatures are not uniform for manual-defrost models, the temperature in any part recommended for storage of frozen foods is likely to be satisfactory. If one has an accurate thermometer and can identify the warmer sections, it may be wise to use this space to store foods that have a fast turn-over. When freezing foods, on the other hand, as many packages as possible should be placed to make contact with freezing coils.

Management

Good use of a freezer calls for good packaging of foods that are to be frozen and a working philosophy that the freezer is a cold-storage pantry for foods that will be used in a reasonable length of time.

Probably freezers do not save money for most families when all the costs associated with operating them are considered—electricity, packaging materials, service calls, depreciation, etc. They may save money if a great deal of home-grown food is frozen or if a family practices extremely good management in purchasing large quantities of foods that the family likes when prices are low. They may also save money if fewer market trips are made,

[3] Franklin Manufacturing Co., *1970 Freezer Service Manual*, sec. 2–3.

especially if such trips are limited to planned shopping with little impulse buying.

On the other hand, families that own freezers can have better and more varied meals throughout the year; they can also have many foods on hand that are ready to be prepared. Management of the freezer therefore should be planned toward better meals and an assortment of foods on hand. A well-used freezer might be used daily. The convenience a freezer provides is not measured in monetary terms alone.

Freezer use is simplified by organizing the foods in sections for meat, vegetables, fruits, desserts, and breads. Some complete meals are stored as a matter of convenience, even though this is likely to use more space than storing the components of the meals.

As new foods are added to the freezer, some rearranging may be desirable to place already frozen foods near the front or on top. Packages should, of course, be labeled with date and content. Keeping a record of the foods in the freezer helps prevent overlong storage of items. Individual users can generally work out the type of records easiest for them.

Several books have been written on meals, baked goods, and other individual categories of food to be frozen. A prospective purchaser of a freezer or an owner might be interested in examining such books against her own interests or family food patterns. Menus might be suggested for meals to be served from the freezer in different lengths of time from 20 minutes to 2 1/2 hours. Different types of appetizers, main dishes, breads, and desserts are suggested. Special items are suggested: individual portions for a family member on a diet, quantities of ice cubes in plastic bags, "blobs" of whipped cream, homemade food gifts such as fruit cakes, sandwiches, and many others.

After food has been defrosted, treat it as though it were fresh food. For example, ground beef frequently is cooked from the frozen state. If, however, it is defrosted and not used immediately, refrigerate it as you would fresh ground beef. Do not *plan* to refreeze thawed foods. Certain foods that have thawed only partially can be refrozen safely, but this procedure does not give the best final product.

Freezing

The extension departments of many state colleges and the U.S. Department of Agriculture distribute bulletins on freezing that contain information on what foods to freeze; instructions on preparation for freezing, such as scalding times for various vegetables; suggestions on varieties of fruits and vegetables that freeze successfully; and suggestions on preparing meat for freezing.

General rules to follow in freezing foods include the following:
1. Freeze good-quality products when possible.
2. Prepare food for freezing when it is at its best. For example, vegetables and fruits should be firm and fully ripe.
3. Use good packaging materials, and package tightly to exclude as much air as possible.
4. Package with minimum area; for example, stack meat patties one on top of another rather than side by side, with sheets of freezer paper between individual patties.
5. Follow reliable instructions on scalding and cooling of vegetables.
6. When freezing liquid or semiliquid foods, leave 1/2 inch of free space, or more, at the top of the container to allow for expansion during freezing.
7. Prepare foods in quantity and freeze meal-size amounts of meat, vegetables, etc.

8. Have an experimental attitude about foods for which freezing instructions are not available. Freezing is easy but it is important to realize that: "Although bacteria are destroyed by freezing, frozen foods are not sterile in the same way as canned products. During freezing, microbial growth and enzyme action are only slowed down, and can continue to grow when the food is thawed."[4]

9. As far as possible, place foods to be frozen so they contact freezing coils. If more food is to be frozen at one time than can be placed against freezing coils, interchanging the positions of the packages during the freezing process may speed freezing for some models.[5]

Packaging Materials

The word *vaporproof* as used for freezing materials usually implies that there will be no significant loss of moisture from foods so wrapped. Loss of moisture causes drying out of fruits and vegetables and "freezer burn" in meats and poultry. In addition to providing a barrier to movement of water vapor out of foods, good packaging materials provide a barrier to movement of oxygen from the freezer air into the package. This characteristic is especially important in packaging materials used for meat and poultry to prevent rancidity of the fat in these foods. No-frost freezers' greater air circulation makes good packaging even more important.

Packaging materials include wrappings and containers. Heavy freezer-weight aluminum foil and polyethylene wrappings furnish excellent protection and are reusable. Sheets of both can be molded to the food. Polyethylene bags can be twisted at the ends and then secured with small pieces of wire sold for that purpose. Laminated sheets combine paper and one or more other materials such as cellophane, aluminum foil, or plastic coating. Ordinarily, when these wrappings are used the paper should be on the outside. Waxed locker papers usually cost less and are recommended for shorter storage periods than the wrappings just discussed.

Glass freezer jars provided with screw-on rustproof metal caps, polyethylene containers, and heavyweight aluminum boxes offer excellent protection and are reusable.

Boil-in-the-bag is a relatively new packaging used for precooked food. Carlin, in the report just cited, cautions: Be sure the air is pressed out of the package and no food is on the surface to be sealed. She also gives the following brief summary on some types of packaging to be used for different categories of food.

Vegetables: Folding boxes with inner bags of a moisture-vaporproof material such as polyethylene.

Fruits and precooked foods: Rigid containers, made of plastic or metal, freezer glass jars or wax-treated cardboard cartons.

Meats, poultry, fish: Heavy moisture-vaporproof paper or plastic wraps such as Saran, polyethylene, pliofilm, and heavy aluminum foil. Also, a laminated paper for which several materials have been combined is used. Use a freezer tape for sealing.

Care

No-frost models require only periodic cleaning. Other models require defrosting and cleaning. The rate of frost accumula-

[4] Francis Carlin, "Goals and Guidelines for Frozen Food," *Report of the National Home Appliance Conference*, 1969. Published by the Association of Home Appliance Manufacturers.

[5] Florence Ehrenkranz, Jeanne Banister, and Reba Smith, "Capacity Load Freezing Time in Home Freezers," *Refrigerating Engineering*, June 1954, p. 63 ff.

tion depends on use condition. Frost should be removed when it interferes with placement of food in the freezer and probably should be removed in any case when it has hardened to a layer of ice. Light and fluffy frost may be removed without disconnecting the freezer motor by using a rubber, plastic, or wooden scraper. Strategically placed paper or towels can collect the frost as it is scraped down.

Ice or dense frost must be melted. When this is necessary, a conventional upright or a chest should be disconnected and the frozen foods removed and wrapped in several thicknesses of paper so that they will not thaw while they are out of the freezer. Some chests as well as some conventional uprights have provisions for draining water which makes defrosting and cleaning easier.

In ordinary use, the interior of a freezer does not need to be cleaned as frequently as the interior of a refrigerator. The interior usually is cleaned with a baking soda solution, rinsed, and wiped dry. As for refrigerators, rubber or vinyl gaskets are cleaned with a cloth dipped in a soapy solution, rinsed, and dried. Exterior surfaces may be polished or waxed with a product recommended by the dealer or manufacturer.

For models that have an exposed condenser attached to the rear of the freezer, the user may be instructed to clean the condenser once or twice a year with a brush or vacuum cleaner attachment. The freezer should be disconnected when the cleaner attachment is used. Where the condenser is enclosed or partially enclosed, the user is usually not instructed to clean it.

If a freezer is to be unused for a long period of time, it should be disconnected, emptied, cleaned, and the door or lid propped open.

The proper procedure in case of failure of the power supply or mechanical failure of the freezer depends on the probable duration of power failure, construction of the freezer, interior temperatures of the frozen foods, amount of food in the freezer, ambient temperature, and other factors. Freezers that contain a large quantity of food will maintain safe temperatures longer than those that contain only a small quantity. Some manufacturers now guarantee to the purchaser a limited cash refund in case of food spoilage for a certain number of hours of power failure, usually 48. Dry ice—not ordinary ice—placed next to or on top of the frozen foods may be a worthwhile precautionary measure in some cases. A freezer thermometer is helpful in deciding whether and when special measures are necessary.

Note: Some packages of food in a freezer remain in a thawing range (about 25° to 31° F) for many hours, while other packages thaw quite rapidly.

FEATURES

Size is one feature. It was estimated that food freezer sales in 1970 would be as follows: 10 to 14 cubic feet, 17 percent; 15 to 19 cubic feet, 64 percent; 20 cubic feet and over, 19 percent.[6] The same source also estimated that of the total one and one-fifth million food freezers that would be sold in 1970 more than three-fifths would be uprights. These estimates suggest how families in general appraise their needs. An individual family, of course, should relate size and upright versus chest to its own situation—food pattern, financial resources, space, freezer capacity in its refrigerator, and other factors.

Since an upright occupies less space

[6] Rossie Ann Gibson, "Cool it—Past, Present and Future," Report of the National Home Appliance Conference, 1969. Published by the Association of Home Appliance Manufacturers.

than a chest of similar capacity, adequate space might be available for an upright when it is not available for a chest. The chest type has the advantage that, since cold air is heavier than warm air, less moisture-laden warm air enters the chest when the lid is raised than enters an upright when the door is opened for the same period. For this reason, an upright may cost slightly more to operate than a chest of the same capacity. A no-frost model would be expected to cost more to operate than a conventional model of the same capacity if a fan runs constantly (except when the door is open) and a heating element operates during automatic defrost.

Some specific features of different models of food freezers are illustrated in Figures 14-6 through 14-8 and described in the figure captions.

Figure 14-6. No-frost freezer with solid shelves for contact freezing. Cold air circulates from the back across top of each shelf, ceiling, and above basket and between front of shelves and door. An ice maker with a 10-pound capacity ice bin is optional. (Amana Refrigeration, Inc.)

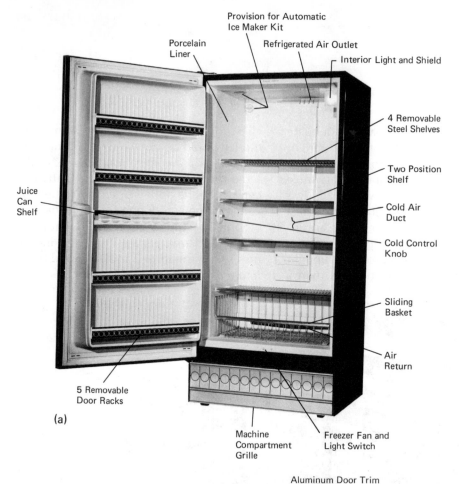

Provision for Automatic
Ice Maker Kit

Porcelain
Liner

Refrigerated Air Outlet

Interior Light and Shield

4 Removable
Steel Shelves

Two Position
Shelf

Juice
Can
Shelf

Cold Air
Duct

Cold Control
Knob

Sliding
Basket

Air
Return

5 Removable
Door Racks

(a)

Machine
Compartment
Grille

Freezer Fan and
Light Switch

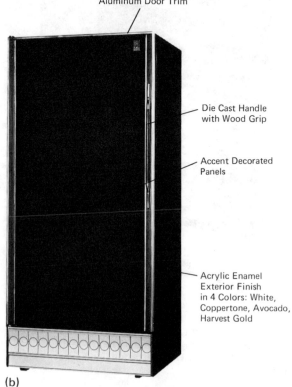

Aluminum Door Trim

Die Cast Handle
with Wood Grip

Accent Decorated
Panels

Acrylic Enamel
Exterior Finish
in 4 Colors: White,
Coppertone, Avocado,
Harvest Gold

(b)

Figure 14-7. (a) No-frost freezer with removable shelves. An automatic ice maker that makes and stores up to 5 pounds of half-sphere shaped "ice quicks" is an optional feature. (b) Picture-frame door styling allows homemaker to slip a decorative panel into trim around door. Overall styling is intended to match "all" refrigerator of the same manufacturer. Note left-hand door opening of freezer. (Gibson Refrigeration Sales Corporation)

Figure 14-8. Chest freezer with defrost drain provision at lower left corner. (Frigidaire Division of General Motors Corporation)

STANDARDS

The current freezer standards are part of the UL and ANSI refrigerator and freezer standards which were considered in Chapter 13. Of special interest is the definition of a household freezer as given in the definitions section of the ANSI standard: "A household freezer is a cabinet which is designed for the extended storage of frozen food at a recommended temperature of 0° F (−17.8° C) in a 90° F (32.2° C) ambient and with inherent capability for freezing of food, which has a source of refrigeration and which is intended for household use."[7]

[7] American Society of Heating, Refrigerating and Air-Conditioning Engineers, Inc., *American National Standard Methods of Testing for Household Refrigerators, Combination Refrigerator-Freezers, and House-Hold Freezers. ANSI B38.1, 1970.*

EXPERIMENTS

Experiment 1. Food Storage Volume

Measure interior dimensions and compute volume to nearest 0.1 cubic foot. How does the measured volume compare with the manufacturer's rated volume?

Experiment 2. Door or Lid Closure

1. Check tightness of door or lid gasket seal by one of the methods described in the experiment on electric refrigerators (Chapter 13).
2. Measure pounds of force needed to open door or lid with a spring balance as described in the experiment on refrigerators.

Experiment 3. Electrical Characteristics

1. Starting current: Use an ammeter with an adjustable pointer-stop. (See experiment on electric refrigerators.)
2. Power factor: Use an ammeter, voltmeter, and suitable wattmeter or use a watt-var meter. (See experiment on electric refrigerators.)
3. On-off cycling and kilowatt-hours per 24 hours for normal setting only. (See experiment on electric refrigerators.)

   Note: A recording instrument usually is needed because on and off times may be too long to "clock" in a two- or three-hour laboratory period.

Experiment 4. Time to Freeze Containers of Water and Warm-Up-Time when Freezer Is Disconnected

Time to freeze four containers of water placed at different locations in the freezer: Place a metal tube in a central hole in the lid of each water-filled container (a plastic freezer container is useful) and slip a thermocouple through the metal tube so that the junction is approximately in the center of the container. Bend the thermocouple so that the junction is out of the tube. Cover the end of the metal tube on the outside of the container with masking tape. Connect the free ends of the thermocouple to a temperature-recording device. The time to reach freezing temperature can be determined from the chart speed of the recorder in inches per hour and the inches of chart used. Record also the temperature of the air in one or more locations in the freezer. Compare fluctuations of temperature, if any, in the containers of water.

   Note: If a recording temperature instrument is not available, an indicating instrument can be used with a rotary switch and several thermocouples. With an indicating instrument, take measurements at 15-minute intervals.

Experiment 5. Effect of Packaging

1. Prepare two sets of patties from a single lot of ground beef. For one set, shape the beef into patties about 4 inches in diameter

and 1 1/2 inches thick. For the other, shape it into patties about 2 inches in diameter and 1/2 inch thick.

2. Package the larger patties in stacks of four, using paper disks to facilitate separation after freezing. Wrap different groups of four patties in different types of freezing paper.
3. Package the smaller patties in one layer, four or six to a package. Wrap different groups of four or six in the same types of freezing paper that were used for the larger patties.
4. Freeze all the packages of meat in the freezing section or storage section of one freezer for two weeks, or a longer interval if that is more convenient. Remove the packages from the freezer, unwrap them and note the following:
   a. Which packaging materials, if any, are associated with desiccated (freezer-burned) meat? (Look for freezer burn at edge of patties.) If *any* freezer burn is observed after as short a time as two weeks, the packaging materials probably are poor risks for longer storage periods.
   b. How do the 4-inch patties in the stack wrap compare with the 2-inch patties that were packaged in a single layer with its relatively large surface?

BUYING GUIDE

The first part applies to upright and chest freezers. However, some points are handled differently for the two types; for example, dimensions for an upright should include width with door open and for a chest height with lid open. The second part applies to uprights, and the third part to chests.

Upright and Chest Freezers

1. Does the appliance have the Underwriters' Laboratories seal of approval?
2. What are the dimensions of the appliance and the space needed for it? Space needed for a chest includes the thickness of the lid if the lid opens to a full vertical position.
3. What is the AHAM (Association of Home Appliance Manufacturers) rated total capacity in cubic feet? What is the capacity of the fast-freeze section if one is provided?
4. Is the nameplate accessible? Does it give the manufacturer's name and address, model, voltage rating, amperes, serial number, and other information?
5. Does it have an adjustable cold control? What is the average temperature corresponding to the normal control setting? Does the manufacturer's literature specify that average temperature is in the 0-degree range?

6. What is the manufacturer's warranty against food spoilage? What is the warranty on the appliance?
7. What provision does the door or cover have to insure closure? A cover might have a flexible or "floating" inner part that aids for secure closing. A door might have a special arrangement of magnets.
8. Does the compressor motor have a protector against electrical overload?
9. In the case of a conventional upright or chest is a drain system provided for defrost water?
10. What type of insulation is used?
11. How well constructed does the freezer seem to be? Are leveling feet or casters (upright only) provided?
12. Is the design of the freezer reasonable to you—from a functional-use viewpoint and aesthetically?
13. Do interior appointments seem useful and does the appliance seem easy to clean?
14. Does it have interior light(s)?
15. Is there a signal light on the exterior to show whether power is reaching the appliance?
16. Does the user's folder seem adequate in terms of how to use and maintain the appliance?

Upright Freezers

1. Is it a conventional or a no-frost type?
2. How many shelves have evaporator (freezing) coils? What additional parts—walls, ceiling, floor—have freezing coils?
3. If it is a no-frost upright, does it have a moist-cold section? Does it have an automatic ice maker? How many heating elements does it have and how frequently does it defrost?
4. What provision is made for large or irregularly shaped packages? If all the shelves have freezing coils are some spaced appropriately for odd-size packages?
5. What additional special features does the upright have and how do you rate them?

Chest Freezers

1. Is the lid counterbalanced to stay open at one or more positions?
2. Is the bottom rear portion of the interior accessible for a short person?
3. If a special fast-freeze section is provided, how many sides of the section have freezing coils?
4. What additional special features does the chest have and how do you rate them?

CHAPTER 15
ROOM AIR
CONDITIONERS

Total home air conditioning includes heating, purifying, humidifying, and air circulation in cold weather as well as cooling, filtering, dehumidifying, and air circulation in hot weather. This chapter deals with one type of equipment for air conditioning in hot weather: namely, room air conditioners. These are units which are designed primarily to cool, filter, dehumidify, and circulate air in an enclosed space such as a room or living zone and are usually mounted in a window or through a wall (Figs. 15-1, 15-2). Some models are designed in addition for supplementary heating of the enclosed space when the outside air is moderately cool, 45° F to 60° F or so.

Room air conditioners are rated in Btu/hr capacity (British thermal units per hour they remove from an enclosed space) and in electrical characteristics—amperes, watts, volts, and power factor. Most are used on 60 cycles, though models are available for use on 50 cycles.

According to ASHRAE,[1] 115-volt units are generally limited to a rating of 12 amperes which is the maximum allowable load of a single outlet 15-ampere circuit in compliance with the National Electrical Code. The same source also notes that a very popular 115-volt model is one which is rated at 7.5 amperes; this rating allows the unit to be plugged into any standard 115-volt multiple-outlet 15-ampere circuit. Models for use on 115 volts are available in ca-

---

[1] American Society of Heating, Refrigerating and Air-Conditioning Engineers, Inc., *ASHRAE Guide and Data Book—Systems and Equipment*, 1967, p. 511.

Figure 15-1. Through-wall air conditioner. (Hotpoint Division of General Electric Company)

Figure 15-2. Window-model air conditioner with a rating of 30,000 Btu/hr designed for multiroom cooling. An optional kit permits through-wall installation. (Frigidaire Division of General Motors Corporation)

pacities from approximately 5,000 to 14,000 Btu/hr. Models with higher capacities (14,000 plus to 36,000 Btu/hr) are designed for use with 230 or in some cases 208 volts. Units with quite high capacities, above 18,000 or so, are used in commercial applications, such as small restaurants and beauty shops, as well as in homes.

For a limited capacity range—about 12,000 to 14,000 Btu/hr—some models are available for use on 115 volts and others on 230 volts.

CONSTRUCTION, CAPACITY CONSIDERATIONS, AND INSTALLATION

CABINET CHARACTERISTICS

Overall cabinet dimensions vary from about 10 inches high, 14 inches wide, and 25 inches deep for small casement models (Fig. 15-3) to 20 inches high, 27 inches wide, and 39 inches deep for the very large-capacity models. Window-mounting spacers are available for windows wider than the unit. The depth which the unit extends into the room and out of the house may vary for different models even if the overall depth is the same (Fig. 15-4).

The cabinet is usually a steel or polycarbonate resin shell in or on which are mounted the functional components described in the next section. A louvered steel panel may be provided on the exterior to cover the condenser. A plastic grille is usual on a portion of the front part of the cabinet that is inside the room.

Acoustical installation is usual on the interior of the cabinet for the part inside the room.

FUNCTIONAL COMPONENTS

A room air conditioner absorbs heat from the room at the evaporator which is inside the room and gives up heat at the condenser which is outside. Principal components include the motor-compressor, evaporator, evaporator blower which causes warm room air to circulate across the evaporator coils, condenser, condenser fan, refrigerant, and capillary or restrictor between condenser and evaporator (Figs. 15-5 through 15-7).

In the schematic shown in Figure 15-5 warm room air pulled into the air conditioner by blower C is cooled by circulating

Figure 15-3. Casement air conditioner designed for operation on 115 volts, 7.5 amperes, and available in capacities of 5,000, 6,000, and 7,000 Btu/hr. (Fedders Corporation)

Figure 15-4. Window model designed with noise-producing parts placed in an insulated cabinet below window height and below cool-air duct. (Westinghouse Electric Corporation)

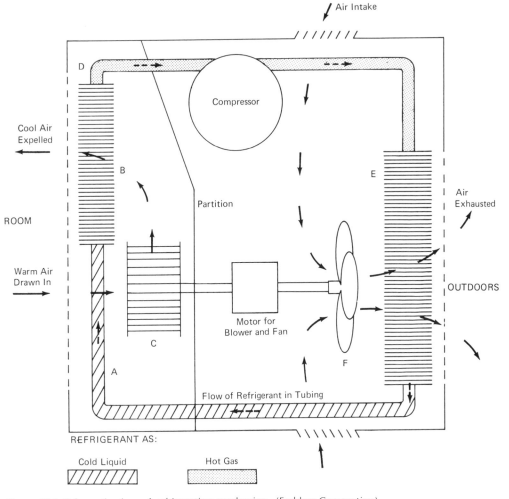

Figure 15-5. Schematic view of refrigerating mechanism. (Fedders Corporation)

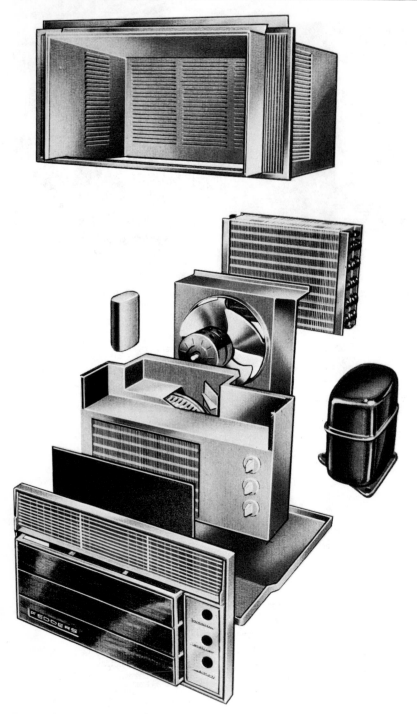

Figure 15-6. Exploded view of actual components. Housing with accordion-fold expanding side panels is at top; component at right is motor compressor. (Fedders Corporation)

across evaporator B and gives up heat to the refrigerant in B. The gaseous refrigerant that leaves the evaporator is compressed in the compressor and condensed in the condenser E. The heat of condensation is expelled to the outside air partly by the action of the condenser fan in moving air across the condenser. (This sketch shows the usual case of one motor for blower and fan; some models, however,

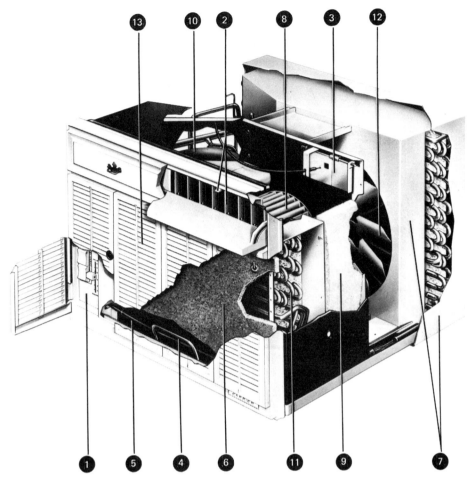

1. Touch control comfort center
2. Front and side air flow controls
3. Positive, separate exhaust and ventilation systems
4. Convenient, slide-out chassis
5. Automatic "Even-Temp" thermostat also prevents freeze-up
6. Extra large, washable, germicidal-treated filter
7. "Weather Armor" casing, internal parts
8. Specially designed blower moves more air
9. Lo-Sound, extra thick insulation
10. Long life compressor and fan motors
11. Staggered rows, copper tubing, traps heat
12. Slinger-fan prevents condensate build-up
13. Highly functional grille design

Figure 15-7. Cutaway view of a deluxe model. This model has a Canadian Standards Association seal as well as an American Home Appliance Manufacturer's (AHAM) seal. (Carrier Air Conditioning Company)

use separate motors.) Moisture from the room air condenses on the evaporator coils, drips into a drain pan, and is carried off by atmospheric air.

Figure 15-6 shows an exploded view of actual components. The cabinet assembly at the top has accordion-fold expanding side panels for installation in windows of widths greater than that of the cabinet. The compressor is the separate part at the right. Other components are a wood-grain front panel with directional air-flow controls, polyurethane filter, bulkhead partition with controls mounted on it and

evaporator and blower in it, condenser fan, and condenser.

CONTROLS

Room air conditioners may have mechanical and electrical controls. Mechanical controls are used to change the air-flow pattern of the room air near the evaporator. The air conditioner shown in Figure 15-7 has front and side air-flow controls.

In a stripped model the electrical control is an on-off switch. In a deluxe model, the electrical controls may be push buttons, rotating parts, or parts that move in

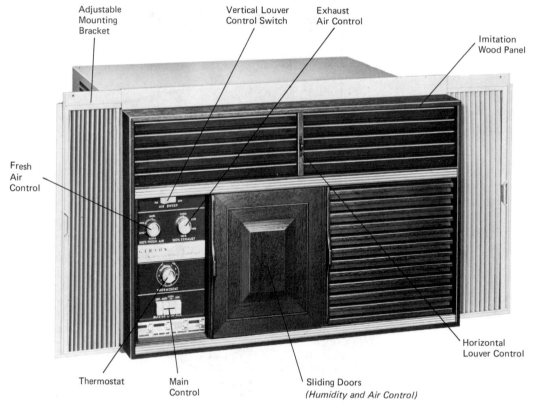

Figure 15-8. Deluxe model with several special controls. In cold weather interior is closed by sliding doors and louvers. On high humidity days partial covering of evaporator maintains a preset temperature *and* dehumidification. Motor-driven vanes provides an "adjustable air sweep." Controls also permit unit to be used as a window fan that brings outside air in or removes inside air. (Gibson Refrigeration Sales Corporation)

a line. The control panel may have, in addition, a switch for operating the evaporator blower at high or low speed and a mechanical or electrical control to regulate ventilation and exhaust dampers if the unit also is designed for ventilation and exhaust without cooling.

Figure 15-8 shows a unit with adjustable humidity control accomplished by sliding front panels. The manufacturer notes that on some high-humidity days normal operation of the cooling system overcools the room. By sliding the panels so as to partially cover the cooling coils, cooling action is decreased without affecting dehumidification.

All room air conditioners cool, dehumidify, and filter the room air. The cold-

control provision is especially appreciated by users who operate an air conditioner for long periods at a time. Without such a control, the room will tend to become increasingly colder if outside temperature remains constant or decreases. The cold control does not govern anything except the on-off cycling of the motor-compressor unit.

For maximum cooling only room air is circulated, cooled, dehumidified, and filtered. Cooling and ventilating permits introduction and cooling of outside air. Fresh air is introduced from the outside, mixed with the recirculated room air, cooled, and dehumidified. For the vent and exhaust settings, air is filtered but not cooled or dehumidified.

HEATING AND REVERSE-CYCLE OPERATION

As indicated earlier, room air conditioners may be designed to provide some heat on cool days. The usual source of heat is an electric resistance element or *reverse-cycle* operation of the refrigerating mechanism, that is, cooling *or* heating. In a cooling cycle, heat from the room is absorbed at the evaporator and given up to the outside air at the condenser. In a heating cycle, heat from the outside air is absorbed at the condenser (acting as an evaporator) and given up to room air at the evaporator (acting as a condenser).

Reverse-cycle operation will be more readily understood by tracing the path of the refrigerant in a cooling cycle and a heating cycle. Figure 15-9 shows schematically how reverse cycling works.

A characteristic feature of a refrigerating system designed for reverse-cycle operation is a reversing valve. For cooling, the solenoid plunger of the reversing valve is in position to the left. Refrigerant discharge gas from the compressor passes through the reversing valve to the condenser. The condensed refrigerant next passes through a restrictor into the evaporator. The vaporized refrigerant then moves to the reversing valve and leaves the reversing valve to enter the compressor inlet. This is a standard cooling cycle.

For heating, the solenoid plunger of the reversing valve is in position to the right. In this case refrigerant discharge gas from the compressor passes through the reversing valve to the evaporator, where it is condensed because the evaporator is now serving as a condenser. The condensed refrigerant then passes through the restrictor to the condenser where it is vaporized because the condenser is now serving as an evaporator. The vaporized refrigerant next passes through the reversing valve to the compressor inlet. The outside air gave up heat to vaporize the refrigerant in the condenser, and this heat was absorbed by room air when the refrigerant was condensed in the evaporator.

CORRECT SIZING

The appropriate capacity window unit to use for a given space depends on several variables: amount of window area and directions the windows face, height of ceiling, type of roof on the house, floor area, number of people who customarily use the room, wattages of lights and other electrical equipment used in the room, and the total number of lineal feet of doors and arches continuously open to unconditioned space.

A cooling load estimate form is available from the Association of Home Appliance Manufacturers.[2] Alternately, a dealer may use an even shorter device supplied to him by the manufacturer for determining the correct capacity, provided the prospective purchaser supplies information on the variables listed above.

Correct size is important for adequate cooling and dehumidification. A size larger than appropriate may not dehumidify satisfactorily even though temperature is controlled adequately during on-off cycling. A too small size will not control temperature or humidity for optimum comfort.

Whenever possible in house planning, all the usual means of maintaining comfort should be utilized in addition to air conditioning. Outside awnings over windows, as well as insulation of walls and roof, decrease the cooling load on the air conditioner.

[2] AHAM, 20 North Wacker Drive, Chicago, Ill. 60606.

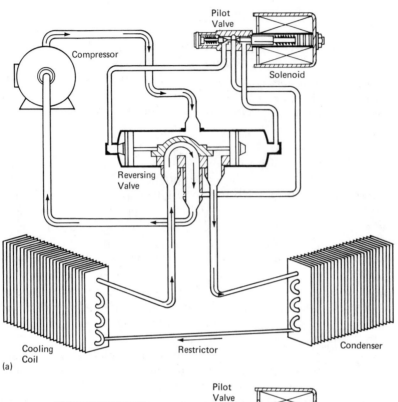

(a)

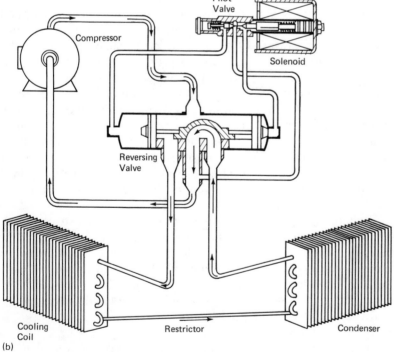

(b)

Figure 15-9. Refrigerating mechanism designed for reverse-cycle operation. (a) Cooling cycle. (b) Heating cycle. (Frigidaire Division of General Motors Corporation)

INSTALLATION, USE, AND CARE

Installation

A room air conditioner should be installed in a window on the cooler side of the house, if the room has windows on more than one side and if installation on the cooler side is practical. Furthermore, the location should be such that free air circulation is possible at both the front or room part and the rear or outside part of the air conditioner.

Because the starting current is high, a time-delay type of fuse is recommended for the electric circuit to the air conditioner.

Plugs supplied on 115-volt models should be three-prong type. Plugs for 208- or 230-volt models are one of several three-prong types. (A special outlet may be needed.)

Care and Use

Suggestions for care of specific models are given in user's booklets. Also, the dealer will often offer suggestions, especially on preventive maintenance.

Generally, the filter is cleaned or replaced by the user according to instructions in the user's booklet. Polystyrene filters may be rinsed in lukewarm water. Aluminum ones may be washed in hot, sudsy water. The air conditioner should not be operated with the filter removed.

Preventive maintenance at the beginning of each cooling season by a service man may include the following operations: cleaning the condenser and evaporator, if required; installing a new air filter, if required; oiling the fan motor and, for some models, the fan bearing; checking tightness of seals around the window in which the unit is installed.

If the room or space is allowed to become excessively hot, more time will be required to obtain a comfortable tempera- ture because of heat stored in walls and furnishings. For maximum comfort, the unit should be turned on before the space becomes excessively hot so that the unit need only *maintain* comfort conditions.

During extremely hot weather it is better to operate the air conditioner at the cool rather than the cool-vent setting to avoid placing the unit under a heavy overload. Ventilating brings in warm outdoor air.

The members of the family should recognize the *usual* operating sound of the unit. If any unusual operation is noted, such as continuous "on" and "off" cycling, the air conditioner should be turned off until the reason for the unusual operation is ascertained and corrected, either by the user or by a service man. One trouble may be too high or too low voltage; another may be that the air conditioner is installed in a hot area with poor ventilation. In the latter case the location of the air conditioner may have to be changed.

When an air conditioner is moved in a truck from one house to another, the motor-compressor assembly of some units should be bolted to the supports (cross members) on which it rests. A dealer will know whether this is necessary for models he sells. *Caution:* The shipping bolts should be loosened or removed before the air conditioner is turned on in the new location.

Perhaps the best suggestion on noise associated with an air conditioner is to listen to the operating sounds of several models before purchase and to choose one with the least objectionable noise. Generally it is not possible to estimate noise in a store and the prospective purchaser should try to appraise the sound insulating parameters that are built into the unit. A *Consumer Bulletin* article makes the following pertinent comments:

A quiet running conditioner will usually have (1) large evaporator and condenser coils, (2) large, low-speed circulating fans which move as much air as smaller (and noiser), high-speed fans and, in some models, (3) a larger-than-usual compressor. . . . It is to be expected that, in a particular manufacturer's line, a model in a larger cabinet will be quieter in operation than a unit of like cooling capacity in a smaller cabinet.[3]

AHAM CERTIFICATION AND ROOM AIR CONDITIONER STANDARDS

The Association of Home Appliance Manufacturers has a certification program for window and through-the-wall air conditioners. Since 1963 the program assures a purchaser of a certified model that the nameplate cooling capacity rating in Btu's per hour and the electrical input in amperes and watts have been independently verified in accordance with Standard CN-1 of AHAM. Electrical Testing Laboratories, Inc., administers the program.

Certified units have a distinctive AHAM seal which is affixed at the time and place of manufacture of the unit and are listed in AHAM directories published three times per year. The certification program provides that samples of models representing at least 50 percent of the participating manufacturer's basic models are tested.

North American standards for room air conditioners are:

Underwriters' Laboratories, Inc., UL 484, Standards for Safety—Room Air Conditioners.

AHAM Room Air Conditioner Standard, No. CN-1.

AHAM Room Air Conditioner Sound Rating Standard No. RAC-2SR. (This is a uniform laboratory means of measurement for arriving at a sound rating. It is not intended as a consumer standard.)

Canadian Standards Association, Specification C 22.2 No. 117, Construction and Test of Room Air Conditioners. (Cited in ASHRAE Guide and Data Book, 1967).

The AHAM Standard No. CN-1 has sections on definitions, rating standards, testing standards, and a cooling load estimate form. Some of the definitions are as follows: Recirculated air is the air discharged by the unit into the enclosed space when all ventilating dampers are closed. Ventilating air is the air introduced by the unit into the enclosed space from outside. Exhaust air is the air removed by the unit from the enclosed space to the outside. Standard air is air having a density of 0.075 pounds per cubic foot and is equivalent to dry air at a temperature of 70° F and barometric pressure of 29.92 inches of mercury.

The rating standards specify the cooling capacity ratings. They include total cooling capacity in Btu's per hour and moisture removal capacity in pints per hour. Heating capacity rating, where appropriate, shall be stated in Btu's per hour. Ratings for recirculated air quantity, ventilating air quantity, and exhaust air quantity shall be stated in cubic feet per minute of standard air. Electrical ratings shall be stated in watts, amperes, and power factor in percent.

Also in the rating standards section is the following requirement on minimum information to be included on the nameplate: (1) Electrical data in accordance with Underwriters' Laboratories, Inc., "Standard

[3] "Room Air Conditioners," Consumer Bulletin, July 1970, p. 25. Published by Consumers' Research.

for Room Air Conditioners," Publication 484. (2) Electrical power input in watts at AHAM rating conditions. (3) Total cooling capacity in Btu's per hour or, if heating is also provided, the combined total of cool-ing and heating capacity ratings in Btu's per hour.

The cooling load estimate form is the same as the AHAM form referred to on page 277.

BUYING GUIDE

General Considerations

1. What are the relevant characteristics of the space to be air-conditioned? Might a larger space than the one initially planned for be conditioned by an adequately sized and well-located window unit—for example, might a unit of suitable capacity in a window at a floor landing serve for two floors? Alternately, instead of a single large unit for a given space that is cut up by architectural barriers, might two smaller units give better zoning of the cooled and dehumidified air?

2. Is it reasonable to consider total house air conditioning for a house heated by forced air and with adequate ductwork? Indeed, is it reasonable to consider ductwork?

Specific Considerations

1. What cooling capacity in Btu's per hour is needed for the space to be conditioned?

2. Are models with approximately this capacity available in the voltage supplied to your home? Will an individual equipment circuit and/or improved house wiring be needed? (In addition a special receptacle to receive the plug of the air conditioner may be needed. This is best installed after the unit has been selected.)

3. What are the AHAM cooling capacity ratings in Btu's per hour and in pints of water per hour for the model under consideration?

4. What are the electrical ratings in watts, amperes, and power factor?

5. What are the physical dimensions: How many inches will the unit extend into the room? What is the extension out of the house? How much space is needed below the window? Will the appearance be satisfactory inside and outside?

6. If the unit is described as portable, what is meant by portable? (Generally the word seems to mean the electrical characteristics of the unit are such that it can be installed in any 115-volt, 15-ampere, multiple-outlet circuit—not that it is designed to be moved like a portable phone.)

7. What are the construction characteristics relative to quietness in operation?
8. What are characteristics relative to control of movement of room air?
9. Is the unit designed so that year-round installation in the climate where it will be installed is practical in terms of winter comfort *and* heating energy costs?
10. Does the unit have the special features of exhaust, ventilate, and heat? If so, and if you want these features, what are the ratings in cubic feet of air per minute or Btu per hour, as appropriate?
11. What additional special features does the unit have and how do you rate them? For example, is the air conditioner designed to accommodate an electronic air cleaner and are you interested in purchasing such a cleaner?
12. What does the manufacturer's warranty promise? Will the dealer exchange the unit if it proves inadequate for the space for which it was purchased? Does the quoted price include initial installation? If the unit needs to be removed and stored each winter, what are the costs likely to be?

CHAPTER 16
ELECTRIC DEHUMIDIFIERS, HUMIDIFIERS, AND ELECTRONIC AIR CLEANERS

As noted in Chapter 15, home comfort involves removal of moisture from the air in the summer and addition of moisture to the air in the winter. Separate, portable dehumidifying and humidifying appliances, as well as fixed appliances, may be used for these purposes in some enclosed spaces.

Air cleaning by electrostatic components is an additional means of filtering air.

ELECTRIC DEHUMIDIFIERS

Electric dehumidifiers are self-contained appliances designed to remove moisture from the air. Generally, they are used in below-ground level spaces such as basement workshops, laundry areas, family rooms, or others. Removal of moisture retards rusting of metal parts (such as tools or vises), mildew on surfaces, and warping of wooden units (such as furniture).

An electric dehumidifier contains an electric refrigerating mechanism similar to that used in an electric compression-system refrigerator or room air conditioner. The condenser, fan and fan motor, and compressor are located behind the evaporator. The fan draws

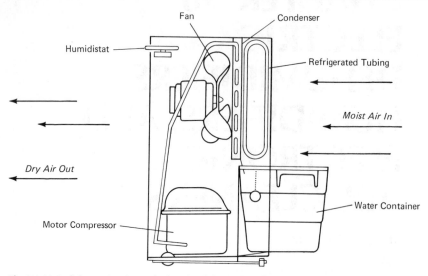

Fan

Condenser

Humidistat

Refrigerated Tubing

Moist Air In

Dry Air Out

Water Container

Motor Compressor

Figure 16-1. Schematic view of electric dehumidifier. (General Electric Company)

air over the evaporator or cooling coils and the moisture in the air condenses on the coils. This moisture drains either into a pan under the coils or through a house drain.

Figure 16-1 shows major components schematically.[1] The compressor consists of a closed shell which houses a pump powered by an enclosed motor containing lubricating oil. Thermal-overload protection is provided so that the unit will shut off automatically if unusual operating conditions occur. Copper tubing forms the connecting passageway between the compressor and condenser. The condenser sketched is tubing with a serpentine configuration which has a fin construction superimposed on it to increase its efficiency in giving up heat to the room. The capillary is tubing of very small diameter which acts as a restrictor to control the flow of the refrigerant from condenser to evaporator. The evaporator or refrigerant tubing is a coil wound in oval configuration. The schematic view also locates a humidistat which maintains the relative humidity in the enclosed space at the control setting dialed.

The deluxe model shown in Figure 16-2 has a 2 1/2-gallon container at the rear

[1] General Electric, *Dehumidifiers—Product Review*, publication no. 57-199.

which receives the water removed from the room or other space. The automatic shut off responds to a float valve which extends from a bead chain and hangs freely inside the container. The purpose of the automatic shut off is to prevent overflow of water from the container. The AHAM capacity of the model shown is given as 22 pints. This is the amount of water that will be removed in 24 hours of continuous operation under specified conditions. (See the section on additional design features later in this chapter.) Exterior dimensions are approximately 11 1/2 inches square by 21 1/2 inches high. The compressor has a capacity of 1/4 horsepower and input electrical characteristics are given as: 115 volts, 6.0 amperes, 60 cycles, AC only, and 550 watts.

Two other models of the same manufacturer have water removal capacities per 24 hours of 17 pints and 13 pints, respectively.

LOCATION, OPERATION, AND CARE

A dehumidifier should be installed in an enclosed space (windows closed) since it is not designed to dehumidify the outdoors. Also, the location in the room should be such that air flow *through* the dehumidifier is not restricted. As for other refrigerator-type mechanisms, the appli-

Automatic Humidistat—Turns dehumidifier on and off at selected humidity level.

Signal Light—Indicates when water
container needs emptying.

Figure 16-2. Deluxe (automatic) dehumidifier. (General Electric Company)

ance should be positioned level. If the appliance is placed in a location where there is a floor drain it is possible to use a garden hose on a threaded hose fitting above the container, thus eliminating the need for emptying the container.

Dehumidifiers that meet current recommendations have a three-wire grounding cord and three-prong plug. The grounding wire and plug are designed to protect the user from electric shock should internal dehumidifier insulation fail *provided* the plug is installed in a properly grounded receptacle. The appliance plug should not be used with an adapter in a two-wire outlet unless it is known that the receptacle is grounded. Also, a dehumidifier should not be used in a space where standing water might accumulate around it.

For models without humidistat the user manually turns the appliance on and off to maintain the desired dehumidification. For models with humidistat the desired level is maintained automatically.

The manufacturer's instructions to the user may state that the humidifier is designed to operate most effectively at room temperatures above 65° F or so. Also the instructions may state that the dehumidifier should be turned off in periods of low humidity to avoid damaging the mechanism. One manufacturer (General Electric) recommends that the container be emptied when the dehumidifier is not in use to avoid unpleasant odor that may be associated with standing water. The same manufacturer recommends that the dehumidification water not be used for any domestic purpose.

Care of the dehumidifier during the use season and before storage involves dusting surfaces, including the grille, and the refrigerant tubing. The appliance should, of course, be disconnected when dusting with an electric cleaner.

ADDITIONAL DESIGN FEATURES

Additional design features relate to a control that cycles the compressor off and on under frosting conditions (below design operating temperature and/or low relative humidity) and air movement capacity.[2]

[2] American Society of Heating, Refrigerating and Air-Conditioning Engineers, Inc., *ASHRAE Guide and Data Book—Systems and Equipment*, 1967, p. 515.

The *Guide and Data Book* states that domestic dehumidifiers will maintain satisfactory humidity levels in an enclosed space with an air flow equal to approximately one air change per hour. For a space 30 by 30 by 8 feet, this corresponds to a fan capacity of 120 cubic ft./min.

STANDARDS

Standards are: Underwriters' Laboratories, Inc., Standards for Safety—Dehumidifiers, UL 474, and the AHAM Dehumidifier Standard No. DH-1.

The AHAM standard pertains to "self-contained, electrically operated, mechanically refrigerated" devices for removing moisture from the surrounding air. Rating standards are as follows.[3]

Voltage rating shall be 115, 208, or 230 volts.

Frequency shall be 60 cycles per second.

Rated capacity shall be stated in pints of water per 24 hours of continuous operation.

Electrical input rating shall be stated in watts.

Nameplate data shall include the following minimum information: electrical data in accordance with Underwriters' Laboratories, Inc., *Standard for Dehumidifiers*, Publication No. 474.

The AHAM rated capacity is based on tests at the following ambient conditions: dry-bulb temperature of 80° F, wet-bulb temperature of 69.6° F, and relative humidity of 60 percent.

ELECTRIC HUMIDIFIERS

As the name indicates, humidifiers add humidity to an enclosed space. This is desired in the winter in many locations to increase comfort, decrease drying out of wooden furniture and other wooden objects, diminish frequency of static electric shocks, and decrease heating bills. (With adequate relative humidity, 25 to 60 percent or so, most healthy people are comfortable at temperatures near 72° F; on the other hand, when the relative humidity is very low in the winter, people at rest or only moderately active may prefer a temperature higher than 72° F.)

COMPONENTS AND CHARACTERISTICS

The moisturizing action of the humidifiers shown in Figures 16-3 through 16-5 depends on a "water wheel" and a fan. The filter belt fits around a drum which rotates through a water reservoir. Dry air enters the appliance through the lattice covering in the back and filtered, humidified, and cooled air is discharged from the top.

Water is added to some models at the back and to others at the front.

The model shown in Figure 16-5(a) is described as having a drum type "evaporative mechanism," individual controls for drum and fan motors, a three-speed fan, refill signal light, water-level indicator, humidistat, shut off (when reservoir is nearly empty), ball-bearing casters, and a capacity of 20 gallons per 24 hours at 10 percent relative humidity and 8 gallons at 40 percent relative humidity. Dimensions are 26 inches high, 27 1/2 inches wide, and 12 1/2 inches deep.

The model shown in Figure 16-5(b) has a solid-state control, a capacity of 24 gal-

[3] Association of Home Appliance Manufacturers, *AHAM Dehumidifier Standard No. DH-1*, pp. 3-4.

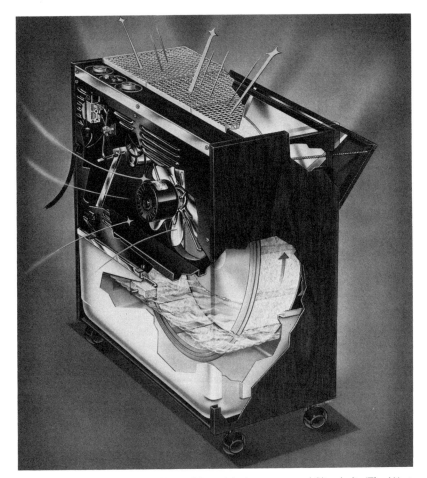

Figure 16-3. Schematic view of humidifier with drum-mounted filter belt. (The West Bend Company)

Figure 16-4. Filling one model of humidifier. (West Bend Company)

(a)   (b)

Figure 16-5. (a) Humidifier with styrene front panels and vinyl-on-steel side panels in Mediterranean style. (Albion Division, McGraw-Edison Company) (b) Humidifier in traditional style. (Albion Division, McGraw-Edison Company)

lons per 24 hours at 10 percent relative humidity and 10 gallons at 40 percent relative humidity, an optional summer top cover, and other features of the model described previously. Capacity of the reservoir is 10 gallons for both models.

The solid-state sensor is described as a modulating system that holds the humidity close to the preset value.

The only routine care required is cleaning and, when necessary, replacement of the filter.

ELECTRONIC AIR CLEANERS

An electronic air cleaner in a housing that suggests an end table was put on the market in 1970.

The basic part is a two-stage electronic cleaning cell which catches and collects microscopic particles of airborne dirt and pollen (Fig. 16-6). In operation, room air is drawn by a fan into the air intake at the bottom of the unit. As the air moves upward it passes first through an aluminum mesh filter which catches lint, hair, and other visible particles. The air then moves into the two-stage electronic cleaning cell where the micron-size particles are ionized and collected. (One micron is one-millionth of a meter.)

In the first stage, the air passes through a high-voltage (5,400 volts) electrical field created by a series of electrodes and ion-

izing wires and the micron-size particles take on a positive electric charge. In the second stage the air enters an electrical field associated with a series of metal plates that have a negative charge. The positively charged particles are attracted by and collect on the ground plates, remaining there until they are washed off. (The ionization process causes the appliance to generate some ozone.)

The air that leaves the electronic cell passes through an activated charcoal filter that absorbs odors and vapors not affected by the electrostatic precipitation in the cell. Finally, the cleansed air enters the room through the front grille. The manufacturers state in their literature that the cleaner removes 99 percent of all pollen and "up to 90 percent" of all airborne dust,

Figure 16-6. Electronic air cleaner. (West Bend Company)

smog, and smoke as measured by the National Bureau of Standards Dust Spot Test using atmospheric dust.

The appliance is listed by Underwriters' Laboratories. It operates on a 60-cycle, 120-volt circuit. Power consumption is 98 watts, 1.25 amperes at the high setting of the fan and 60 watts, 0.82 amperes at low setting. Dimensions of the cabinet are approximately 13 inches deep, 26 1/2 inches wide, and 21 inches high.

Maintenance involves washing the aluminum mesh prefilter and the electronic filter and replacing the disposable charcoal afterfilter. A safety interlock is provided so that access to the filters requires that the appliance be disconnected from the power supply.

BUYING GUIDE

Electric Dehumidifiers

1. Is the appliance UL approved?
2. What is the dehumidification capacity in pints of water per 24 hours?
3. What are the electrical input ratings? The compressor horsepower rating?
4. Is an overload protector provided against damage due to high or low voltage?
5. What are the dimensions?

6. Does the appliance have a humidistat? A signal light to indicate when the container needs to be emptied? An automatic shut off to prevent water overflowing the container? Other features that are important to you?
7. How sturdy do the casters seem? Is the container easy to take off and put on?
8. Does information in the user's booklet recommend use where you plan to use it? If the booklet is not available, try other sources for this information.
9. Will out-of-season storage present a problem?
10. What does the warranty promise?

Humidifiers

This guide compares with that for a dehumidifier, with some exceptions.

1. Is the appliance UL approved?
2. What is the humidification output in gallons of water per 24 hours at different ambient relative humidities?
3. What is the capacity of the water reservoir? Does it seem easy to fill?
4. Does the appliance seem to move easily so that it will not be a chore to move it to the water supply?
5. What are the electrical input ratings?
6. Does the appliance have a humidistat? What type? Does it have a signal light that indicates when water should be added? Does it have other features that are important to you?
7. Will out-of-season storage present a problem?
8. Is the user's booklet helpful?
9. What does the warranty promise?

Electronic Air Cleaners

Perhaps the first most important consideration is: Will the appliance fill the need for which it is purchased? For example, if you wish to use it in a living room in the summer, are you prepared to have the windows closed? Can the living room be closed off from other rooms?

Specific guide points of the type outlined for the other two appliances in this chapter derive from the specific characteristics of this appliance.

CHAPTER 17
HOME CLEANING AND VACUUM CLEANERS

Currently, electric vacuum cleaners and their tools or atachments are the most labor-saving devices available for removing dust and other soil from many of the large and small surfaces in the home.

Electric floor polishers are a most effective home tool for polishing and buffing waxed floors and hard floor coverings. By changing brushes on the polishers some can be used for shampooing rugs either with liquid or dry cleaners.

One reason for using vacuum cleaners and floor polishers is to decrease the amount of human energy expended in cleaning. An additional means of decreasing the work of cleaning is, of course, to keep as much dirt as possible *out* of the home. Dust and other soil enter through open windows, filter through openings around windows and under doors, filter through screens, and are carried on shoes and clothing of persons and paws and other parts of pets.

Effective mats at all entrances decrease the amount of soil carried into the house on shoes. Good stripping around windows and doors cuts down the amount of soil that gets through openings. Some airborne dust can be removed by installing a furnace filter in home-heating systems. In air-conditioned homes or rooms some dust collects on the filters in the air conditioner rather than on surfaces that have to be vacuumed frequently.

ELECTRIC VACUUM CLEANERS

Vacuum cleaners and their attachments are used to clean carpets, rugs, wood floors, cement floors, hard floor coverings, ceilings, walls, and some furniture. The standard attachments also are used to remove dust from lamps, lighting fixtures, pictures, drapes, books, and bric-a-brac.

Vacuum cleaners and their attachments are not ordinarily used for counter tops or other surfaces on which food is handled.

Every electric vacuum cleaner has one or two fans or blowers assembled at one end of a motor. As the fan(s) rotate, air is thrown outward, creating an area of low pressure around the fan hub. This low pressure produces a vacuum or suction effect. In most upright cleaners the overall effect of the high-speed rotation of the fans is to cause air to enter the floor nozzle or other cleaning attachment, move through the fan chamber, through the bag or other filtering device, and out into the room.[1] In some uprights and most cylinder-type cleaners the air is filtered before it moves through the fan chamber. This entering air picks up soil from the surface under the nozzle and leaves it in the filtering device.

As dirt collects in the cleaner, the filtering action is impeded and the efficiency of the cleaner decreases. Comments on frequency of removal of accumulated dirt from the cleaner are given in the section on care and maintenance.

[1] In the upright cleaner, air usually leaves through the bag. In the cylinder-type cleaner, air leaves through blower ports after passing through the bag.

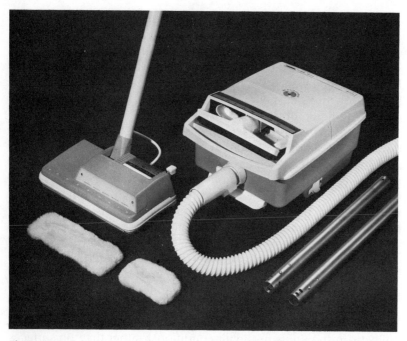

Figure 17-1. Cylinder-type vacuum cleaner with a motor-powered rotary beater in floor-cleaning attachment; acrylic dusting pads can be laundered. (Sears, Roebuck and Company)

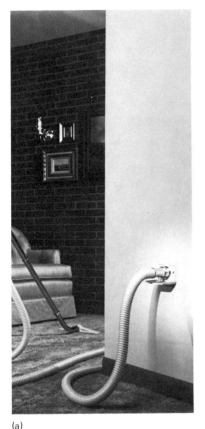

(a)

(b)

Figure 17-2. (a) Automatic wall inlet turns on when lightweight vinyl hose assembly is inserted. (b) Six-gallon soil bag receptacle installed in a closet. (NuTone Division of Scovill Manufacturing Company)

In addition to cleaning by rapid movement of air, modern vacuum cleaners make use of brushes. The type of brushes used varies with the design of the cleaner. Rotating brushes usually are found in upright cleaners; stationary brushes usually are used in cylinder-type cleaners.

TYPES OF VACUUM CLEANERS

It becomes more and more difficult to classify vacuum cleaners into discrete types. Canisters, tanks, and special-purpose types depend primarily on rapid movement of air through the cleaner for their cleaning effect. Some types which look similar to an upright vacuum cleaner actually operate more like the tank type. However, the rug tool of the cylinder-type usually has a free-floating brush and/or a comblike device in the nozzle. Some

cleaners have a rotating brush which may be powered by air-flow or an electric motor (Fig. 17-1). This device can be positioned in two ways according to the type of soil on the rug or carpet to be cleaned. If the brush is positioned to make contact with the rug, it picks up threads and similar light litter as the nozzle is moved across the rug. If the brush is retracted into the nozzle, it is inoperative. This position is used when no more light litter is present on the rug.

Built-in vacuum cleaners, which are cylinder-type cleaners, are most efficiently installed when the house or apartment is built. The power unit and dirt receptacle usually are installed in a garage, utility room, or basement (Fig. 17-2). The ease with which the cleaning tool and hose can be maneuvered depends in part on the location of the wall inlets. The floor

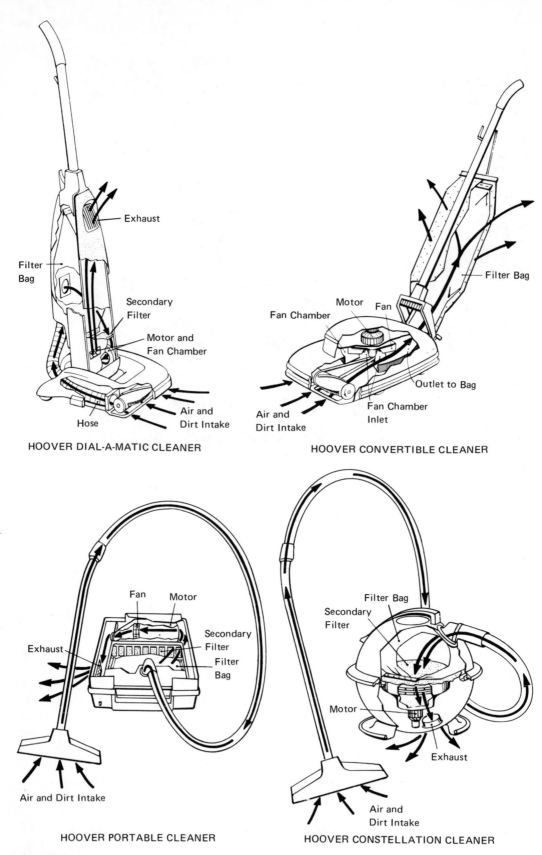

Exhaust

Filter Bag

Secondary Filter

Motor and Fan Chamber

Hose

Air and Dirt Intake

HOOVER DIAL-A-MATIC CLEANER

Motor

Fan

Fan Chamber

Filter Bag

Outlet to Bag

Fan Chamber Inlet

Air and Dirt Intake

HOOVER CONVERTIBLE CLEANER

Fan

Motor

Secondary Filter

Filter Bag

Exhaust

Air and Dirt Intake

HOOVER PORTABLE CLEANER

Filter Bag

Secondary Filter

Motor

Exhaust

Air and Dirt Intake

HOOVER CONSTELLATION CLEANER

Figure 17-3. Phantom view of two upright cleaners and two cylinder-type cleaners. (Hoover Company)

tool is similar to that found in other cylinder-type cleaners.

Upright cleaners use a motor-driven rotating brush for cleaning carpets and rugs, in addition to air movement. A cylinder or roll is mounted inside the part of the cleaner that makes contact with the rug. Different manufacturers use different names to describe the cylinder such as brush roll or agitator. The roll is belt connected to the motor used for the fan. The belt passes over the motor shaft just ahead of the fan and over a recessed portion of the roll. Brushes are mounted in or on the roll, either parallel to its axis or in a spiral around it. The brush roll rotates at high speed when rugs are vacuumed.

In addition to brushes, there may be a curved bar mounted spiral fashion on the exterior of the brush roll. Its function is to beat the rug and loosen embedded dirt that is then picked up by the air current and carried into the cleaner (Fig. 17-3).

CYLiNDER-TYPE CLEANERS

Appearance and convenience features of cylinder cleaners have been changing almost continuously since 1953. Some of the housings approximate vertical cylinders or rectangular boxes; others are the shape of a sphere. The suction opening is in the top or the front of the cleaner.

Interior Components

The movement of air through the cylinder cleaner is shown in Figure 17-3. The exhaust opening in one example is on the side; on the other it is on the bottom. The filtered air leaves through the exhaust opening. The motor is protected by two filters—the filter bag and the secondary filter.

A paper bag may fit into the bottom part of most cleaners of this type (Fig. 17-4). On some models the motor is protected by a filter cap.

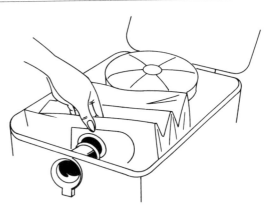

Figure 17-4. Schematic view of cylinder-type cleaner showing placement of disposable bag. (Sears, Roebuck and Company)

Other models of cylinder cleaners have similar functional parts though the actual physical parts are different. In the case of a horizontal cylinder cleaner, the filter and the filter bag are installed at the suction end of the cleaner ahead of the fan.

Hose and Extension Tubes

A hose and extension tubes to connect the cleaner with the tools are provided with cylinder cleaners and many uprights. A metal part at one end of the hose is attached to the cleaner, and another metal part at the other end is connected to an extension tube or directly to a dusting tool. The hose is usually about 8 feet long. It generally is made of vinyl and is flexible. At least one manufacturer uses a hose that stretches twice its apparent length, its advantage being that the cleaner needs to be moved less frequently when a large area is cleaned.

Two hollow extension tubes or wands, each about 20 inches long, are usual. A friction fit may be used to connect hose, wands, and tools, or mechanical locking devices may be provided. In general, mechanical devices are considered more satisfactory than a friction fit for locking tool to wand, one wand to another wand, and wand to hose.

One manufacturer provides wands that telescope; that is, one wand slides into and out of the other. Their advantage is that they are handled as one piece instead of two.

Most cylinder cleaners have a variable

vacuum or suction control. This is a small hole or set of small holes with a sliding ring or cover usually located near the place where the hand grips the wands. Moving the cover off the hole or holes permits air to enter the wand, thereby decreasing the vacuum effect of the cleaner. Some cleaners have the vacuum control in the cleaner itself or in one of the metal parts of the hose, rather than in a wand. Maximum suction (opening covered) is used when cleaning heavy rugs and carpets. Less suction (uncovered or partially uncovered opening) is used when cleaning lightweight scatter rugs and dusting drapes and certain other lightweight furnishings. Too much suction will draw drapery materials into the nozzle.

Tools

Usually, tools are provided with the cleaner for five types of use, and additional attachments are sometimes available as accessories. Standard models of cylinder cleaners may have a rug tool, a floor-and-wall tool, a dusting tool, an upholstery tool, and a crevice tool. Many models have one dual-purpose, rug-and-floor tool instead of separate tools for rugs and bare floors. A dual-purpose, dusting-and-upholstery tool may also be provided. Effective dual-purpose tools simplify cleaning since fewer tools need to be removed and replaced in the wand.

As indicated earlier, the rug tool or the rug side of a rug-and-floor tool often has a two-position floating brush or comblike device inside the nozzle for picking up light litter. The comblike device has narrow openings to concentrate air flow through the nozzle.

The floor-and-wall tool or the floor side of a rug-and-floor tool usually has a brush around the nozzle, which provides a sweeping action when walls and hard floors are cleaned. The small dusting tool also has a brush around the nozzle; its bristles are generally softer than those of the brush on the floor tool. The bristles may be trimmed at an angle which makes it easier to clean in corners. The crevice tool has no brush and the upholstery tool generally has no brush.

Because of its narrow opening, greater suction is obtained with the crevice tool than with any other tool. It is used for cleaning hard-to-get-at narrow spaces and drawers by means of high suction. In addition, it may be attached, by means of the hose, to the blower end of the cleaner when a strong, directed blast of air is desired—the only practical means of dislodging small particles and dirt from narrow enclosures.

One of the convenience features of some models of cylinder cleaners is provision for storing the tools on the cleaner

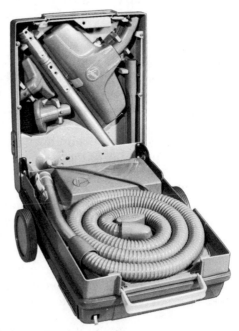

Figure 17-5. Storage of attachments and hose within vacuum cleaner case makes for easy portability. (Hoover Company)

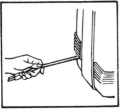

Automatic cord return.

Inside tool storage.

Super capacity quick change filter bag.

Full bag indicator.

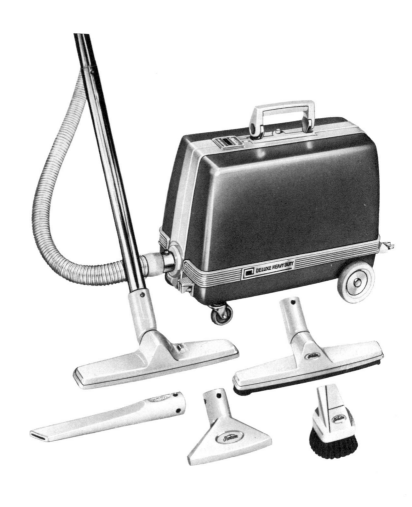

Figure 17-6. Cleaner with automatic cord return, inside tool storage, quick-change filter bag, and full-bag indicator. (Sunbeam Appliance Company)

or in a container that holds the cleaner and the tools. Various arrangements are used to make the tools easily available when the cleaner is in use; between uses, cleaner and tools can be stored as a unit (Figs. 7-1, 17-5, and 17-6).

Cylinder cleaners can be stored with hose, wands, and rug or other tools connected. The hose will store vertically by looping it over a hook in a storage closet or other storage space. Usually two hooks are more satisfactory than one because the weight of the hose is not concentrated in one place on the hose.

Accessories

Attachments for demothing and for spraying wax, paint, and insecticides are not commonly included as accessories. When they are available they are attached by means of the hose to the blower end of the cleaner. Extra storage space needs to be provided for these accessories.

On some models the crevice tool can be used for demothing. Instead of adding demothing crystals to a special attachment, they may be added to a clean disposable filter bag inside the cleaner. Hose and crevice tool then are attached to the blower end of the cleaner and the motor is turned on.

A cord winder is another accessory. It has a window-shade action that rewinds the cord automatically. This device keeps the cord out of the way when the cleaner is in use and provides convenient storage for it when the cleaner is not in use.

Specifications and Features

Specification sheets of different manufacturers differ in completeness of information supplied, and specification sheets are not always available to the consumer. The type of information found on detailed specification sheets is summarized below.

Dimensions usually are stated. These are important to the user, since they determine amount of storage space required. Weight in pounds is often specified, with or without tools, hose, and wands. For example, one cleaner is described as weighing 25.6 pounds including all tools. Weight of cleaners is not an important consideration when they are equipped with wheels for movability.

Maximum water lift sometimes is given. For cylinder cleaners, water lift is usually quoted at some value between 50 and 100 inches. This is a measure of suction effect; it is the height of a column of water that can be supported by the difference between atmospheric pressure and the air pressure in the cleaner. Working suction which is measured at the end of the hose, using a standard tool, is approximately one-third of that measured at the motor.

Input wattage is sometimes given on specification sheets and always on the nameplate of appliances that carry the Underwriters' Laboratories seal of approval. Current models of cylinder cleaners usually have an input rating between 550 and 850 watts, but some have higher ratings. Sometimes a model rated at 750 watts is described as a 1-horse-power cleaner. This description tends to be confusing, because horsepower ratings of motors of other appliances refer to output rather than input.

Length of cord often is noted; a usual length is 20 feet.

Reference may be made to extra filters on the intake or discharge side of the motor, if provided.

Special features vary with different manufacturers. Several cleaners have a special recess in the shell around which the cord may be wound. The motor circuit may have a condenser to minimize radio and television interference while the cleaner is operating. A series of blower holes around the bottom of the cleaner allows the air to be spread as it leaves the cleaner. When the blower end of the cleaner is used the holes are closed automatically so that the force of the air is more concentrated.

Two wheels or one long roller near the rear end of the dual-purpose, rug-and-floor tool tend to insure correct fit of the nozzle to the rug or floor, as well as ease of motion of the nozzle.

The specifications will usually note provision for making the operating noise of the cleaner less, if such provision is made.

Finally, the specifications may give information on the disposable bag and/or the cloth bag. Some cylinder-type cleaners have an indicator on the outer part of the cleaner to tell the user when the level of dirt accumulation is such that the bag should be changed (Fig. 17-6). Disposable bags make the cleaning task easier, but add to upkeep cost.

UPRIGHT CLEANERS

Fewer companies manufacture uprights than manufacture the cylinder-type cleaner. Uprights are designed primarily for effective cleaning of carpets and rugs.

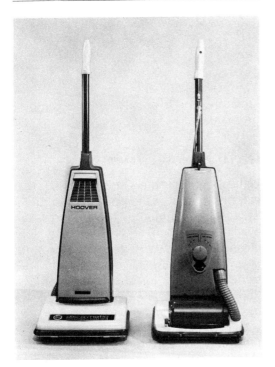

Figure 17-7. Upright cleaner can be changed to canister-type operation by setting of switch. (Hoover Company)

They are sold with or without attachments for cleaning bare floors and furnishings, although combination models are available (Fig. 17-7).

Construction

The main parts of an upright that are visible when the cleaner is in operating position are the handle assembly, the bag, and the motor hood. The switch is usually in the upper part of the handle. The outer bag is made of cloth or plastic, and in some models a disposable bag is inside it. The motor hood may have a light bulb at the front—"to seek out the dirt"—and a narrow plastic strip or furniture guard that extends almost completely around the bottom.

The motor and fan are assembled in a housing or case underneath the motor hood. The nozzle for the brush roll is located at the front of the cleaner under the motor hood and a rubber belt connects the brush roll with the motor shaft (Fig. 17-8).

Two front wheels are located under the hood just behind the nozzle, and two smaller wheels are located under the rear part of the hood. The wheels help make the cleaner easy to move.

A toe-operated handle control or pedal is located at or near the back of the motor hood. This is depressed to lower the handle from the vertical used for storage to the slanting position that is usual for cleaning carpets. The handle can also be lowered to an almost horizontal position for cleaning under furniture. Usually the height of the motor compartment determines the clearance needed under furniture. This is approximately 4 to 7 inches.

Attachments for Bare Floors and Furnishings

The floor-and-wall tool, upholstery tool, dusting tool, crevice tool, hose, wands, and connector or converter that may be used with an upright generally are called the attachments. The connector is a device used to attach the hose to the cleaner.

On some models the hose may be attached to the cleaner without removing

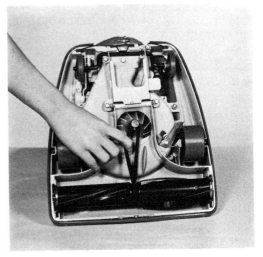

Figure 17-8. Bottom view of upright showing rubber belt that connects motor shaft and brush roll. (Hoover Company)

the belt from the brush. The connector is placed in the opening provided for it and the hose is assembled to it. On other models, the belt is disconnected before the connector is installed in the cleaner, and the brush roll is not rotated by the motor when the attachments are used. The model shown in Figure 17-7 requires only the flip of a switch and the completion of the hose connection to make the change.

Specifications and Features

The information on specification sheets for uprights is somewhat similar to that for cylinder-type cleaners. Some characteristic differences are summarized here. Wattage input varies but may be less than that for cylinder-type cleaners; on some models it is the same or even greater. Maximum water lift may not be specified, but is less than that for cylinder-type cleaners. The relatively low water lift or suction is adequate for cleaning carpets and rugs with the motor-driven brush, but is sometimes disadvantageous for above-the-floor cleaning.

Some uprights provide for two speeds of rotation of the brush roll, according to the setting of the on-off switch. Automatic adjustment of nozzle height to rug thickness or pile may be provided by use of a spring mounting in the support for the rear wheels. (Ideally, the nozzle lips *just* clear the top of the pile and the brushes make contact with the pile.)

Uprights may also have a mechanical device for lowering the roll to compensate for wear of the brush bristles. When mechanical adjustment is no longer possible the entire brush roll is replaced, or the brushes only are replaced for models that have replaceable brushes.

If positive means are provided to prevent spillage of dirt when the bag is removed from the cleaner, this feature will probably be noted in the manufacturer's literature. This is one feature that can best be evaluated by use.

SPECIAL-PURPOSE CLEANERS

Light-weight cleaners are designed that can be hung over the operator's shoulder by a strap or carried by hand. Models are available that can be converted to floor cleaning, others are designed for floor cleaning, or these cleaners may be designed primarily for above-the-floor use. Portable cleaners are also useful for cleaning automobile and boat interiors.

Lightweight cleaners pick up surface litter and are easy to manage. For quick cleanup tasks such as cleaning the kitchen carpet after meals, they are most useful. They are not as effective in removing imbedded soil from carpets and rugs as the regular upright. The cleaner usually is designed for vacuuming rugs, bare floors, and furnishings. The cleaner shown in Figure 17-9 has a bag that does not have to be removed to empty accumulated dirt. Instead, dirt that collects in the filtering bag is shaken down into a dirt cup just below the bag, and only the dirt cup is emptied.

USE OF VACUUM CLEANERS

Specific suggestions for the different tools or attachments are given in the user's booklets and other literature distributed by manufacturers. The discussion here covers general procedures for using a vacuum cleaner to clean a family room, living room, or den.

Without a plan for room cleaning, one is likely to take many unnecessary steps. Planned cleaning of a room, on the other hand, can be a satisfying activity. You know in advance how you are going to do the work. You know where you are in your work after unplanned interruptions, such as telephone calls. You know what you

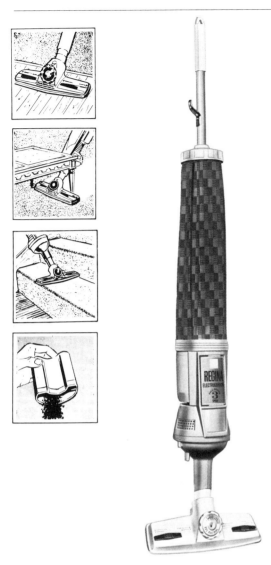

Figure 17-9. Lightweight cleaner adjusts for bare floors, low, medium, or high pile carpet by turning a dial; dirt cup empties like an ash tray. (The Regina Corporation)

have accomplished when you are through. Finally, you may learn effective short cuts to improve your plan.

Two types of room-cleaning plans have been suggested. The first and probably the better known one is to use one tool for all the cleaning in the room for which that tool is appropriate, then to use a second tool for all the cleaning for which that tool is appropriate, and so on. The second plan is to clean a portion of a room completely, changing tools as necessary, then to clean another portion of the room completely,

and so on, until the entire room is cleaned.

These two plans evolved before the introduction of dual-purpose tools. With such tools available, a modified form of the second plan seems most logical. The modification consists of using *both* parts of a dual-purpose tool in cleaning portions of a room.

In the general work plan outline given below some of the suggestions are adapted from the Hoover cleaning manual.[2]

1. Pick up magazines, clothes, toys, etc.
2. Empty ash trays and set them aside for washing. Move wastebaskets to a hall or elsewhere for later emptying and replacement.
3. Divide the room, mentally, into several areas. For example, a corner might be one area. The space associated with one-half of a wall might be another area.
4. Move furniture slightly away from the wall, provided it is not too difficult to move.
5. Start in one area and clean the wall or walls, moldings, drapes, lamps, etc. Clean the front, back, top, bottom, and sides of the furniture in that area. Clean the floor register. Clean the carpet, rug, or wood floor between the back of the furniture and the wall. The floor brush may be the reasonable tool to use for the portion of the carpet or rug near the walls.
6. Move to the adjacent area and repeat step 5. Then move to other areas until you have gone all around the room.
7. Move the furniture back into place against walls.
8. Clean the uncleaned portion of the floor, including areas under the furniture. (There is no reason to assume

[2] Hoover Home Institute, *House Cleaning and Home Management Manual*, 1950.

that moth larvae boycott the parts of rugs that are under furniture.)

9. Empty and replace wastebaskets. Wash and replace ash trays.
10. Clean the dusting tool and the floor tool with the vacuum cleaner itself. Empty the fabric bag or replace the disposable bag, if necessary.

CARE AND MAINTENANCE OF VACUUM CLEANERS

Care of vacuum cleaners is rather simple and consists essentially of four parts. One is regular emptying of the cloth bag or replacement of the disposable bag. The second is replacement of filters when necessary. The third is regular cleaning of brushes and replacement when necessary. The fourth is replacement of the belt, when necessary, on vacuum cleaners that have a belt-connected brush roll.

As noted earlier, the efficiency of a vacuum cleaner decreases as dirt accumulates in the bag, because the dirt in the bag and that covering the interior surface of the bag offer a mechanical resistance to the flow of air through the cleaner. For maximum efficiency, the bag should be empty and the interior surfaces of cloth or vinyl bags should be clean, that is, free of clogged dirt, at the start of each use. Many homemakers are unwilling to empty the cleaner after each use. (Emptying after use insures that the bag is empty before the next use.) A practical compromise probably is to empty a cloth bag or discard a disposable bag when it is about half full. Actually, the size of the bag may make some difference, since the larger the bag, the larger the filtering surface available. If cleaning is routine, one can learn from experience about how much cleaning can be done before the bag is half full. On those cleaners which have a full-bag indicator the user does not have to decide when the bag should be changed.

Brushes on the tools should be cleaned after each use and during use for some types of cleaning. Use the suction end of the hose for this purpose.

As the brushes on the roll wear, the height of the roll should be adjusted according to instructions in the user's booklet. A test for correct length of bristles of the brush roll is to hold the edge of a small, stiff card perpendicularly across the nozzle lips and rotate the roll by hand. If the length of the bristles is correct, the bristles will just brush the edge of the card.

The belt should be replaced when the roll slips as it is driven by the motor. Another indication of need for replacement is elongation of the belt. This can be checked by laying a new belt alongside the old one or by measuring the length of the belt when it is new and again when it looks worn. Instructions on replacement are given in the user's booklet. Always check the model number of the vacuum cleaner when buying a new belt.

If any periodic maintenance by a service man is necessary, this fact would usually be mentioned in the user's booklet. Such maintenance is not necessary for all cleaners. On the other hand, if a cleaner does not pick up threads and lint with a very few strokes, it probably is desirable to have a factory-authorized service person check the cleaner, after you have made sure that the bag is not too full, that the nozzle is adjusted properly and the hose, if one is used, is clear of obstructions.

SELECTION CONSIDERATIONS

Frequent model changes benefit the consumer in that early incorporation of improvements is possible. On the other hand, frequent model changes make it difficult for a consumer to know which models have proved themselves in use.

The Buying Guide at the end of the

chapter outlines some use characteristics that can be checked without a home trial.

One type of cleaner probably is not best for all homes. Some homemakers may find it desirable to have a lightweight cleaner as well as a standard cleaner. The size of the house as well as the type of cleaning done are important factors to be considered. The best procedure at present for selecting a new cleaner seems to be to try a few models that you have inspected in your home. If a cleaner performs satisfactorily for a few days and is convenient to use for your type of housecleaning, it is likely that it will be adequate for the length of time you might reasonably expect to use it. If home trials are not practical, try to visualize the cleaning tasks that you do in your home when you make your choice of model.

ELECTRIC FLOOR POLISHERS, SCRUBBERS, AND WASHERS

As is the case with practically all other household appliances, more than one type of electric floor polisher is available. A floor polisher is seldom "just" a polisher. The twin-brush floor conditioner uses a pair of brushes (Fig. 17-10). The brushes alone are used for scrubbing, waxing, and polishing, and pads are used under them for buffing. With special brushes the floor polisher can be used for shampooing rugs. The single-brush type may have one brush for waxing and polishing, one for shampooing rugs, one for scrubbing hard surface floors, and a disposable buffing pad.

Brushes are assembled to the motor through a pulley and gears. When the motor is turned on, the brushes rotate at high speed.

A motor-driven floor polisher is one of the household appliances that people seem to enjoy owning. Probably, the enjoyment is really associated with the superior results obtained when the appliance is used to polish and buff hard floors and hard floor coverings.

USE AND CARE

The floor polisher is used for five types of operation: scrubbing and rug shampooing, waxing, polishing, and buffing.

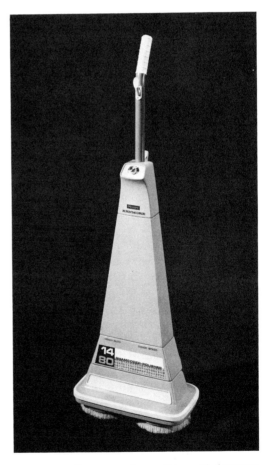

Figure 17-10. Shampooer-floor polisher with detergent dispenser for scrubbing or shampooing; also polishes floors. (Sears, Roebuck and Company)

Scrubbing and Rug Shampooing

The scrubbing brush or brushes may be used with detergent and water to scrub hard floor coverings or concrete floors or with cleaner for wood floors. Follow the procedure described if the appliance does not have a dispenser for detergent solution and a container for the soiled water.

1. Vacuum the floor to remove dust, litter, and other surface dirt.
2. Use a sponge mop to sponge a convenient area of floor, perhaps 6 square feet. Note that the recommendation is to *sponge* an area, not to flood it with water.
3. Scrub the sponged area with the polisher.
4. Rinse with the sponge mop. Then pick up as much water as possible with the mop.
5. For wood floors, a solvent cleaner or a liquid cleaning-and-polishing wax is used instead of detergent in water. Cleaner or cleaning wax can also be used for sealed concrete floors and linoleum. No extra mop is necessary.

The cleaner or cleaning wax is applied with fine steel wool, such as 00 grade. Stainless steel pads which snap onto the brushes of floor polishers are available with some models. If not, a roll of fine steel wool may be purchased in a hardware store. The steel wool can be cut into 4-inch by 4-inch strips and pressed or embedded into the scrubbing brushes of the polisher. Small puddles of cleaner are poured onto the floor, or for many models the cleaning solution is dispensed directly from a container on the appliance. The polisher with scrubbing brush is used to scrub. The steel wool strips are discarded as they become soiled.

An appliance is available which can be used to scrub hard surface floors and vacuum the dirty water from the surface. It has two containers for detergent solution —one clean and one used. Other appliances are designed for only carpet shampooing. Most have rotary brushes, but at least one has straight-line brushes that have a back-and-forth action (Fig. 17-11).

Waxing, Polishing, and Buffing

Floors and coverings are waxed to enhance their appearance, to increase their life, and to minimize staining. To achieve these results, it is necessary to use the correct wax. So, read the label on the package, and use a type of wax appropriate for the material of the floor or floor covering. Generally, a wax that is not self-polishing and that has a naphthalike odor is used for wood, sealed cork, and sealed concrete. Self-polishing wax only is recommended for asphalt and rubber tile. Either a wax that is not self-polishing or a self-polishing one may be used for linoleum or vinyl floor coverings.

Liquid as well as paste waxes can be applied with some electric polishers, although a hand applicator sometimes is recommended. It is usually not recommended that self-polishing waxes be applied with an electric polisher or that the floors be polished after the self-polishing wax has dried. Cleaning and waxing can be done in the same operation with a cleaning wax and disposable or washable pads.

A suggested procedure for waxing, polishing, and buffing with a floor polisher is as follows:

1. Vacuum the floor or scrub it, if necessary. Scrubbing is necessary only when the floor is very soiled or when a thick coat of wax has been allowed to build up on it. (One indication that a coat of wax is too thick is that the wax peels.)
2. Paste wax usually is spread on brushes, special waxing pads, or steel wool pads. Liquid wax is poured directly on the

Straight-Line
Power Stroke

Figure 17-11. Carpet shampooer with straight-line, back and forth, cleaning stroke, 1,500 strokes per minute. (Bissell, Inc.)

floor. Only a thin coating of either wax should be applied.

3. Let the wax dry thoroughly on the entire floor. This will require 20 to 30 minutes or longer.

4. If the polisher was used to scrub the floor, wait until the brushes are dry before proceeding with step 5. If the

polisher has an extra polishing brush, replace the scrubbing and waxing brush with the polishing brush and continue with step 5, as soon as the wax on the floor is dry.

5. Polish the entire floor with the polisher.

6. Buff the floor with wool pads or other pads provided with the polisher to give the floor a high luster. The bright look obtained by buffing means good polishing to many homemakers.

Caution: Users' booklets are likely to note that two or more very thin coats of wax give a harder, smoother, and more lasting finish than one coat. However, floors that are too slippery may cause falls. Hence, a practical compromise is to use only one thin, polished coat of wax and to renew it as necessary. Polished wax floors are less slippery than poorly polished ones, as wax can stick on the soles of shoes and cause slipping. This is especially important for households with aged members. The appearance of a floor should not take precedence over safety in the home. Wax may be sticky, dull, or conducive to falls if all the detergent is not removed from the brushes or from the floor before it is waxed.

Care

Polishers require little maintenance. Occasional, once a year or so, oiling of a motorshaft bearing may be specified in the user's booklet. Brushers are cleaned with water and detergent or a cleaner, as suggested in the user's booklet. The one care procedure always emphasized is to store the polisher so that the bristles do not rest on the floor. Either hang the entire polisher, using the hook provided on the handle, store the polisher and brushes with brush bristles pointing up, or remove the brushes and store separately from the polisher.

EXPERIMENT

Vacuum Cleaners

1. Use several cleaners in a living room or family room, if possible.
2. Follow the plan outlined in the section on use of vacuum cleaners.
3. Evaluate the cleaners, as well as you can, in terms of dirt-removing ability and ease of use.

Do not assume that you can rate cleaners on the basis of a single use in one part of one room.

BUYING GUIDE

Vacuum Cleaners

1. What is the model number? Who is the manufacturer?
2. What is the wattage? Amperage? Voltage?
3. What seals of approval does the appliance carry? For what qualities do these seals of approval stand?
4. Is the appliance easy to move? Check by moving the cleaner over a rug, if possible. If the cleaner is a cylinder-type, move it with the hose, wands, and rug tool attached. Is the cleaner stable or does it have a tendency to tip over when it is moved? Can you lift it fairly easily?
5. Do the rug and floor tools of the cylinder-type appear to be designed so as to insure correct fit to the floor when the cleaner is handled naturally?
6. Is the cloth bag easy to empty or the disposable bag easy to replace? If possible, check by removing and replacing a bag. Does the cleaner have a positive means for preventing spillage of dirt when the bag is emptied or replaced?
7. Are positive mechanical devices provided for locking tools, wands, and hose?
8. If the cleaner is an upright, note the height of the top of the motor hood from the floor and estimate whether it will fit under furniture for under-furniture cleaning.
9. What provision is made for storing the tools? How is the cord handled when the cleaner is not in use?
10. Will the cleaner and its tools or attachments be easy to store in a closet? Approximately how many square inches of floor area will be needed for storing cleaner and tools?
11. Does the air leave the cleaner in a single stream or is the discharge-air distributed over a broad area?
12. Are a specification sheet and an instruction booklet supplied

with the cleaner? Does the manufacturer's literature indicate that periodic servicing is necessary? If so, where is this servicing available? Where will you obtain additional disposable bags, filters, and replacement brushes? How much will these cost?

13. Does the manufacturer's literature indicate that some provision has been made to decrease the noise of operation? Check the effectiveness of this provision by listening while the cleaner is operated. (Remember that noise is relative and may seem different at home.)

14. Does the manufacturer's literature indicate that the electrical circuit of the motor has some provision for minimizing radio and television interference? (If no such provision is made, radio and television reception may be poor in your neighbors' sets as well as your own when the cleaner is in operation.)

15. Precisely what does the manufacturer's warranty promise you?

Polishers and Scrubbers

In addition to points 1, 2, 3, 12, and 15, given for vacuum cleaners, consider the following:

1. Is the appliance easy to guide while it is in operation?
2. Will the handle remain upright or does it have to be propped?
3. Are the brushes easy to remove and replace?
4. Do the buffing pads stay in place when the machine is used for buffing?
5. Does splattering accompany scrubbing or application of wax?
6. Is it possible to polish floor corners and space next to the walls?
7. What provision is made for storage?
8. How often will you use the polisher? What does it cost to rent a polisher? Would it be convenient for you to do so?
9. What upkeep is necessary? Does it involve much time? Much cost?

CHAPTER 18
WATER, LAUNDRY SUPPLIES, AND MECHANICAL WATER CONDITIONERS

Laundering soiled articles with modern equipment involves use of water, water heaters, mechanical water softeners, detergents, and other aids such as fabric softeners, washers, dryers, and irons.

WATER AND LAUNDRY SUPPLIES

Hard Water

Water that contains calcium and magnesium salts is considered hard and water that does not contain the salts of one or both of these minerals is soft. It is common in a discussion of water and home laundering to consider hardness of the *supply* water only. However, the soil removed in a family laundry load introduces hardness minerals into the wash water.

Calcium and magnesium carbonates produce temporary hardness in water; calcium and magnesium sulfates and chlorides produce permanent hardness in water. The relative amounts of temporary and permanent hardness in a water supply depend on the relative amounts of the minerals producing the two types of hardness.

Temporary hardness in the water supply is removed when water is heated, because the acid carbonates are decomposed and ordinary or normal carbonates are precipitated. The precipitate, which is mostly calcium carbonate, is observed as a white or gray deposit on the utensil used for heating the water. Permanent hardness is not removed when water is heated, because the sulfates and chlorides are not decomposed by heat.

Water hardness is expressed in grains per gallon or parts per million. One grain-per gallon hardness is equal to 17.1 parts-per-million hardness. Water with a hardness of 1 grain per gallon contains 1 grain or 1/7000 of a pound of hardness-producing minerals per gallon of water. Water with a hardness of 1 part per million contains 1 part of hardness-producing minerals per 1 million parts of water.

The average hardness of the water supply varies in different regions of the United States. Some regions—most of the eastern seaboard, for example—have soft water. Others have very hard water. A particular location in a soft-water region may have hard water and a particular location in a hard-water region may have soft water. For example, most of the midwest has very hard water, but some locations have a water supply of approximately 0 grains hardness.

Though many municipalities now soften water before delivering it to homes through the mains, such water is not completely softened; in fact, a hardness of 5 to 7 grains per gallon is left in the water because water of 0 hardness sometimes causes corrosion problems in city water mains.

Hard water, even that with a hardness of 5 to 7 grains, creates some problems in laundering, as well as in dishwashing and general cleaning. These problems are primarily associated with the use of soap in hard water. Whenever soap is used with such water, some of it combines with the calcium and magnesium sulfates and chlorides to form a gray curd or scum. As far as cleaning is concerned, this part of the

soap is lost. The scum thus formed in hard water also adheres to laundered fabrics and washed surfaces, such as washed walls.

When synthetic detergents are used in hard water, formation of curd is uncommon.

Although only calcium and magnesium salts contribute to water hardness as usually calculated, other minerals dissolved in the water supply may also cause problems in home laundering and dishwashing. For example, a 0.3 part-per-million concentration of iron may cause rust stains on washed fabrics.

DETERGENTS AND SOAPS

A detergent is an agent that cleanses and technically, therefore, a soap is a detergent. A soap is a cleansing agent made by reacting an alkali such as lye with a fat or fatty acid; a detergent is a cleansing agent synthesized from various materials derived from petroleum, fatty acids, and other sources.

The particular fat used by manufacturers in the commercial production of soap depends partly on world prices of different sources. In the home, suet is the fat traditionally used in soap production. In the United States, commercially made soap packaged in a box is likely to say soap on the box.

Detergent Issues in the 1960s and 1970s—Biodegradability, Enzymes, Eutrophication

For centuries soap was the accepted cleaning agent for laundering. A shortage of fats led to the introduction of synthetic detergents about the time of World War II. The synthetic products were improved over the years and increasingly used. In the late 1950s public concern developed relative to foaming and sudsing in streams due to detergents that had not been broken down in sewage treatment plants and detergents that had not passed through sewage plants before entering the streams. During 1965 the detergent industry in the United States completed the changeover of the type of organic surfactants (active agents) in laundering and cleaning products to the "soft" or readily biodegradable type.[1] The biodegradable material is altered by bacteria in surface waters and in soils and by treatment in "standard" sewage-treatment plants so that it loses its wetting and sudsing characteristics.

The next issue in the detergent saga was the introduction in the late 1960s of enzymes into laundry presoak products and detergents. According to Dunn and Dowlen,[2] three major types of enzymes are used for laundry compounds: (1) proteolytic—which act on protein substances, (2) amylolytic—which act on starchy substances, and (3) lipolytic—which act on fatty substances. These investigators state also that enzymes normally are most effective in the removal of protein-based and starch-based stains; but they also influence the removal of nondigestible soils since proteins and starches can serve as binding agents for such soils.

Enzymes in laundry products should be considered separately from phosphorus in laundry products, even though detergents with enzymes tend to be high-phosphorus products. The concern with enzyme products is due to possible allergic side effects on the user. By 1972, some manufacturers of widely sold laundry products were promoting enzyme detergents for stain removal in laundering.

At the same time some manufacturers stopped including enzymes in the formulations of their best-known name-brand detergents.

Also in the early 1970s concern developed relative to eutrophication of lakes, and to a lesser extent of rivers and streams,[3] by the phosphorus used in detergents—in various forms (compounds) and percentages. Eutrophication is a process by which lakes evolve and eventually change from pure, clean water to swamps and meadows. This is a natural process. The concern in the early 1970s came about because addition to lakes of phosphorus from sewage treatment plants and directly by runoff of surface waters appeared to be at least partly responsible for speeding up the eutrophication process from one that seemed to require centuries or millenia to one that only required a generation.

In response to this concern, manufacturers of detergents are pushing research and development of substances other than phosphorus to supply alkalinity in the wash water, to emulsify greasy soil, and so on. Hopefully any new detergents that are developed will be effective in cleaning and not harm the environment or the user.

Built and Mild Detergents and Soaps

Mild soaps and detergents are considered less harmful to the skin than built products and are used for hand laundering lightly soiled articles. The mild products contain surfactant, inactive ingredients, and little or no builder. Built products are used for laundering in washers and for general cleaning. Specially formulated (built) detergents are used for dishwashers.

[1] The Soap and Detergent Association, "The Facts about Today's Detergents," April 1971.

[2] Mary Ann Dunn and Rowena P. Dowlen, "A Comparison of Enzyme and Detergent Presoaking," *Journal of Home Economics*, Feb. 1971, vol. 63, no. 2, p. 104.

[3] Lowell Hanson, Wanda Olson, and Roger Machmeier, "Detergent decision guide to minimize water pollution," Agricultural Extension Service, University of Minnesota Fact Sheet, Environmental Control No. 1, 1971.

Built detergents are mixtures of surfactants which wet the fabric and the soil and remove soil; builders to soften water and to aid in cleansing by providing alkalinity in the wash water and emulsifying greasy soils;[4] a small quantity, less than 1 percent of sodium carboxymethylcellulose (cmc) to help suspend the emulsified soil; some silicates to inhibit corrosion of susceptible metal parts of the washer; fluorescent dye to aid in making the washed articles look whiter or brighter; perfume; and some inactive ingredients. Additional materials may be provided in detergents for specific purposes.

Built soaps generally will be alkaline but will not have the polyphosphates commonly used in built detergents.

Type and Quantity of Washing Product to Use

The type and amount of detergent to use should be conditioned by a concern for oneself and the environment, as well as the cleanliness and "hand" or feel of the washed articles. (A harsh hand of washed articles suggests unrinsed detergent or soap.) As far as practical, a concerned individual might use soap with a water supply that is naturally soft or with water that has been softened by a mechanical water softener.[5]

A user can observe cleanliness of the washed articles after repeated washings with minimum amount of product (detergent or soap) and increase the amount of product if that is necessary. Until amounts of ingredients are clearly stated on packages it will be difficult for the consumer to

[4] Mary E. Purchase, "Phosphates and Detergents in Water Pollution," *Information Bulletin No. 12*, March 1971. Extension publication of the New York State College of Human Ecology.

[5] Florence Ehrenkranz, "Water Quality Requirements in the Home," Proceedings Fifth International Water Quality Symposium, Aug. 1970, pp. 56–58. Water Quality Research Council.

judge how much of different types of builders or other additives she is using. Since products vary in density, a concerned person will relate volumes of product used to weight of product used—both in judging cost of product and in determining amounts of additives.

ADDITIONAL LAUNDRY AIDS

Laundry products used in addition to detergents and presoaking products include: packaged water softeners, bleaches, fabric softeners, starch, and other sizing agents.

Packaged Water Softeners

Packaged water softeners are of two types —the precipitating type which does not contain phosphorus compounds and the nonprecipitating type which does. Washing soda (Na_2CO_3) is an example of the former. This type combines with the hardness-producing salts to form water-insoluble particles. When the softener is added to hard water in a container such as a laundry tub, some of the precipitate floats to the surface and some disperses throughout the water. The best procedure in laundering would be to skim off the surface scum before clothes are placed in the wash or rinse water. The surface precipitate, but not the dispersed precipitate, can be removed when separate wash and rinse tubs are used as with wringer-type washers. Removal of surface precipitate with an automatic washer before wash and rinse would, of course, detract from the automaticity of the washer.

The nonprecipitating type of packaged water softener combines with the hardness-producing salts in water to form water-soluble compounds. The water does not become cloudy and no scum forms on the surface. Also, when enough non-precipitating softener is used, enough active ingredient is available both to soften the water and combine with hardness minerals in detergent built up in the fabrics—that is, unrinsed detergent from previous washings. Thus cleanliness of the load is helped by use of the softened water and the removal or stripping of unrinsed detergent from the load.

A drawback is that the current non-precipitating type is very high in polyphosphates; also it is not useful for water that is harder than 30 grains or so.

Bleaches

Home laundering bleaches are oxidizing type and utilize chlorine or oxygen to cause the bleaching process. Liquid chlorine bleach has been the most widely used home bleach for many years. However, it should not be used if the water supply contains iron. It is an effective aid in laundering white and colorfast cottons and linens and white nylons, Dacrons, and orlons. It should not be used on silk, wool, spandex, blends of these fibers, or most resin-treated materials. The chlorine bleach aids in retaining or restoring whiteness to fabrics and in soil removal. Also, chlorine bleaches that contain 5.25 or more percent of sodium hypochlorite as the active ingredient are effective sanitizers in home laundering.[6]

Follow instructions in using the bleach. In particular take care that undiluted chlorine bleach does *not* come in contact with clothes. Some washers are designed to insure that liquid bleach is diluted before it contacts the load. (A bleach injector mixes water and bleach before the bleach enters the main part of the tub.) Also, some washers provide for delayed injection of bleach added at the beginning of the wash cycle to insure that the bleach

will not interfere with the action of the fluorescent dye in the detergent or soap.

Dry chlorine bleaches are available which release the active ingredient more slowly than the liquid product. These may have lower bleaching effectiveness but they require less care in use to insure against fabric damage for washers that do not mix water and bleach before the bleach enters the wash tub.

Diluted liquid hydrogen peroxide may be used with silk and wool. Note that the compound decomposes on standing.

Granular bleaches that have sodium perborate or potassium monopersulfate as their active ingredient are mild. The instructions on the box usually indicate that they are safe for fabrics and colors that hot water will not harm. They are most effective in maintaining whiteness and brightness if used each time articles are washed.

Fabric Softeners

Fabric softeners provide a lubricating film on fibers.[7] Because the chemically active portion of fabric softeners is charged oppositely from that of most detergent surfactants, most fabric softeners are not designed for use in the wash water; rather they usually are used in the final rinse. However, one manufacturer of a built *soap* included fabric softener in the soap formulation.

The liquid fabric softeners make fabrics softer and fluffier, minimize wrinkling, and eliminate static cling. The last effect is especially useful for man-made fibers that do not have much natural moisture content. The fluffing effect is useful for terry towels. The soft hand imparted by the softener is advantageous for sweaters and for baby diapers.

[6] "Sanitation in Home Laundering," *Home and Garden Bulletin No. 97.* Washington, D.C.: U.S. Department of Agriculture, July 1970.

[7] *The Maytag Encyclopedia of Home Laundry,* 3rd ed. Popular Library, pp. 101–102.

Starch and Other Sizing Agents

Laundry starch is used chiefly to give body and crispness to laundered fabrics. *Dry* starches mostly must be cooked or mixed with hot water before use. One manufacturer, however, provides a dry starch ("instant starch") that dissolves in cold water. *Liquid* starch is diluted with water before use. *Aerosol* starches are sprayed directly onto damp or dry fabrics.

A spray fabric *finish* is used to give a *light* body. It is described as especially suitable for synthetics.

Both starch and fabric finish may improve appearance and restore body lost in laundering.

MECHANICAL WATER CONDITIONERS

For some water supplies, total water conditioning might involve: (1) removal of hardness-producing minerals; (2) removal of all ferrous and ferric iron; (3) clarification of silt, sand, dirt, and other materials causing turbidity by filtration through a suitable filtering agent; (4) removal of some objectionable flavors and odors by filtration through activated charcoal; (5) neutralization of acidity.

WATER-SOFTENING APPLIANCES

These appliances "soften" water by an ion-exchange process *and* may remove very minute amounts of ferrous iron and other impurities. Ferrous iron sometimes is called "clear" iron because ferrous salts are almost colorless in solution. Ferric iron, sometimes called "red" iron, is not removed by the softening agent.

Basically the appliance softens water as follows: Hard water flows through a bed of resin which exchanges the calcium and magnesium ions in the water for sodium ions. Thus the water that leaves the softener has mineral in it, namely sodium, but does not have hardness-producing minerals. The exchange material is a synthetic resin, a natural zeolite, or a mixture of natural zeolites and synthetic resin.

The resin or zeolite will exchange its sodium ions for calcium and magnesium ions until it becomes saturated with the hardness-producing minerals. Then the exchange material is regenerated in a multistage regeneration cycle which at a minimum includes backwash, brining, and rinsing. In backwash, water is caused to flow upward through the resin bed to compensate for compacting during down flow. In brining, a sodium solution is added at or near the top of the appliance. As this solution flows down through the bed, the magnesium and calcium ions in the bed are exchanged for sodium ions. In rinsing, water is flushed through the appliance to remove brine which now contains calcium and magnesium salts rather than sodium salt. Some softeners have more than one backwash and rinse.

Figures 18-1 through 18-3 show fully automatic models of two manufacturers. Fully automatic refers to regeneration at the intervals set by the user on the control. Manually operated regeneration controls may require the user to operate valves for backwash, brining, and rinsing, or one of the steps may be automatic.

Figure 18-4 shows a model with a special automatic feature—namely, a sensing device ("Aqua-Sensor") that "detects the need for recharging to provide more softened water" and thus eliminates the need for setting the frequency with which recharging is to take place.

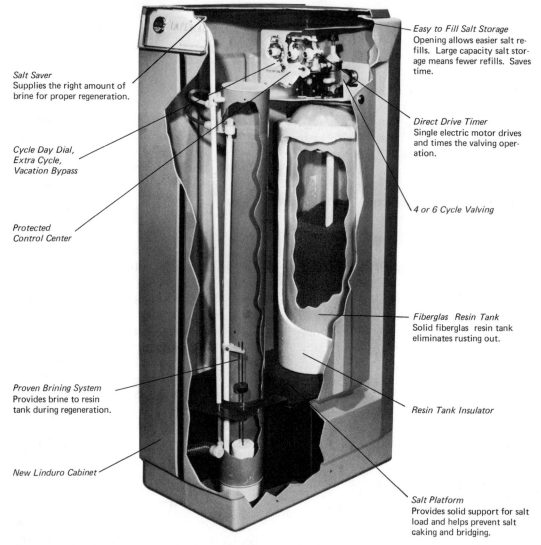

Salt Saver
Supplies the right amount of
brine for proper regeneration.

*Cycle Day Dial,
Extra Cycle,
Vacation Bypass*

*Protected
Control Center*

Proven Brining System
Provides brine to resin
tank during regeneration.

New Linduro Cabinet

Easy to Fill Salt Storage
Opening allows easier salt refills. Large capacity salt storage means fewer refills. Saves
time.

Direct Drive Timer
Single electric motor drives
and times the valving operation.

4 or 6 Cycle Valving

Fiberglas Resin Tank
Solid fiberglas resin tank
eliminates rusting out.

Resin Tank Insulator

Salt Platform
Provides solid support for salt
load and helps prevent salt
caking and bridging.

Figure 18-1. Interior components of rectangular-design completely automatic water
softener. (Ecodyne Corporation, The Lindsay Division)

Complete specifications on a fully automatic softener might include the following:

1. Softening capacity in grains—that is, the total hardness in grains that can be removed before regeneration is necessary. This might be some value between 5,000 and 33,000 grains. (Models might also be rated according to basic grains and low setting grains. The latter capacity rating corresponds to more frequent regenerations with lower salt input per regeneration.)

2. Salt usage in pounds for rated capacity —for example, 12 pounds for a 33,000-grain model.

3. Number of regenerations that can be set on the control—for example, one every other day, daily, or some other interval—or a sensor which initiates regeneration when needed.

4. Flow rate from the softener in gallons per minute at a specified pressure drop —for example, 9 gallons per minute at a 15 pound per square inch pressure drop.

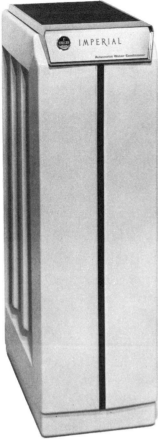

Figure 18-2. Exterior view of softener shown in Fig. 18-1. (Lindsay Division, Ecodyne Corporation)

(a) (b)

Figure 18-3. Automatic softener of cylindrical design. (a) Exterior view. (b) Interior view. (Water Refining Company, Inc.)

5. Maximum ferrous iron in parts per million (ppm) that will be removed— for example, 0 ppm or 1 ppm for different softeners.

6. Maximum water hardness in grains per gallon that can be handled at the rated flow rate. This might be as low as 15 or as high as 65 to 100 for different softeners.

7. Total time required for regeneration— for example, two hours.

8. Water in gallons used for regeneration. One manufacturer quotes 21 gallons for a 6,000-grain model and 32 gallons for a 14,000-grain model.

9. Salt storage capacity—for example, 100 pounds, 200 pounds, or some other number of pounds.

10. Dimensions; materials.

11. Water-Conditioning Foundation Validation Seal.

12. Manufacturer's warranty.

Not all the above information pertains for mechanically operated softeners.

WATER-FILTERING EQUIPMENT

A separate, free-standing iron filter handles more ferrous iron than softeners and some ferric iron which softeners do not handle at all. The appliance may filter the ferric iron directly and the ferrous iron after it is oxidized. Regular backwashing is needed to remove accumulated iron particles. One model removes 20 ppm of ferrous or ferric oxide. This same model has a seven-day electric clock mechanism to provide up to seven automatic backwashes per week.

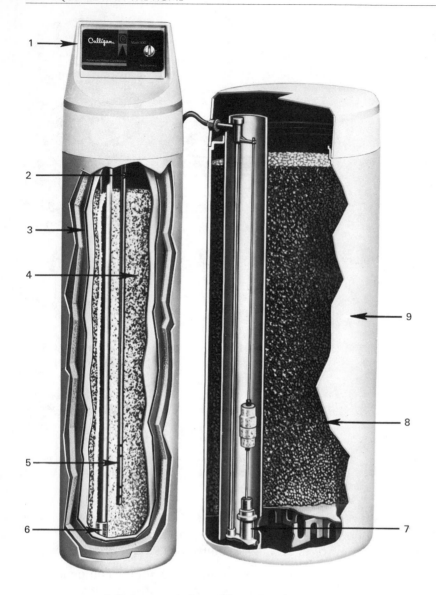

1. Recharge controller and five-cycle recharge valve
2. Backwash freeboard
3. Quadruple-hull insulated tank with plastic liner
4. Ion exchange water-softening resin
5. Electronic sensing device
6. Collector for treated water
7. Brine-collecting valve
8. Dry salt storage
9. Salt-storage container.

Figure 18-4. Automatic water softener with electronic sensing device. (Culligan, Inc.)

Accessory-type iron filters also are available. These are attached to a water softener to maintain the efficiency of the resin bed in the softener in removing rated amounts of ferrous iron.

A free-standing clarifying filter may be described as removing "large" quantities of foreign materials before backwashing is necessary. An accessory-type clarifying filter with replaceable cartridge is available for water supplies with smaller amounts of foreign particles.

A taste and odor filter may depend chiefly on activated charcoal.

BUYING GUIDE

Water Conditioners

For automatic and nonautomatic models get complete information on installation requirements: water supply, water drain, 120-volt outlet. Try to get approximate information on installation cost with hard water bypass during regeneration; softened water to all water-bearing equipment; softened water to most outlets but hard water for drinking, outside faucets (for watering the grass), and for toilets.

For fully automatic models get information of the following type:

1. What is the softening capacity in grains; that is, the total hardness in grains that can be removed before regeneration is necessary?
2. What is the salt usage in pounds for rated capacity?
3. What is the number of regenerations that can be set on the control—for example, one every other day, daily, automatic as needed?
4. What is the flow rate from the softener in gallons per minute at a specified pressure drop?
5. What is the maximum ferrous iron in parts per million (ppm) that will be removed?
6. What is the maximum water hardness in grains per gallon that can be handled at the rated flow rate?
7. What is the total time required for regeneration?
8. How much water (in gallons) is used for regeneration?
9. What is the salt storage capacity?
10. What are the dimensions? Materials?
11. Does the appliance have a seal specifying that it was "tested and validated under industry standards by the Water Conditioning Foundation"?
12. Does it have a warranty?

For semi- and nonautomatic models get appropriate information using as a guide the points outlined for fully automatic models.

CHAPTER 19
GAS AND ELECTRIC WATER HEATERS

GAS WATER HEATERS

Water heaters give long service. Some models carry a ten-year guarantee to replace the heater without cost if the tank should leak within that period (Fig. 19-1). Prorating the costs of replacement is no longer common. Various combinations of time and costs are available in guarantees depending upon the quality of the heater and company policy.

CONSTRUCTION

The main parts of a gas water heater are the tank, the burner with its thermostatic control, the flue, the insulation and exterior shell, and the relief valves.

It has been stated that the life of the tank is the life of the water heater. The material from which the tank is made should be able to resist corrosion and withstand the water pressures to which it may be subjected. Some of the materials in use for tanks of gas water heaters are galvanized steel, Monel metal, copper, aluminum alloy, and glass-, copper-, or ceramic-lined steel. Practically all gas water heaters are now made with glass-lined tanks.

The useful life of a water heater depends not only on the material of the tank, but also on the kind of water heated in it, and the temperature to which the water is heated. With use, tanks of all materials become lined with carbonates if the water supply is hard. A small deposit of hard-water scale serves essentially as an extra finish on the interior of the tank; a thick deposit may be expected to in-

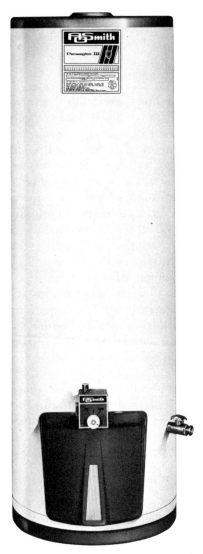

Figure 19-1. Automatic storage water heater changes heating speeds to meet varying demands. (A. O. Smith Corporation, Consumer Products Division)

crease the operating cost of the heater. Tanks of certain materials corrode when the water supply is corrosive. Water that has acid in it or a high dissolved oxygen content is likely to be corrosive. Dissolved solids, too, may contribute to the corrosive effect.

Galvanized steel is more likely to corrode than the other tank materials listed above. For this reason, in a water heater that has a galvanized steel tank a magnesium rod is sometimes used to "direct" the corrosive action to the rod rather than to the zinc lining of the tank, as discussed below. The magnesium rod is not effective for all types of water.

A magnesium rod coupled with a dissimilar metal and placed in water will develop electrical energy. The magnesium will corrode and destroy itself while protecting the other metal. The protection a magnesium rod supplies in a water heater tank takes place in three ways: (1) The corrosion takes place on the magnesium rod rather than on the metal of the tank; (2) the tank walls are covered with a protective film; and (3) the water in the tank is chemically conditioned so that less corrosion will take place.

Magnesium protective devices may be installed on new equipment, or they may be added to water heaters already in use. The rod may be solid and rigid or a link type. Link anodes may disintegrate at the joint. When the anode is no longer grounded to the tank it does not serve the purpose of protecting the interior of the tank.

The life of a water heater is prolonged by using a thermostat setting only as high as needed for the hot water demand in the home, since the reactions that wear the tanks are accelerated at high temperatures. The walls of glass-lined tanks, however, are not corroded, as are most metal tanks without anticorrosion rods.

Most gas water heaters use the Bunsen type of burner, which has a primary air inlet, a mixing tube, and a burner head with ports. There may be one burner or a cluster of burners. Smaller size, automatic storage water heaters have less gas input than larger ones. The orifices used may be of the fixed or adjustable types. The burner flame should not touch the bottom of the tank or any other surface within the burner compartment. Flame control should be so designed that it cannot be changed accidentally. Burners for use with liquefied petroleum gas have fixed orifices.

The door or doors to the burner assembly should fit tightly. If a door is warped or is not closed as tightly as it should be, excess air and drafts are admitted, which may cause the pilot to go out. It may also cause a loss of heat when the burner is not in operation. Doors should be large enough so that the pilot is easily accessible and the burner can be removed for servicing, if necessary.

The automatic pilot, now standard equipment on automatic storage type water heaters, is a safety device that stops the flow of gas to the burner if the constantly burning pilot should go out. In the 100 percent shut-off type, gas supply to both the pilot and main burner is shut off if the pilot goes out. In many water heaters the operation of the automatic pilot depends on a thermomagnetic effect.

In the thermomagnetic type of safety cutoff, two dissimilar metals are joined together, and the junction is placed in the pilot flame. This produces a small amount of electricity. When the flame goes out, no electricity is produced, the magnet controlling the gas valve is deenergized, and the flow of gas is stopped.

The pilot should remain on at all times. The flame should be high enough to maintain heat to operate the automatic gas shutoff. It should be adjusted so that the main burner will ignite rapidly. The pilot burner should be located so that the flame can easily be seen even if the main burner is in operation.

The thermostat controls the temperature of the water in the storage tank by controlling the flow of gas to the burner. The heat-sensitive part of the thermostat is located in the water tank. Many thermostats for gas water heaters are actuated by a copper tube (expanding element) over an Invar rod (nonexpanding element).

The thermostat may operate on a snap-acting, graduating, or quick-acting principle. In snap-acting thermostats the valve controlling the flow of gas moves from open to closed, or vice versa, very rapidly. In graduating thermostats the valve is wide open, allowing a full flow of gas to the burner when the water in the storage tank is cold. As the water is heated, the main valve gradually closes until the thermostat setting is reached, at which time the main valve is closed completely. A bypass or minimum flame is used on this type because of the problem of ignition on slow-opening valves. Quick-acting thermostats react much more rapidly to changes in water temperature. This type of thermostat is much like the graduating type, but because the action is more rapid, little or no bypass gas is required.

Tank corrosion usually does not affect an immersion type of thermostat, but the copper part of the thermostat can cause galvanic corrosion on the tank adjacent to the control. The copper may be covered with a plastic sheath to prevent this reaction.

The gas burner in the automatic storage type of water heater is placed below or underneath the water tank (Fig. 19-2). The heat from the burner and from the gases surrounding the tank must be kept close to the tank. A liner may be used around the tank that will allow the passage of the hot gases at the correct speed. Flue liners must be made of heavy enough material so that they will not lose their shape. It is important that there be equal flue space around the tank.

A center flue may be used in the tank, with baffles in it to regulate the rate of flow of the hot gases.

Good insulation is necessary in an automatic storage type of water heater to keep the stored water from cooling too rapidly. In the type of heater using the outside flue, insulation is placed between the liner and the outer shell. In the center flue type,

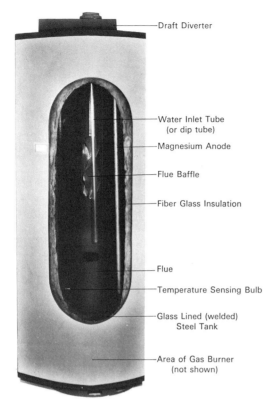

Draft Diverter

Water Inlet Tube
(or dip tube)

Magnesium Anode

Flue Baffle

Fiber Glass Insulation

Flue

Temperature Sensing Bulb

Glass Lined (welded)
Steel Tank

Area of Gas Burner
(not shown)

Figure 19-2. Automatic storage water heater, glass-lined tank, underfired, high-low automatic flame. (Sears, Roebuck and Company)

insulation is placed between the tank and the outer shell (Fig. 19-2). Spun glass or rock wool are the insulating materials usually used.

Relief valves for a gas water heater are a pressure-relief valve, a temperature-relief valve and a vacuum-relief valve. A vacuum-relief valve is recommended in the cold-water line when cold water enters the tank externally at the bottom. It functions as an antisiphon device provided it is higher than the tank.

Water heater tanks may be subjected to pressures much higher than ordinary city water pressure. A sudden closing of a faucet can cause a momentary high pressure, or there may be increased pressure in the street main.

The pressure-relief valve should preferably be located on the cold-water line. This valve unseats slightly when excessive pressure occurs within the tank and allows

water to run out, thus relieving the strain on the tank.

A pressure-relief valve is not always required on open-water piping systems. It is, however, a very wise precaution, as an open-water piping system can become a closed system without warning. It is quite possible to have excessive pressure within the water tank and not have high temperatures. The reverse may also be true. Pressure-relief valves in a closed system may drip continuously during heating periods.

The temperature-relief valve is designed to prevent excessively high temperatures of water within the storage tank. It does not relieve high pressures nor does the pressure-relief valve relieve high temperatures.

The temperature-relief valve usually acts at about 205° F: a fusible plug melts and the overheated water can escape. The plug must then be replaced. Another type of valve reseats itself. Manufacturer's instructions and city plumbing codes govern the location of these valves.

One manufacturer (Sears) states that a combination pressure- and temperature-relief valve is the only means to effective protection. This valve should be installed in the top six inches of the tank. Manufacturer's instructions and local ordinances should always be checked.

TYPES OF GAS WATER HEATERS

Gas water heaters may be manually operated or automatically controlled. Water heaters in use today include the automatic storage, nonautomatic storage, adjustable recovery, automatic circulating tank, and automatic instantaneous types. Instantaneous heaters and the larger circulating tank heaters generally are used for multi-family homes, institutions, recreational centers, and industrial buildings; the internal tankless instantaneous heater is used in some single-family dwellings.

The internal tankless instantaneous heater consists of a coil of copper tubing with fins placed in the boiler of hot water where the temperature is usually 180° to 200° F. The operating control is located in the boiler near the cold-water inlet to the tankless heater. Tankless water heaters are long-lasting because the copper tube construction is not readily susceptible to corrosion. However, these heaters are not recommended for use in hard-water areas unless the water is softened. The heat for the house is maintained independently of the hot water for the house. These gas boiler and tankless heater units may require less space than a water heater and separate furnace.

Water is heated and stored in both automatic and nonautomatic storage types of water heaters. The nonautomatic storage type does not have a thermostat; the user must regulate the temperature of the water by controlling the source of heat. The automatic storage type is thermostatically controlled and is probably the more common type of water heater used in the home. The heat input rating for these water heaters may not exceed 75,000 Btu's per hour.

The automatic storage type of water heater has the heater, storage tank, insulation, and controls all incorporated into a single unit. The design is such that it can be installed in the kitchen, recreation room, or utility room if desired (Fig. 19-1).

Peak demands for hot water can be met by setting a control higher than needed so that by tempering the very hot water with cold it will serve more purposes. In some instances this excessively hot water is a safety hazard, particularly in homes with small children, elderly, or infirm people. Heaters that operate at a high recovery rate during peak demands, and normal recovery rate for ordinary demands, avoid

this problem (Figs. 19-1, 19-2). If a gas input of 40,000 Btu's for a 30-gallon size will raise the temperature of 33.6 gallons through 100° F per hour the same input will heat 56.0 gallons 60° F per hour.

Most automatic storage water heaters deliver water at one temperature only. However, two-temperature water heaters are available that can supply water to the dishwasher and washer at a higher temperature than that supplied to bathroom and lavatory faucets. Plumbing connections must be planned for use with the two-temperature heater. The hotter water is taken out at the upper level of the storage tank; the tempered (less hot) hot water is a result of mixing hot water with some of the water from the cold-water-inlet line before the tempered water is drawn off.

Natural, manufactured, or mixed gas or liquefied petroleum gas may be used as fuel for gas water heaters. Water heaters designed to be used with liquefied petroleum gas should be plainly labeled. It is possible to change a burner designed for use with other gases for use with liquefied petroleum gas, but this is not recommended.

Sizes and Shapes

Size of water heater should be determined by the needs of the household for hot water. Shape usually is determined by the place where it will be installed and by the size needed.

Sizes commonly available in the automatic storage tank types are 20, 30, 40, 50, and 60 gallons. The recovery rate should be considered as well as size when selection is made.

A water heater that is too small for the use required of it will literally be overworked and will not last as long as an adequate size water heater which is not overworked.

Future uses, anticipated and perhaps not

yet dreamed of, should be considered when deciding upon the size water heater to install. According to the Gas Appliance Manufacturer's Association, there are 140 different uses for heated water in the home and the hot water may be turned on 100 times or more each day.

The American Gas Association lists the following factors as some of the more important points to be considered in choosing the correct size of water heater. The number, sex, and age of people in the house, including the number of servants, is one consideration. How the dishes are washed, how the laundry is done, the number of baths or showers taken daily, and the number of times hands are washed—all these factors influence the demand for hot water. Leaky hot-water faucets and the distance from the hot-water tank to the faucet influence the amount of water that needs to be heated. The size of the faucet and the water pressure are also factors. How the water was heated previously is considered important in the correct sizing of a water heater for the home.

The more people there are in a home, the more baths, the more clothes to wash, and the more dishes there are to do. Many families do not use 1,000 gallons of hot water per month; other families use much more. On the basis of many tests conducted in various parts of the country it is estimated that the average hot-water consumption is between 7 and 12 gallons per day per individual.[1]

Water heaters may also be sized according to the number of rooms in the home, the number of bathrooms, and the appliances that use hot water. The FHA minimum requirement for a four-bedroom home with two bathrooms is a 40-gallon gas heater with 33,000 Btu input, or a 66-gallon electric storage-type heater with a 3 kw input, or an IWH[2] rated tankless heater with a rating of 3.25 gallons per minute per 100° F rise.[3]

The temperature rise through which the water must be heated also helps to determine the size water heater to purchase. In northern sections of the United States, the lowest temperature at which water is supplied to the water heater might be around 40° F. If water is heated to 140° F this would be a 100° F rise. Water delivered at 60° F would require no more fuel to be heated to 160° F.

The common shape of a gas water heater is a cylinder approximately 60 inches high and 14 to 25 inches in diameter. Some water heaters have an all-over exterior covering that fits to the floor (Fig. 19-2). Others are supported on legs several inches off the floor. Some models have the exterior shell in a rectangular shape. Cabinet models are also available. Cabinets have the appearance of kitchen base cabinets and may fit nicely into the kitchen cabinet installation.

INSTALLATION AND CARE

An insulated, automatic storage water heater that is underfired and approved by the American Gas Association may be located as close to the walls of a room as 2 inches. Others which have one or more flat sides designed to fit adjacent to wall surfaces may be safely installed if the temperature on the adjacent wall surface does not go over 90° F above room temperature. The AGA approval seal is important because many tests such as this that a homemaker could not make are made before

[1] Gas Appliance Service Water Heater Manual, 4th ed., American Gas Association 1953, p. 13.

[2] Indirect Water Heater Testing and Rating Code.
[3] Minimum Property Standards, Federal Housing Administration FHA No. 300. Washington D.C.: U.S. Government Printing Office, 1960.

the water heater is given an AGA approval seal. It should be located near a good flue and have plenty of air around it. Gas water heaters should not be located in closets or cubbyholes, unless one side is permanently open to provide good ventilation. Water heaters located in any confined space require special venting. The American Gas Association describes the specifications for these openings.[4] Water heaters should not be installed in bathrooms or bedrooms.

Preferably the water heater should be not over 15 or 20 feet from the place where hot water will be used most often—usually the kitchen sink. The correct diameter of the hot-water pipe depends partly on the number of faucets it will supply. A larger diameter than necessary results in unnecessary heat loss. When practical, copper pipe is preferred over iron pipe because it is thinner and therefore has less area to cool. Insulation around the hot-water pipe also helps decrease loss of heat from the pipe. If copper pipes are connected directly to steel or aluminum tanks, an electric current may be set up. To avoid this, nonmetallic gaskets to separate the metals are available. These are sometimes called dielectric unions.

Draft hoods should be used, as they help to regulate the draft or speed with which the heated gases pass around or through the water heater. They also serve to divert any back drafts. Back drafts may cause the pilot to go out or interfere with combustion of the gas. If the heated gases pass over the water heater too quickly, they do not give up as much heat as they should, and the fuel bill goes up.

A water heater should be level and on a solid fixed base. It should be easily accessible for operation, maintenance, and servicing.

---

[4] *Gas Appliance Service Water Heater Manual*, op. cit., pp. 18, 19.

Care

Water heaters actually require very little care. Pressure-, temperature-, and vacuum-relief valves should be checked occasionally by a serviceman. Corrosion or sediment around the seat or fusible metal of the valve may keep it from operating as it should.

If the automatic storage type of water is used in a soft-water system, very little or no flushing of the tank is needed. If the heater is to be used in areas where the water is very hard it should have a hand-sized, clean-out hole, as it is almost impossible to flush out the deposits of precipitated solids. Occasionally, perhaps once a month, the drain cock should be opened and at least a bucket of water drawn off. If the house is to be closed during cold weather, the supply of cold water to the tank should be turned off and the tank completely drained.

The outside of the heater can be kept clean by wiping with a damp cloth and occasionally washing with a detergent, followed by a rinse of clear water.

Leaky faucets should be repaired, as they may be the reason for high fuel bills. A water faucet leaking at the rate of 60 drops per minute will waste about 200 gallons of hot water per month. If it leaks twice as fast it will waste enough water to supply an average family with hot water for 11 days.

Do not attempt to relight a gas pilot or burner that has been extinguished accidentally until after the gas has been turned off, the lighting door opened, and the room adequately ventilated for at least ten minutes.

AMERICAN GAS ASSOCIATION STANDARDS

The American Gas Association, Inc., has set up standards for the construction and performance of gas water heaters with heat inputs of 75,000 Btu's per hour or less.

These standards, which cover the different types of gases that may be used for heating, have been approved by the American National Standards Institute, Inc.[5] Construction and performance of gas water heaters may be better than these standards which are considered basic for safe operation, acceptable performance, and substantial and durable construction of gas water heaters.

ELECTRIC WATER HEATERS

Some construction, use, and installation factors are the same for electric and gas water heaters. Others factors are, of course, quite different.

CONSTRUCTION

Electric water heaters, whether of cylindrical (Fig. 19-3) or table-top form, contain an interior tank and an exterior shell with insulation between tank and shell. Materials used for tanks of electric water heaters include galvanized steel, steel with glass fused on the inside surface, Monel, and copper. The material of the shell is steel and the exterior surface of the shell is usually baked-on synthetic enamel. The finish on the tops of table-top heaters may be porcelain enamel.

The cold-water inlet may be at the bottom or top of the tank. When the inlet is at the bottom, a cold-water baffle may be welded to the interior of the tank above the cold-water inlet. Its function is to direct the incoming cold water to the bottom of the tank, thereby minimizing mixing of cold and hot water when hot water is drawn off. When the opening is at the top, a water-inlet tube carries the cold water to the bottom of the tank. An anti-siphon hole located near the top prevents water from draining out of the tank should the water in the main lines be shut off.

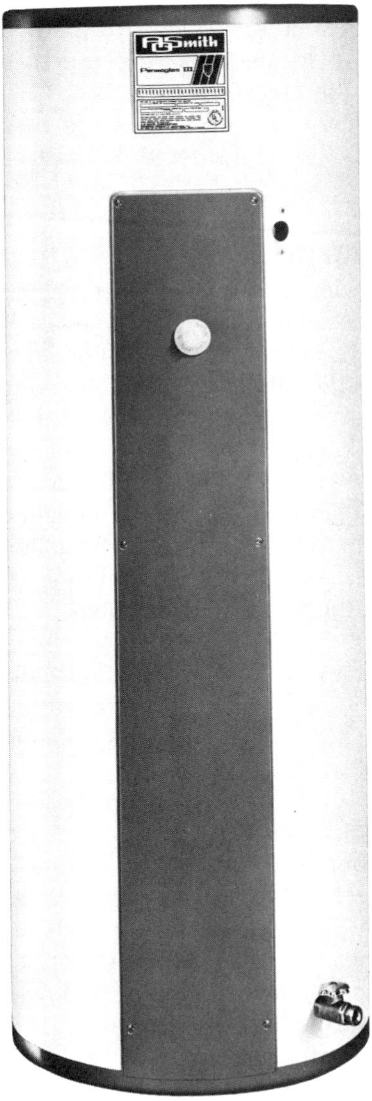

Figure 19-3. Automatic storage electric water heater, glass-lined tank. (A. O. Smith Corporation, Consumers Products Division)

[5] American National Standards Institute, Inc., *Z21.10.1-1971 Gas Water Heaters*, vol. I, approved March 10, 1971.

The hot-water outlet is at the top of the tank. The outlet pipe usually has a curved or gooseneck shape. This curve prevents cool pipe water from moving down into the tank when hot water is drawn off.

A drain faucet is provided near the bottom of the water heater to enable the user occasionally to drain water from the bottom of the tank into a pail. (Draining from the bottom of the tank removes sediment.) When the heater is installed this outlet should be so located that the user can manipulate the faucet.

Factors that affect the useful life of a water heater were considered earlier (see p. 324).

Tank Capacities and Wattages

Automatic electric storage water heaters are available with gallon capacities of 6, 15, 30, 40, 52, 66, 80, and 120 gallons. For many years the wattage of the heating unit or units was related to capacity; that is, certain tank capacities were associated with certain wattages and the larger the tank, the larger the wattage of the heating unit or units. For example, an 80-gallon tank would have one heating unit of 4,000 watts near the bottom of the tank or a heating unit of 2,500 watts in the upper part of the tank and another of 1,500 watts near the bottom. A 66-gallon tank would have one or two heating units totaling about 3,250 watts.

Quick recovery electric water heaters with units of rather high wattage are also available. For example, one manufacturer's line of electric water heaters includes high wattage, 30-, 40-, 52-, 66-, and 80-gallon cylindrical models. Each has a 4,500-watt upper unit and a 4,500-watt lower unit. Other wattages are available. This same manufacturer states that recovery can be figured on the basis of 4.1 gallons per hour per kilowatt at 100° F temperature rise, or for these heaters, 36.9 gallons per hour. A consumer can get a large quantity of hot water with a tank approximately comparable in size to that used for gas heaters, provided the house is adequately wired and the local power supplier permits use of the high-wattage heater.

The unit or units of most standard models and the units of the quick-recovery heaters are of the immersion type.

Factors to be considered in choosing the size of water heater to be purchased were discussed on page 323. A reliable local dealer is likely to be a useful person to consult.

OPERATING CHARACTERISTICS
Heating Units

The operation of a single-unit heater consists simply in the electric circuit to the unit being open or closed according to the thermostat setting and the temperature of the water in the tank. The operation of water heaters with two units is slightly more complex. The units may be connected separately or in parallel, if the local power supplier permits a parallel connection. Usually, water heaters that have two units are connected so that only one unit is on at a time, and the major part of the water in the tank is heated by the lower unit.

The cycle of operation is as follows. Assume that the tank is filled initially with cold water. The upper unit operates and heats about the top fourth of the tank. When the water in the upper part of the tank reaches approximately 150° F, a double-throw thermostat switch opens the circuit to the upper unit and closes the circuit to the lower unit. The lower unit now operates until all the water reaches approximately 150° F; then a single-throw thermostat switch opens the circuit to the lower unit.

As hot water is drawn from the top, cold

water enters near the bottom, and the single-throw thermostat of the lower unit closes the circuit to the lower unit again. The upper unit ordinarily will operate only when most of the hot water in the tank has been replaced by cold water.

Thermostats may be surface mounted units, held against the tank by a bracket bolted to the element mounting flange.

Relief Valves

Relief valves that will open the tank or the water line to the air in case of unsafe water pressure or temperature in the tank are provided with electric water heaters or are available as accessories. Requirements of the local power supplier and local plumbing ordinances may specify the type of relief valve to be used.

A pressure-relief valve supplies minimum protection in an open system only. The National Plumbing Code requires that a pressure-relief valve be installed in the cold water supply as close to the water heater as possible.

Protection against excessive water temperature is obtained with either a cutout switch or a temperature-relief valve. Underwriters' Laboratories requires a temperature cutout switch, which is essentially a thermostat so installed that it opens the electric circuit to the heater if the outside of the tank reaches a too-high temperature. A temperature-relief valve gives protection against excessive temperature by opening the tank to the air if the water in the tank becomes excessively hot.

A combination temperature-pressure-relief valve provides better protection. It should be installed in the top of the tank.

INSTALLATION

Many of the considerations here are the same for electric and gas water heaters. The heater should be installed as close as

possible to the place where hot water is used most frequently. Ideally, the water heater should be next to the kitchen sink or directly under it in the basement. Frequency of use rather than volume of hot water used is the criterion because after each withdrawal of hot water, the water in the pipe cools due to radiation.

Water pipes to all hot water faucets and water-bearing equipment, such as a washer, should be as short as possible. In fact, if the house has a very large area, it may be better to have two water heaters rather than one in order to have short hot water lines. The considerations that determine the diameter and material of the hot water pipe are given on page 324.

Suggestions on Use

Three recommendations usually are emphasized for electric water heaters. (1) It is important to repair leaking faucets. (2) Remove sediment from the bottom of the tank occasionally. This is done by placing a pail under the drain faucet and turning the faucet on. If a clean-out plate is provided, it may be possible to clean sediment from the inside of the tank with a brush. (3) Use a thermostat setting only as high as needed for the hot water demand in the house. (The factory setting of 150° F can be changed; for some heaters this is done most conveniently when the heater is installed.)

AMERICAN STANDARDS ASSOCIATION STANDARD

The existing standard that applies to electric water heaters is C72.1.[6] This standard applies to electric storage type heaters of not less than 30-gallon capacity. The purpose of the standard is to establish a

[6] American Standards Association, C72.1, *American Standard Household Automatic Electric Storage-Type Water Heaters,* 1949.

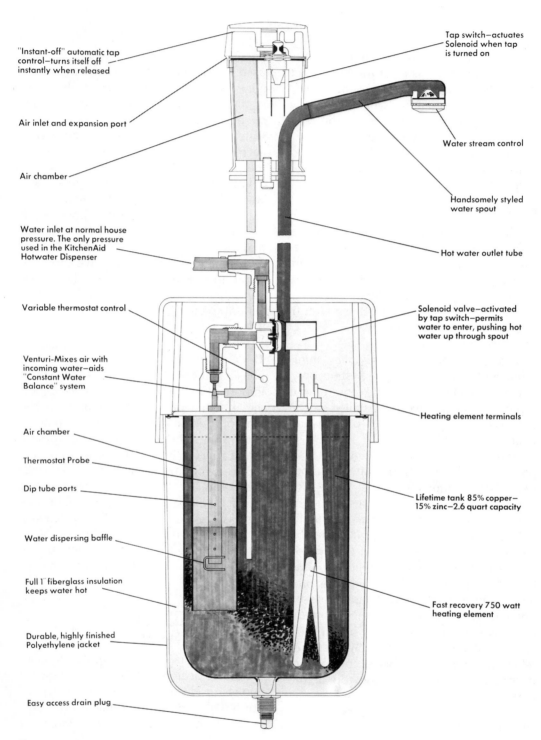

"Instant-off" automatic tap control—turns itself off instantly when released

Air inlet and expansion port

Air chamber

Water inlet at normal house pressure. The only pressure used in the KitchenAid Hotwater Dispenser

Variable thermostat control

Venturi-Mixes air with incoming water—aids "Constant Water Balance" system

Air chamber

Thermostat Probe

Dip tube ports

Water dispersing baffle

Full 1" fiberglass insulation keeps water hot

Durable, highly finished Polyethylene jacket

Easy access drain plug

Tap switch—actuates Solenoid when tap is turned on

Water stream control

Handsomely styled water spout

Hot water outlet tube

Solenoid valve—activated by tap switch—permits water to enter, pushing hot water up through spout

Heating element terminals

Lifetime tank 85% copper— 15% zinc—2.6 quart capacity

Fast recovery 750 watt heating element

Figure 19-4. Hot-water dispenser is installed in hot water line. (KitchenAid Division, Hobart Manufacturing Company)

uniform procedure for determining performance under specified test conditions and to establish certain minimum requirements.

Introductory material includes definitions and ratings. The fourth section deals with "general standards," some of which are summarized below.

Parts intended to be serviced shall be reasonably accessible. Water service connections to the heater shall permit use of standard pipe or tube fittings. Provision shall be made to install temperature- and pressure-relief devices effectively. A data nameplate conveniently accessible for observation shall be attached to the heater.

The normal factory setting is 150° F. The upper limit of the range of adjustment may be considerably higher than this. The secondary unit of a two-unit heater is located so as to heat effectively the top 25 percent of the actual tank capacity. Under normal operating conditions, a water heater should not cause a temperature higher than 170° to 190° F or so to be attained at any point on surfaces near or upon which it may be mounted in service when operated at the highest thermostat setting.

Section five prescribes test conditions and procedures. These cover sturdiness of the complete heater, strength of tank, recovery characteristics, rate of delivery of hot water, and other performance aspects. The user is interested, of course, in durable construction. On the performance side, the following two recommendations are of particular interest to the user: (1) The efficiency of heating water in the tank, as determined by the specified test procedure, shall not be less than 90 percent. (2) At least 90 percent of the actual tank capacity can be withdrawn before the temperature of the delivered water drops 30° F.

More recent standards for electric storage type water heaters for household use include specifications that are designed to give the consumer "better quality and performance at no higher cost." The types of heaters, tank sizes, and number of heating elements are reduced. Suppliers of equipment conforming to the specifications are not required to state that the equipment meets the standards specified.[7]

SPECIAL-PURPOSE HEATERS

An instantaneous heater is available in a small-size model known as the "hot water dispenser" (Fig. 19-4). This is thermostatically controlled to provide water up to a temperature of 190° F which is hot enough for the preparation of instant foods and instant hot drinks. It is connected to the hot water line at the kitchen sink and will provide approximately 15 quarts of 190° F water per hour. The brass container has a 2-quart capacity. It operates on 115 volts and uses 750 watts. It can be installed on a counter adjacent to the sink or through the opening for the spray attachment on the kitchen sink. If the latter, the spray attachment cannot be used. A higher wattage model is designed for use with water coolers that are located in offices. Other instantaneous heaters are designed for use in heating water for home swimming pools.

[7] *The electric utility industry's specification for electric storage-type water heaters for household use,* approved and adopted by Edison Electric Institute Marketing Division and Southeastern Electric Exchange Marketing Division, August 1969.

BUYING GUIDE

Refer to Chapter 8 for general factors to consider when buying major appliances. The "right answer" to the following questions may sometimes vary for different family needs but suggestions to help in making individual decisions will be found in the preceding pages.

1. How much hot water does the family need? What are peak demands? Number and ages of family members, use of automatic washer and dishwasher, and the number of baths in the home influence this decision greatly.
2. What capacity is the heater? Capacity includes size and recovery rate and size needed depends upon:
   a. Amount of hot water needed each day
   b. Peak demands
   c. Water inlet temperature and hot water temperatures
   d. Level of income and living standards of family
3. Would a two-temperature model be advisable? Water for laundry and dishwashing is piped off before water for other uses. Would this involve the need for additional plumbing to be installed in the home?
4. What is the lining of the tank? How hard or soft is the water in area? Does the tank have an anodic rod? Is it replaceable? Can the average homeowner do this or is it a task for a service man? How much would the rod and replacement cost?
5. What is guaranteed? By whom? For what length of time? Who services it? Water heaters seldom require service, but they are not necessarily service-free.
6. What approval seals does it have? What do the seals really mean?
7. Will this heater require a new installation of the utility in the home or moving of an installation to another location in the house? How much would such an installation cost?
8. If gas, how will it be vented?
9. How much room space is required for this model?
10. If gas, is the pilot and burner the 100 percent shut-off type?
11. What provisions are made to make the heater rust resistant? Will it be easy to maintain the finish?
12. What is the estimated cost of operation?
13. What is the purchase cost? For how many years is the heater guaranteed? What is the average yearly cost?

CHAPTER 20
LAUNDRY AREA,
WASHERS, AND
WASHING GUIDES

LAUNDRY AREA

The appliances and accessory equipment used for washing, drying, and ironing household textiles may be assembled either in a single laundry area or in several places in the home.

A complete laundry area is one planned for prelaundering tasks and for washing, drying, and ironing.[1] But family needs or preferences for use of space may be such that a complete, single laundry area is not provided. For example, a homemaker may prefer to do such prewashing tasks as sewing torn articles in a sewing room or in a living room; she may wish to vary the room in which she irons according to other activities taking place in the home at a given time.

Whether a complete center is planned or not, the washer and dryer should be located close to each other, since they usually are used together. Cabinet or shelf space should be provided near the washer for storage of laundry supplies. Some well-lighted sink space should be provided for prewashing treatment of heavily soiled parts and counter space for folding articles after removal from the dryer.

---

[1] Helen E. McCullough, *Laundry Areas, Space Requirements and Locations,* University of Illinois Bulletin, Circular Series C 5.4, December 1957.

Figure 20-1. Laundry center planned for left-to-right work procedure includes space to hang permanent press items as they come from dryer, space to store and sort soiled clothes, and space for ironing and sewing supplies. (Maytag Company)

Figure 20-2. Corner sink installation between washer and dryer. (Elkay Manufacturing Company)

Figure 20-3. Washer and dryer in a left-to-right arrangement. Dryer basket is 10 inches higher than former models. (Westinghouse Electric Corporation)

LOCATION

If a laundry area is provided, available space in the home tends to determine its location. For many years, part of the basement has been used for the laundry area. In ranch-style houses with no basement, the laundry area is, of course, on the first floor. Split-level houses are likely to have the laundry area in the basement level.

Laundry equipment sometimes is installed in the kitchen. A washer only or a washer and dryer may be placed along the "fourth wall," that is, the wall not used for the kitchen work area. Alternatively, if the kitchen includes a peninsula of cabinets and appliances, a washer and dryer can be installed on the side that faces away from the food preparation and cleanup area. This use of space in the kitchen can be one way of meeting the recommendation that an area be provided in the kitchen for activities appropriate to the family.

A somewhat unusual location is near the bedrooms. This is convenient from the point of view that soiled bed linens, towels, and clothes are "picked up" in bedrooms and bathrooms. But if the bedrooms are on the second floor and the kitchen is on the first, the homemaker is likely to have to do some walking up and down stairs, just as she does when the laundry is in the basement.

Figures 20-1, 20-2, and 20-3 show arrangements of laundry equipment. Note that the dryer can be located on either side of the washer. Most dryers have the door hinged on the right. However, the dryer door (Fig. 20-1) may be hinged on the left. Placement of the dryer at the left of the washer permits the right-to-left work sequence that is normal for a right-handed person.

A separate sink, as shown in Figure 20-1 and Figure 20-2, is most useful in a laundry area.

WASHERS

Currently, the most widely used washers may be classified in three ways. One is to classify them as wringer washers, spinner washers, or automatics. A second is to classify them as agitator-type or cylinder washers. The Association of Home Appliance Manufacturers classifies washers as automatic, wringer, and portable types.

The first classification differentiates between washers partly on method of extraction of water. In wringer washers washed clothes are passed between two rolls. In spinner washers and automatics water is extracted from washed clothes by centrifugal action; that is, rotation of a spin basket or tub causes excess water to squeeze through and spin off the clothes. In the spinner washer, washing takes place in one tub and extraction of water and rins-

ing in another. Since the user must transfer clothes from washtub to spin tub, operation of the spinner washer is not automatic. Most portable washers use this method of water extraction. In automatics, one tub is used for the entire washing process. Fill, wash, extraction, and rinse operations are all controlled by moving one or more dials or levers at the beginning of the fill part of the cycle.

The second classification differentiates washers according to washing method. Agitator-type washers include wringer washers, spinner washers, and automatics that have an agitator or pulsator. Cylinder washers are automatics that have a cylinder with interior projecting baffles instead of an agitator or pulsator. This cylinder rotates about a horizontal axis (Fig. 20-4).

Figure 20-4. Stacked washer and dryer uses 27 inches of floor space. (Westinghouse Electric Corporation)

The washing action of agitator-type washers is due primarily to water currents produced by back and forth movement of an agitator or up and down movement of a pulsator. The water currents flex the clothes and cause them to turn over and move around in the detergent-water solution.

In cylinder washers the washing action consists in clothes tumbling through a detergent-water solution. They are carried up from a bottom position by the rotating cylinder, fall and tumble through the detergent-water solution, and the process repeats.

WRINGER WASHERS

In 1970 the number of wringer and spinner washers sold was approximately one-third of the number sold ten years earlier.[2] Consumers may choose this type of washer for varying reasons. One consideration is availability of water. With wringer washers wash and rinse waters can be reused for several loads. Another consideration is laundering work habits. In a wringer washer eight or more loads can be washed in 2 hours, whereas in automatics the

[2]*Merchandising Week*, Billboard Publications, Inc., Feb. 22, 1971, pp. 22, 23.

complete cycle usually requires 30 minutes or longer for a normal load. Some homemakers who have large quantities of laundry each week prefer to do all the washing one morning per week rather than do two or three loads every few days, even though time used for washing with an automatic is partially free time in that the user need not stay at the washer. A third consideration in determining choice of model is that stripped models of wringer washers cost less than stripped models of automatic washers.

Construction Characteristics

Wringer washers are illustrated in Figures 20-5 and 20-6. The tub in which the clothes are placed is separated from the outer shell by an air space, which acts as a thermal insulator and helps maintain the temperature of the water in the tub. The inner tub is either finished in porcelain enamel or is made of a rustproof material such as aluminum or stainless steel. The exterior finish on the outer shell is usually synthetic enamel.

Capacities may be stated in pounds of clothes—8, 10, 12, or more pounds—and/or gallons of water required to fill the inner tub. Washers rated at 10 to 12 pounds usually take about 17 to 18 gallons of water when filled to the fill line. Low-priced or stripped models are likely to have the lower capacities. However, the actual weight of clothes that should be washed at one time in a washer depends on the bulkiness of the articles as well as the pound-capacity rating of the washer. (See

Figure 20-5. Wringer washer with instinctive wringer release and safety-release bar. (Speed Queen Division, McGraw-Edison Company)

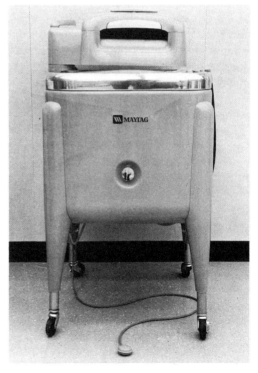

Figure 20-6. Wringer washer with wringer-control foot pedal; when pressure from foot is released, rolls stop. (The Maytag Company)

the later section in this book on washing guides.)

Lids of washers that have a square tub may be hinged to the body of the washer. Round lids are not hinged to the tub. A hook for the lid on the side of the washer is a convenient feature.

Most of the agitators are polypropylene. These usually have a "positive fit" to the agitator shaft by splines and a locking cap. The agitator is formed with vanes. Vane designs differ but often extend farthest from the axis near the bottom, curve in toward the axis, and extend out from it again near the top of the agitator, but not as much as at the bottom. Most agitators can be removed from their shafts by removing the locking cap, if one is provided, and lifting the agitator straight up.

A sediment trap may be provided in the bottom of the tub to catch insoluble bits of dirt. A water-discharge pump for rapid emptying of the tub after use is a convenient feature. Wringer washers with pumps usually have some provision to prevent their clogging—sometimes a strainer at the location where the water is pumped from the tub.

Power-driven wringers can usually be swung through a complete circle and locked in one of several operating positions. Wringers have self-adjusting pressure mechanisms so that there is a tight fit of the rolls when thin fabric is to be wrung and a looser fit for bulky items.

Underwriters' Laboratories require that "power to the wringer rolls shall be controlled by a device or system that (1) must be continuously actuated by the operator or (2) will stop the rolls when a 20-pound force is applied opposite to the direction of infeed to an object in the rolls. . . . (Effective October 1, 1968.) Unless the roll pressure is released automatically when the rolls are stopped, the wringer shall be provided with a safety release having an operating member plainly marked to indicate its function and operation."[3]

Tilt or feed "boards," usually aluminum, are provided on both sides of the rolls to permit the extracted water to drain back into the tub or laundry tray. Safe and correct usage requires that the user flip clothes onto a feed board, whence they will be drawn onto the bottom roll.

Controls provided on the washer vary with price and with manufacturer. A control such as a lever or button for starting and stopping the agitator is usual. Some manufacturers also provide a motor on-off switch. A time control may be provided that stops the agitator automatically when the preselected washing time is up. A minute indicator and bell may be provided that indicate elapsed time but do not control length of wash period.

Operating Components

Washers that use a gasoline engine are available for homes not wired for electricity. The usual type, however, is the washer that operates on house current and has an electric motor rated at 1/3 or 1/4 horsepower. A gear mechanism or transmission is used to change rotary motion of the motor shaft to the oscillating (back and forth) motion of the agitator shaft. This mechanism also provides for rotation of the extension shaft for one of the wringer rolls. Usually, the gear parts that drive the agitator shaft and the wringer roll extension shaft are permanently lubricated and sealed in a metal gear case housing. The gear mechanism and motor may be connected by direct-drive rod linkage or by pulley and belt.

The pump for removing water is separate from the gear mechanism of the washing machine.

[3] Underwriters' Laboratories Standards for Safety, *Electric Home-Laundry Equipment UL 560,* 1968.

Special Use and Care Recommendations for Wringer Washers

Particular instructions for different wringer washers are, of course, given in users' booklets. Some considerations that should be noted especially are these:

1. Dissolve the detergent in the water before adding the clothes.
2. As far as possible, flip clothes onto the feed board of the wringer. Also, to minimize creases, straighten and pass clothes and other textiles one at a time through the wringer.
3. Release the tension on the rolls after washing. Whenever the washer is not to be used for some time, separate the rolls by striking the safety release.
4. Follow instructions on oiling and greasing. For example, the user's booklet may recommend that every month or so a drop of oil be applied to the bearings for the upper wringer roll.

5. Have a serviceman lubricate the motor at the intervals—three to five years or so—recommended by the manufacturer.

SPINNER WASHERS

As stated earlier, a spinner washer is one in which clothes are washed in one tub and rinsed and extracted in another. After a load of clothes has been washed, they are lifted from the washtub into the basket tub. Excess sudsy water is drained back into the washtub, and rinse water is sprayed over the clothes in the basket tub. Many of the washers designated as portable by the Association of Home Appliance Manufacturers are designed with a separate basket for water extraction (Fig. 20-7).

Portable washers are designed especially for use where space is at a premium and hot water or even water is in short supply. They are favored by those consumers who

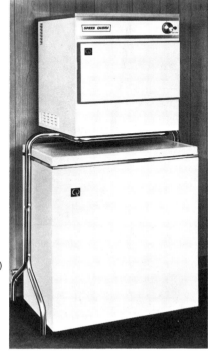

(a)   (b)

Figure 20-7. Portable washers occupy little floor space; require no special plumbing. (a) Clothes are spun damp dry at 2,100 rpm. (Hoover Company) (b) Washer has 210 degree agitator stroke; dryer mounted above. (Speed Queen Division, McGraw-Edison Company)

prefer to invest less money in washing equipment. Normally the washing capacity is less than that for automatic washers. The washer tubs hold approximately 10 gallons of water. The water requirement for a complete wash cycle is approximately 30 gallons.

Washer tubs may be polypropylene, porcelain enamel, or stainless steel. Spinner tubs are usually of aluminum or polypropylene. Separate motors for washing action and for the spinner unit are common (Fig. 20-8).

After the first load is washed, two loads can be handled at a time; that is, one load

can be washed while another is being rinsed. Some care must be taken to load the rinse tub uniformly.

AGITATOR-TYPE AUTOMATICS

Agitators used in automatics are similar to those used in wringer washers. The parts of one type of agitator are shown in Figure 20-9. The agitator oscillates or has a back and forth movement. It has a lint filter that fits down into the top of the agitator. The fabric softener dispenser cup sets in the top of the lint filter. A type of agitator, not shown, known as a pulsator has flex-

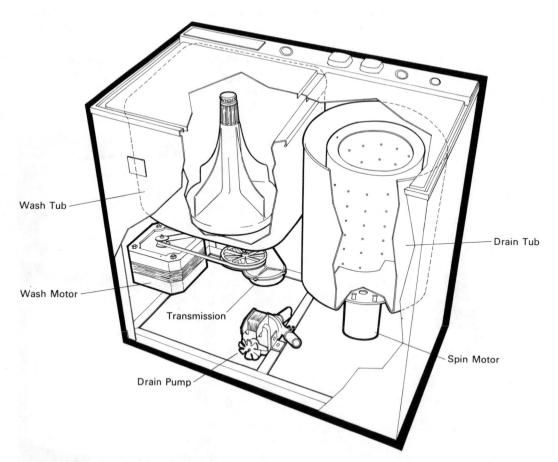

Wash Tub

Wash Motor

Transmission

Drain Pump

Drain Tub

Spin Motor

Figure 20-8. Schematic drawing of common parts of a portable washer. (Adapted from illustrations by Speed Queen Division, McGraw-Edison Company)

ible disks that move up and down on a shaft instead of oscillating about it.

Construction Characteristics and Operating Components

Operating components and construction features of agitator-type automatics are illustrated in Figures 20-10 and 20-12. The construction features one sees on the outside are controls, top with lid assembly, cabinet or chassis, and supply and drain hoses. Inside the cabinet are an agitator, wash-spin tub, outer tub, and a base assembly on which are mounted motor, gear mechanism, pump, valves, and other parts.

A water-level control switch actuates

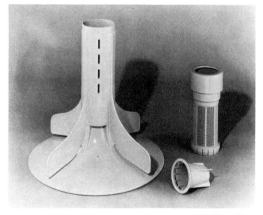

Figure 20-9. "Power Fin" agitator with lint filter and fabric softener cup. (Maytag Company)

the timer to start the washing action when the selected level of water is reached in the tub. The unbalance switch shuts off the motor if the washload becomes unbalanced. The lid hinges are nonrusting

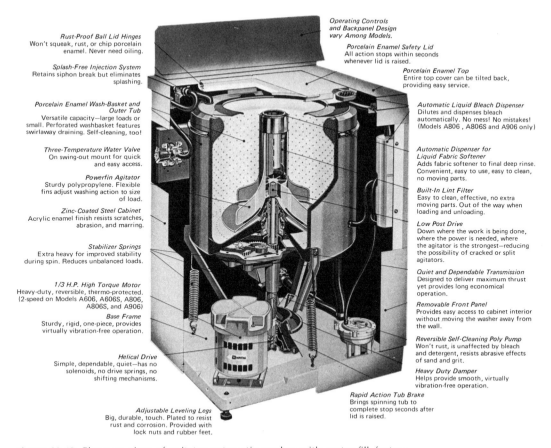

Rust-Proof Ball Lid Hinges
Won't squeak, rust, or chip porcelain enamel. Never need oiling.

Splash-Free Injection System
Retains siphon break but eliminates splashing.

Porcelain Enamel Wash-Basket and Outer Tub
Versatile capacity—large loads or small. Perforated washbasket features swirlaway draining. Self-cleaning, too!

Three-Temperature Water Valve
On swing-out mount for quick and easy access.

Powerfin Agitator
Sturdy polypropylene. Flexible fins adjust washing action to size of load.

Zinc-Coated Steel Cabinet
Acrylic enamel finish resists scratches, abrasion, and marring.

Stabilizer Springs
Extra heavy for improved stability during spin. Reduces unbalanced loads.

1/3 H.P. High Torque Motor
Heavy-duty, reversible, thermo-protected. (2-speed on Models A606, A606S, A806, A806S, and A906)

Base Frame
Sturdy, rigid, one-piece, provides virtually vibration-free operation.

Helical Drive
Simple, dependable, quiet—has no solenoids, no drive springs, no shifting mechanisms.

Adjustable Leveling Legs
Big, durable, touch. Plated to resist rust and corrosion. Provided with lock nuts and rubber feet.

Operating Controls and Backpanel Design vary Among Models.

Porcelain Enamel Safety Lid
All action stops within seconds whenever lid is raised.

Porcelain Enamel Top
Entire top cover can be tilted back, providing easy service.

Automatic Liquid Bleach Dispenser
Dilutes and dispenses bleach automatically. No mess! No mistakes! (Models A806, A806S and A906 only)

Automatic Dispenser for Liquid Fabric Softener
Adds fabric softener to final deep rinse. Convenient, easy to use, easy to clean, no moving parts.

Built-In Lint Filter
Easy to clean, effective, no extra moving parts. Out of the way when loading and unloading.

Low Post Drive
Down where the work is being done, where the power is needed, where the agitator is the strongest—reducing the possibility of cracked or split agitators.

Quiet and Dependable Transmission
Designed to deliver maximum thrust yet provides long economical operation.

Removable Front Panel
Provides easy access to cabinet interior without moving the washer away from the wall.

Reversible Self-Cleaning Poly Pump
Won't rust, is unaffected by bleach and detergent, resists abrasive effects of sand and grit.

Heavy Duty Damper
Helps provide smooth, virtually vibration-free operation.

Rapid Action Tub Brake
Brings spinning tub to complete stop seconds after lid is raised.

Figure 20-10. Phantom view of agitator automatic washer with meter-fill feature. (Maytag Company)

Figure 20-11. Nonrusting ball-lid hinges. Ball shown on left; on right, ball hinge in place. (The Maytag Company)

(Fig. 20-11). The inner washtub has perforations over its entire surface—sides and bottom. A balancing ring is provided at the top of the double tub to minimize vibration when the tub spins.

Water enters at the top of the double tub through valves. Whether it is hot, cold, or warm is determined by which control lever the user has pushed. Temperatures for the hot and cold settings are the temperatures of the hot and cold water supplies. The warm setting provides a mixture of cold and hot water.

Water entering the washtub is controlled by electrical and mechanical action. As the tub fills with water, the air in the pressure chamber or air dome is compressed. This compressed air acts against a diaphragm in the pressure switch. The spring tension in the pressure switch varies according to the water level for which the washer is set.

When the preselected minutes of washing time are up, the agitator stops oscillating and the washtub starts to rotate. Water is drawn through an opening into the outer tub as the inner basket or tub is spun. Pump action pulls the water from the outer tub into the household drain system or, in the case of a suds-saver model, it may be directed to a separate storage container for later use.

Automatic operation of the timer starts the flush rinse while the washtub is still rotating and water is being extracted from the clothes. Next, the washtub stops rotating and is filled with water for the deep rinse. This rinse fill is the same as that selected for the wash; for example, if a partial fill was selected for washing, the tub will automatically have a partial fill for rinsing. The deep rinse is a rinse with agitator operating. After it, the tub spins again, water is expelled, and the motor turns off. All parts of this cycle are automatic after the user has turned the timer control to a specific number of minutes for wash.

The function of the tub brake is to stop the rotation of the tub quickly.

CYLINDER AUTOMATICS

Figure 20-4 shows a front-opening cylinder automatic. The cycle of operation is controlled by a timer. Baffles are provided

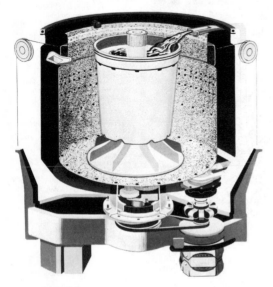

Figure 20-12. Phantom view of agitator washer with mini-basket in place. (General Electric Company)

on the inner surface of the rotating cylinder or tub. During the wash part of the cycle, the clothes are carried between and on the baffles from a bottom to a top position and then drop back. This tumbling action in the detergent-water solution washes the clothes. During the spin part of the cycle, the rapid rotation of the tub throws excess water off the clothes. Tumbling action rinses the clothes.

Washing, rinsing, and spinning take place in the perforated washtub mounted in an inner unit. This unit includes a front tub and a back tub that are clamped together and surround the washtub. The bearing in which the washtub shaft rotates extends through the rear surface of the back tub. The complete inner unit is suspended from the washer housing by a heavy-duty suspension system. Friction devices are provided for stabilization.

A transmission is used to rotate the washtub at two speeds: a relatively slow speed for the wash and a faster one for the spin. Belts transfer power from the motor to the transmission and from the transmission to the washtub pulley. The pump for draining water is driven by a friction wheel.

Adjustable feet are provided for leveling the washer.

FEATURES AND WATER REQUIREMENTS OF AUTOMATICS

It is likely to be true for automatic washers as for other appliances that no one model will have every feature a well-informed prospective purchaser might want. Usually, therefore, a purchaser must decide which features are most important to her.

It is difficult to make accurate comparisons between washers relative to size, water used, clothes washed, and so on, because the industry does not have a standard capacity-measuring technique.

A built-in, motor-overload protective device is a desirable feature for the following reason: A 15-ampere fuse is used in the washer circuit to take care of the high starting current, but this fuse is too large to protect the motor *after* it is running.

An effective balancing or stabilizing device for handling moderately unbalanced loads without excessive vibration is, of course, also desirable.

A brake mechanism to stop the rotation of the washtub quickly is a desirable feature. It is also important that the machine stops operation when the lid is opened.

Among the features related especially to operation of the washer are: type of fill control, provision for removing sediment, type of rinses, and speed of rotation of the washtub during spin. A time fill may be provided, or the fill may be controlled by a pressure switch. The latter control insures correct water level in the washer, regardless of the pressure of the water supply.

The effectiveness of the provision for removing sediment, such as sand or grit, from the wash and rinse waters can best be determined by using the washer. Practically, therefore, information on this point must usually be obtained from a person who has used the washer.

Washers may have deep rinses and spray rinses, deep rinses only, or spray rinses only. Some models also have overflow rinses, in which enough water enters to fill the washtub and overflow into the drain basin. The purpose of such a rinse is to float off loose soil at the start of the power rinse. To date no experimental data have been published to indicate superiority of any particular combination of rinses.

Different models of automatics have different speeds of rotation of the washtub during spin. If all other construction characteristics are the same, the amount of water extracted from clothes will be greater, the faster the speed of rotation of the tub during spin. Also, within limits, the longer the spin part of the cycle, the greater the amount of water extracted.

Some washers have a special suds-saver feature.

Simplicity of operation and appearance is desirable in controls. Overall construction that facilitates diassembly for servicing is also desirable. If the washer can be serviced from the front and/or the top, repair is simplified.

Middle-price and top-of-the-line models have provisions for washing and rinsing small loads in smaller than normal amounts of water and for regular and "fine-fabric" cycles. The washer in Figure 20-12 has a separate washtub which is inserted for small loads and/or delicate fabrics. Sixteen gallons of water are used in the complete washing cycle. In the fine-fabric cycle, the speed of oscillation of the agitator and the length of the wash, rinse, and spin operations is usually less than in the regular cycle.

Another feature is provision for built-in installation of matching washers and dryers. A vertical arrangement may be used, with the dryer above the washer, or horizontal arrangements provided for side-by-side installation of washer and dryer.

The total amount of water and the amount of hot water used per cycle vary for different models. Sometimes it is difficult to determine the amount of water used by a washer because the information may not be given on the specification sheet. One manufacturer states that small loads require 20 gallons of water, medium loads use 27 gallons, normal loads use 34 gallons, and large loads use 40 gallons for a complete cycle (Fig. 20-10). Large-capacity models may use as much as 56 gallons. A permanent-press cycle usually requires an additional fill of cold water. Many washers have a soak cycle; this re-quires additional water and detergent. If comparisons are made on the basis of full loads with maximum water requirements, cylinder automatics generally use less than agitator automatics. Among the agitator automatics, those that have overflow rinses are likely to use more water than those without them. The above statements are general and washers might be available that refute both statements.

In the permanent-press cycle there is an automatic cool-down before the first rinse. The suds-saver feature does not combine well with a permanent-press cycle because of the cooler temperature of the wash water that would be saved.

Various dispensers are found on the more deluxe washers—for detergent, bleach, fabric conditioner, and water conditioner. The emptying of the dispensers may be controlled by a solenoid that opens and closes a valve (Fig. 20-13). A solenoid is made by winding wire on a cylindrical form. This is a tubular coil for the production of a magnetic field. When current flows the solenoid, or helix, acts like a magnet. The solenoid is electrically energized at specific times depending upon the action desired. For instance, the solenoid controlling a fabric-conditioner dispenser is energized at the beginning of the rinse cycle. The outlet is closed during other periods of the washer operation.

INSTALLATION AND CARE OF AUTOMATICS

Installation of automatic washers should be in accordance with local electrical and plumbing requirements in communities where such requirements are imposed. For maximum satisfaction in use, it is recommended that an automatic be connected to a 115-volt, individual-equipment circuit that has a 15-ampere time-delay fuse or a 20-ampere circuit breaker. To meet 1968 UL requirements, washers must have a

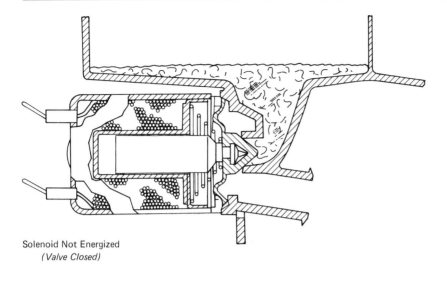

Solenoid Not Energized
(Valve Closed)

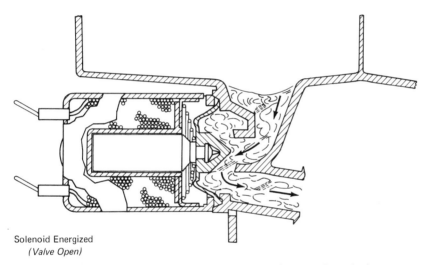

Solenoid Energized
(Valve Open)

Figure 20-13. Schematic drawing of solenoid valve used in laundry aids dispenser. (Frigidaire, Division of General Motors Corporation)

grounding conductor of flexible cord connected to a grounding-type attachment plug.

Washer supply hoses are connected to hot and cold water taps or faucets. The water pressure should be between the limits specified by the washer manufacturer, approximately 20 to 120 pounds per square inch. The type of house drain required may also be specified by the washer manufacturer. Some washers can be used with a floor drain, provided the local plumbing code permits this and provided also that an air gap is left between the drain hose and the back-up water in the drain system. Some washers should be drained into a fixed laundry tray or into a standpipe provided with a suitable trap.

Generally, automatics need not be bolted to the floor; however, they should stand level. Leveling legs usually are provided on the washer. Sometimes the serviceman who installs the washer places it on rubber pads or cups cemented to the floor.

Care

Some general points can be made for care of automatics: (1) To relieve pressure on the hoses, the supply taps or faucets should be turned off when the washer is not in use. (2) The lid or door should be left open after the washer has been used. (3) In freezing weather, if the washer is installed in an unheated area of the house, the supply hoses should be disconnected from the water faucets and all water drained from the washer.

WASHING GUIDES

Users' booklets give instructions on operation of controls. These may include information about short cycle of operation, prewash or soak, starching, bluing, tinting, procedures for laundering such special items as small rugs, pillows, electric bed coverings, and so on.

Some guides that apply, in general, for all washers are summarized here.

1. Wash frequently those garments that come in contact with the skin. Chemical changes may take place in the soil as it ages and what was originally a soiled item may in reality become a stained one.

2. Stains also are less difficult to remove when they are fresh. In any case, articles with certain stains should not be washed in hot water. For example, articles that have blood or meat juice stains should be soaked in cold or cool water before they are laundered and those with egg stains should be treated before they are washed. Some types of stains, on the other hand, are removed by hot water or by hot water plus bleach. These include light dye stains, such as those transferred in a mixed wash, perspiration, mustard, and light scorch. But even for these stains, it is better to treat them when they are fresh rather than let the stained articles remain in the soiled clothes hamper for several days.

3. Excessively soiled areas, such as collars of shirts and blouses, should be pretreated with a brush and a thick or jelly-like solution of detergent or soap.

4. Clothes and household textiles are sorted to make loads that should be washed under the same conditions; that is, at the same water temperature, for the same length of time, and for the same cycle of operation if the washer has more than one cycle. Articles that are not colorfast are washed separately. White nylons are washed only with other white articles. Heavily soiled articles are not washed with lightly soiled ones. Infants' clothing and bedding are washed separately.

5. Loads are made up not by weight alone but with some consideration given to size and bulkiness. If some of the articles are large, sheets for example, the total weight of the load should be decreased in order to insure free motion of the articles in the washer. If overloaded the washer may not wash all the clothes clean. It is also possible that clothes can be damaged and in extreme cases the washer mechanism could be harmed.

6. A prewash or a presoak will usually be helpful when entire articles are quite soiled, as may be true of jeans or denim shorts and gardening clothes. The prewash preferably uses soap or detergent and warm, not hot, water.

7. A wash time of about 15 minutes is the maximum that should be used for cottons, linens, and the like. If the cotton-linen load is not clean in 15 minutes, the water should be extracted and the load washed again. Many washer loads are most satisfactorily washed in 7 to 10 minutes. On the other hand, many syn-

thetic fabrics need to be washed for only a few minutes.

8. To get the *cleanest* washed fabrics, the best wash water temperatures for the soaps and detergents commonly available appear to be at least 135° F for white and colorfast cottons, white nylons, Dacrons, and Orlons, and for some white and colorfast rayons.[4] Certain materials—such as wools, silks, some acetates and all materials that are not colorfast—should be washed at "warm" temperatures, which might range from 100° F for wool to 110°–120° F for some colored materials. Solution-dyed acetates can be washed at higher temperatures, provided a label on the material so indicates.

It is pertinent to note that the water heater thermostat will usually have to be set at 145°–155° F to obtain a temperature in the washer of 135° F for the first wash load. After one load is washed, the washer may be sufficiently warm so that the temperature differential between the water in it and that in the water heater will be less than 20° F.

9. A rinse water temperature of 90° F to 105° F is usual, but some washers provide settings for cold rinses. These actually are likely to be slightly warmer than the cold tap water because clothes and washtub will be warm when the rinse water enters the tub. Up to this time, experimental work has not been reported on the relative effectiveness of 100° F versus cold rinses, though Williams found no significant difference for a 100° F versus a 140° F rinse for cotton washed in soft and hard (25-grain) water at a wash water temperature of 140° F.[5]

[4] Experimental work on laundering cotton treated with a radioactive soil at wash-water temperatures of 120° F, 140° F, and 160° F is discussed in an article by Ehrenkranz and Jebe entitled "Carbon-14 Method Tests Home Laundering Procedures," *Nucleonics*, March 1956.

[5] Velma L. Williams, "Removal of Soil from Fabrics Laundered in Home Washers. I. Soft and Hard Water; Number of Rinses and Temperature of Rinse Water," unpublished master's thesis, Iowa State College, 1951.

EXPERIMENTS

Experiment 1. Wrinkling of Washed Clothes for Different Washers, Different Wash Water Temperatures, and Different Fabrics

Because of the numerous drip-dry fabrics now on the market an experiment on wrinkling is of practical importance. Some of these fabrics may have a few wrinkles if washed in a washer but may not if washed by hand.

If the effect of the washer alone is to be observed at a given wash water temperature and for a particular fabric, the washed fabric must be dried on a line or a wooden horse. Wrinkling of fabrics washed in a washer and dried in a dryer would be a separate experiment.

The experiments should be carried out two or more times before conclusions are drawn on washers, water temperatures, or fabrics.

Additional experiments: The most significant additional experiment would be a test of the effectiveness of different washers getting clothes clean. Such an experiment performed in only one or two laboratory periods might give unreliable data and lead to misleading conclusions. Research on home laundering methods attests to the fact that experiments on the effectiveness of washers have to be repeated several times before valid conclusions can be drawn.

BUYING GUIDE

General

1. What is the manufacturer's name? What is the model number of the appliance?
2. Does it have the UL seal?
3. What are the dimensions in inches with lid fully opened: height, width, depth?
4. Materials:
   a. What finish materials are used for wash basket, outer tub, housing, top?
   b. Is a rust preventive used on base materials?
5. Cost: What is the installation cost? Allowance for old model if user has one?
6. What does the warranty promise the purchaser?
7. What arrangements are available for servicing?

Electrical Characteristics

1. Amperes at rated voltage, motor horsepower rating.
2. Grounding connection, three-prong polarized plug.
3. Is there motor overload protection with reset button to stop motor in case of overload or inadvertent operation?
4. Is there a lid or door shut-off switch that stops the motor in *any* part of the cycle when lid raised and door opened?

Installation Requirements

1. What is the minimum height of drain standpipe, if one is needed? (Provision of standpipe of correct height is necessary for some washers to prevent possible emptying of the washer prematurely when wash action is stopped during agitation.)
2. Water pressure in pounds per square inch: What is the minimum? The maximum?

Construction and Operating Characteristics

1. *For top-opening washers:*
   a. What type of wash action is provided—conventional agitation, pulsation, other (specify)?
   b. What is the agitation speed in oscillations or strokes per minute for regular action, gentle action, other action (specify)?

    c. What is the spin speed in revolutions per minute for regular extraction, gentle extraction, other (specify)?

    d. Type of draining action—does the water spin out of the wash basket over or near the top, through holes in the wall, or does it drain through the bottom of the basket?

    e. Is there a guard to prevent small articles such as children's socks clogging drain pump?

    f. Is there an unbalance device to stop the washer automatically if the load becomes "excessively" off balance? *Note:* This device by stopping the motor prevents damage to the washer due to the wash basket or outer tub hitting the housing and also prevents poor extraction that may be associated with irregular or wobbly spinning. An unbalance shut-off device is *not* necessary if the washer is designed to perform satisfactorily under large off balance—for example, a rug bunched in one section of the basket. If an unbalance device is provided, what type is it—ballast material near top flange of basket, snubbers mounted to inside of housing, other?

2. *For front-loading washers:*

    a. What is the cylinder speed in revolutions per minute for regular wash, gentle wash, other (specify)?

    b. What is the cylinder speed in revolutions per minute for regular extraction, gentle extraction, other (specify)?

    c. Is there provision for alternate acceleration and deceleration of cylinder during extraction?

3. *For top-loading and front-loading washers:*

    a. What type of fill—water level or pressure versus time?

    b. Does the wash basket brake stop spinning within ten seconds or so at the end of each spin period and whenever the lid or door is opened?

    c. What is the tub or cylinder capacity in gallons for complete fill, partial fill, or fills?

    d. What is the total water demand in gallons for each cycle for complete and partial fills?

    e. What is the approximate *hot* water demand in gallons for different cycles?

    f. Is temperature of warm setting thermostatically controlled? Does it depend on temperatures of hot and cold water supply and amount of opening of hot and cold water faucets?

    g. Are there special provisions for quietness in operation? Specify.

Convenience Features for Washers

1. Does it have adjustable leveling feet or glides?
2. Is a lint filter provided? Is it easy to clean? Does the filter operate at all water levels?

3. Is there a device for dispersing detergent?
4. Is there a rinse-conditioner dispenser for packaged water soft-ener or fabric conditioner?
5. Is there a device for delayed injection of chlorine bleach?
6. Is there a special soak setting? With agitation? Does control move automatically from soak through spin to fill for wash?
7. What temperature and time characteristics can be preset by the user: examples, wash water temperature, rinse water tempera-ture, wash time, power rinse time, extraction time?
8. Can the user reset controls during a cycle to get two washes, two deep rinses?
9. Can the user change cycles after one has started? If she can, what happens if the basket is filled with water when the change is made?
10. What characteristics, if any, can the user change *within* a cycle—for example, can she shorten the wash time?
11. During the cycle can the user change the wash, rinse, extraction action from regular to gentle or vice versa?

Accessories

1. Does it have a suds-saver? Is one available? Is a separate tray or tub needed to save the suds water? Is movement of suds water from wash basket to storage tub and return switch controlled or does the user do some mechanical operations?
2. Is a portability kit for nonpermanent installation available?

Special Features

List features not previously noted.

Portable Washers

1. Are the connections for water, electricity, and drainage easy to make? Can water be obtained at the sink when the washer is in operation?
2. Is the washer easy to move? Is it stationary when in operation?
3. What provision is made to keep clothes inside spinner basket during operation? Does the spinner stop when the cover is re-moved?

CHAPTER 21
DRYERS AND IRONS

DRYERS

When removed from a washer, a 10-pound (dry weight) load of clothes and household textiles may weigh 15 to 20 pounds—the precise weight depends on the kind of articles constituting the wash load and the water extraction characteristics of the washer. In the dryer, the wet load from the washer tumbles in a current of heated air and the 5 to 10 pounds of water added in the washer are removed chiefly by evaporation.

Users like dryers for the speed with which washed loads are dried, for the labor-saving aspects (relative to hanging wet loads), and for special results obtained with many articles and textiles. Some special results are: fluffiness of terry towels, "finished" appearance of polyester and other knit garments, finished appearance of fleece bathrobes, ready-to-wear appearance of such articles as men's shirts, childrens' clothing, rain-and-shine coats made of permanent-press fabrics. In addition some users appreciate the tumbling without heat feature of many models for freshening draperies and wrinkled clothes.

CONSTRUCTION AND DRYING MECHANISMS

The construction of a home dryer is illustrated in the labeled cut-away sketch (Fig. 21-1). A centrifugal blower mounted on the motor shaft forces air heated by electrical heating elements or a gas burner

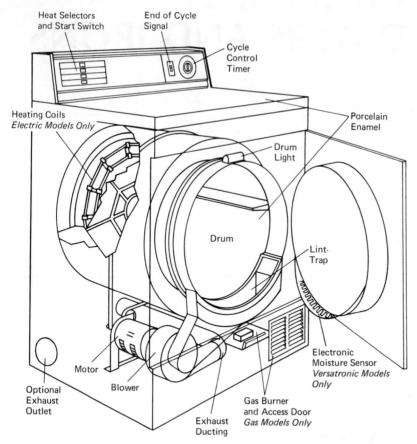

Heat Selectors and Start Switch

End of Cycle Signal

Cycle Control Timer

Heating Coils *Electric Models Only*

Porcelain Enamel

Drum Light

Drum

Lint-Trap

Motor

Blower

Optional Exhaust Outlet

Exhaust Ducting

Gas Burner and Access Door *Gas Models Only*

Electronic Moisture Sensor *Versatronic Models Only*

Figure 21-1. Schematic drawing of deluxe, large-size gas or electric dryer. (General Electric Company)

through a rotating drum. The drum is driven by a pulley from the dryer motor.

As the heated air moves through the tumbling textile load, water in the load vaporizes and is carried through an exhaust duct. Lint in the load mostly collects in a lint trap.

The exterior housing is a welded set of panels finished in enamel. The drum usually is porcelain on treated steel but may be stainless steel.

The controls include a temperature-regulating thermostat, a temperature-limiting thermostat, a time regulator set by the user or an automatic time control or both, a start switch, a safety interlock that causes heating and rotation to stop when the door is opened, and a speed (revolutions per minute) control for the dryer drum. The temperature-regulating thermostat controls the air temperature in the dryer and is either set by the user or is

preset by the manufacturer. The temperature-limiting thermostat is a safety mechanism designed to prevent temperatures that would be high enough on the exterior of the appliance to constitute a hazard to persons or the house and/or so high in the interior of the appliance as to constitute a fire hazard there.

The cubic feet of air moved per minute by the blower is a design characteristic of the blower.

When the user sets a time control, she or he is estimating the length of time the load will require to reach desired dryness. (This actually is not difficult for homemakers who regularly dry the same or nearly the same loads.) The automatic control is of different types in different models. Whatever the type, the user essentially sets a switch for "automatic" rather than a time interval. The most advanced automatic drying-time control

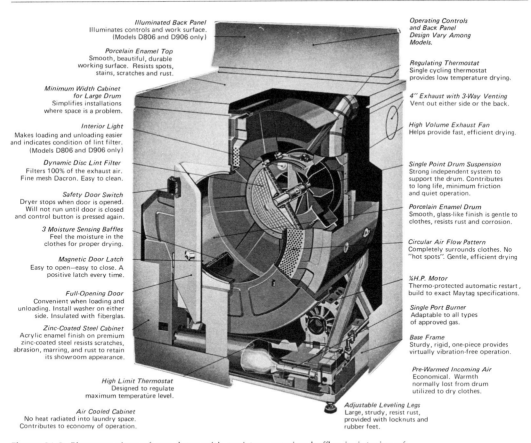

Illuminated Back Panel
Illuminates controls and work surface.
(Models D806 and D906 only)

Porcelain Enamel Top
Smooth, beautiful, durable
working surface. Resists spots,
stains, scratches and rust.

Minimum Width Cabinet
for Large Drum
Simplifies installations
where space is a problem.

Interior Light
Makes loading and unloading easier
and indicates condition of lint filter.
(Models D806 and D906 only)

Dynamic Disc Lint Filter
Filters 100% of the exhaust air.
Fine mesh Dacron. Easy to clean.

Safety Door Switch
Dryer stops when door is opened.
Will not run until door is closed
and control button is pressed again.

3 Moisture Sensing Baffles
Feel the moisture in the
clothes for proper drying.

Magnetic Door Latch
Easy to open—easy to close. A
positive latch every time.

Full-Opening Door
Convenient when loading and
unloading. Install washer on either
side. Insulated with fiberglas.

Zinc-Coated Steel Cabinet
Acrylic enamel finish on premium
zinc-coated steel resists scratches,
abrasion, marring, and rust to retain
its showroom appearance.

High Limit Thermostat
Designed to regulate
maximum temperature level.

Air Cooled Cabinet
No heat radiated into laundry space.
Contributes to economy of operation.

Operating Controls
and Back Panel
Design Vary Among
Models.

Regulating Thermostat
Single cycling thermostat
provides low temperature drying.

4" Exhaust with 3-Way Venting
Vent out either side or the back.

High Volume Exhaust Fan
Helps provide fast, efficient drying.

Single Point Drum Suspension
Strong independent system to
support the drum. Contributes
to long life, minimum friction
and quiet operation.

Porcelain Enamel Drum
Smooth, glass-like finish is gentle to
clothes, resists rust and corrosion.

Circular Air Flow Pattern
Completely surrounds clothes. No
"hot spots". Gentle, efficient drying

¼H.P. Motor
Thermo-protected automatic restart,
build to exact Maytag specifications.

Single Port Burner
Adaptable to all types
of approved gas.

Base Frame
Sturdy, rigid, one-piece provides
virtually vibration-free operation.

Pre-Warmed Incoming Air
Economical. Warmth
normally lost from drum
utilized to dry clothes.

Adjustable Leveling Legs
Large, strudy, resist rust,
provided with locknuts and
rubber feet.

Figure 21-2. Phantom view of gas dryer with moisture-sensing baffles in interior of drum. (The Maytag Company)

depends on electronic moisture sensors which incorporate a solid-state element and are located in the drum or on the interior of the door (Figs. 21-1 and 21-2). When the moisture in the load reaches a level corresponding to dry or damp-dry, as selected by the user, the electronic sensors interrupt the heat output and, after a five- or ten-minute interval, the drum rotation.

Other automatic drying controls depend on the moisture content of the air being exhausted from the dryer. The user sets "automatic" or automatic and type of fabric load.

INSTALLATION, USE, AND OPERATING COSTS

For convenience in use, a permanently installed dryer should be as near the washer as possible. Local ordinances often require outdoor venting of a gas dryer. When a gas dryer is vented to the outside,

the products of combustion of the gas, as well as the moisture and some lint from the loads, pass to the outside rather than to the interior space. Even though not required by ordinance, outdoor venting of an electric dryer is advantageous for keeping moisture out of the house in humid weather. Indeed, the nameplate of the electric dryer shown in Figure 21-3 suggests exhausting to the outside.

A publication on sanitation also recommends exhausting any automatic dryer to the outside to prevent the "atomizing" of bacteria released from fabrics during drying into the room.[1] Manufacturers of both gas and electric dryers supply accessories for venting. Ductwork to the outside of the house should be as short as possible

[1] "Sanitation in Home Laundering," *Home and Garden Bulletin No. 97.* Washington, D.C. U.S. Department of Agriculture, 1970.

Figure 21-3. Air flow in electric dryer of low-pressure type (pressure less inside than outside). Top rear location of heating element plus bottom rear location of fan provide for preheating of air before it enters drum. Dryer is rated at 5,600 watts, 24 amperes, 120/240-volt circuit. Note door-opening foot pedal at lower left. (Speed Queen, Division of McGraw-Edison Company)

and should have a minimum number of 90-degree turns. The drying efficiency is reduced with long ductwork or several sharp turns. Dryers should *not* be vented through a chimney.

Gas dryers require a 115-volt, 60-cycle current to operate the motor. The circuit should be fused with a 15-ampere fuse or a 20-ampere circuit breaker. Electric requirements for electric dryers are given on page 356 in connection with Underwriters' Laboratories requirements.

Good practice is to clean the lint trap after each load is removed. Articles cleaned with any dry-cleaning solvent should *not* be dried in a household dryer. Articles that contain foam rubber or are made of rubberlike materials should not be dried in a dryer. (This injunction does not cover normal amounts of elastic in clothes.)

Fabrics should not be over-dried. To check moisture content remove an article from the dryer. Clothes and household textiles dry more wrinkle free when the dryer is reasonably loaded. It is especially important *not* to overload for a permanent-press cycle since this helps nullify the permanent-press feature of the dryer. For any cycle the load should be such that articles can tumble freely.

Approximate operating costs per load may be figured from the following data: for an electric dryer, 4.4 kilowatt-hours[2] times the cost per kilowatt-hour in your community; for a gas dryer, 1/5 to 1/3 therm (100,000 Btu's) times the cost per therm in your community. (The cost of operating the motor might equal the cost of 0.1 kilowatt-hours in your community.)

FEATURES

The least deluxe models sold for home use in the United States are likely to have at least two cycles—regular load and permanent press. The most deluxe models have four or more cycles. In contrast, dryers made primarily for public use generally can be controlled for time and temperature only. Additional features are described in the captions for Figures 21-4 through 21-6, and are noted in the Buying Guide.

SPACE-SAVING MODELS

Models of large capacity (18 pounds or so dry load weight) are likely to be approximately 28 1/2 to 31 inches wide, although one or more manufacturers may supply a large-capacity model that is less than 28 1/2 inches wide.

Among the earliest and still available space-saving models is a matched washer

[2] Lucile F. Mork, "Figuring the Cost of Doing Laundry at Home," *Agricultural Research Service Family Economics Review,* Dec. 1970.

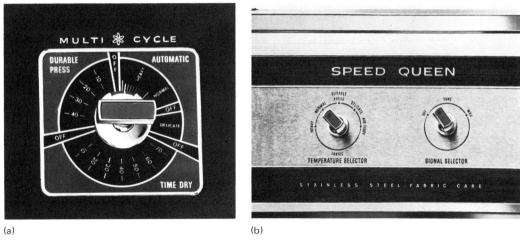

(a)   (b)

Figure 21-4. Controls for electric dryer shown in Figure 21-3. (a) Timer. Timer control provides an automatic cycle for heavy, normal, and delicate fabrics, a time-dry cycle with settings up to 75 minutes, and a durable-press cycle with settings up to 50 minutes. (b) Temperature selector and signal selector. Temperature selector provides suitable temperature for safe drying plus an air-fluff setting. Signal selector regulates volume of buzzer that signals completion of drying cycle. Selector may be turned to off if no signal is desired. (Speed Queen Division, McGraw-Edison Company)

Figure 21-5. Lint screen in door of gas dryer. Pilot is located under small hinged door on top of dryer. An instant ignition burner or a constantly burning pilot is provided. The 200 cfm fan is located at lower left rear part of dryer. (Speed Queen Division, McGraw-Edison Company)

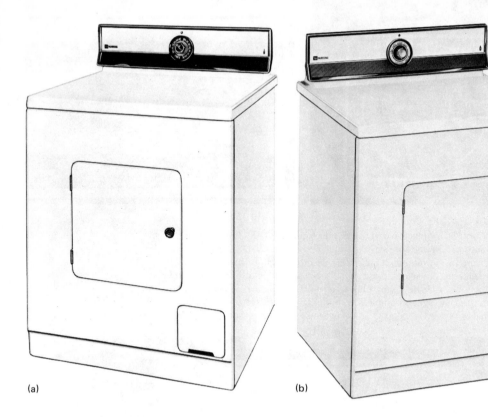

(a)   (b)

and dryer, each 27 inches wide. These may be installed side-by-side or stacked with the dryer above the washer (Fig. 21-7). Electric and gas dryers are available. The electric model uses 5,200 watts at 240 volts and 1,300 watts at 120 volts. The dryer connects to a three-wire, 120/240-volt circuit of 30-amperes capacity or a 120-volt appliance circuit. The gas model (not shown) has a rated input of 20,000 Btu/hr.

Figure 21-8 shows another compact laundry center. The dryer is mounted

Figure 21-6. Three gas models of one manufacturer. (a) Model with timed drying control set by user, permanent-press dry cycle; air-fluff setting, full opening door; flush-to-wall installation, adjustable locking leveling legs, low-temperature drying; and end-of-cycle clothes-conditioning period. (b) Model with electronic drying only; other features as listed for (a), plus end-of-cycle signaling, damp-dry setting, and operating signal light. (c) Top-of-the-line model with following features additional to or in place of those listed for (b): push-button rather than dial control; control panel light; and interior drum light. All models are rated for 18,000 Btu/hr input. (Maytag Company)

(c)

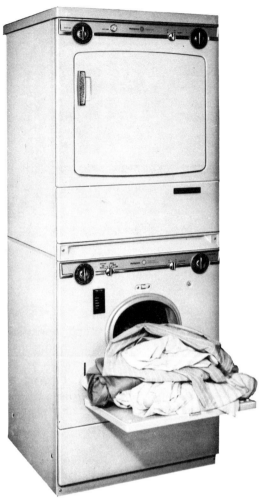

Figure 21-7. "Space-mates"—electric dryer mounted above cylinder (tumbler) type washer. Each appliance is 27 inches wide and 25 inches deep; stacked height is 70 inches. If installed under counter, height is 36 5/8 inches and width 54 inches. Washer has three wash and two rinse temperatures. Dryer has five cycles: regular, low, damp-dry for ironing, air fluff, and "auto-dry permanent press." Last cycle first dries at low temperature, next temperature is raised to 165° F, last is ten-minute cool down. (Westinghouse Electric Corporation)

Figure 21-8. Compact laundry center with dryer in same housing as washer. Dimensions are 24 inches wide, 27 1/8 inches deep, and 65 3/4 inches high. (Frigidaire Division, General Motors Corporation)

DRYER STANDARDS

in the same housing as the washer and is located above the washer. Heating unit wattage at 240 volts is described as 3,000/1,800 watts. Controls for the dryer and washer are shown in Figure 21-9 and the lint trap is shown in Figure 21-10.

Other space-saving models are shown in Figures 21-11 and 21-12 and are described in the captions. Both may be used as portable or fixed appliances.

Separate American National Standards Institute (ANSI) standards are given for electric home dryers[3] and family-type gas dryers.[4] The "family-type" gas dryers are called Type 1 and are covered by the standard just cited. Gas dryers intended pri-

[3] *C33.13-1968 Electric Home-Laundry Equipment, Safety Standard* (same as UL 560, June 1968).
[4] *Z 21.5.1-1968 Gas Clothes Dryers,* vol. I, *Type I Clothes Dryers, including Addenda Z21.5.1a,* 1969.

(a)   (b)

Figure 21-9. (a) and (b) Washer and dryer controls for compact laundry center. Both are mounted on the front of the dryer cabinet. The dryer has timed-cycle drying with special settings for permanent press, delicate, and no heat. The timer may be set for a drying time up to 110 minutes. (Frigidaire Division, General Motors Corporation)

marily for public use are called Type 2 and are covered by a third standard.

The UL and ANSI standard requirements for electric dryers cover construction aspects including power supply connections, location of heating elements, motor-overload protection, automatic temperature controls, grounding; performance characteristics including power

input, starting current, safety relative to becoming a fire hazard when operated under specified abnormal conditions; ratings; and markings.

The construction requirements on the automatic temperature controls specify the number of cycles over or through which the controls shall perform acceptably in endurance tests. Different numbers of cycles are specified for the temperature-regulating thermostat which controls the normal operating temperatures in the appliance and the temperature-limiting thermostat which prevents abnormal temperatures.

Effective December 1970 a new safety requirement was introduced:[5] namely, "a dryer shall be provided with a means to prevent opening of the door while the machine is in its normal operating cycle, or opening of the door to a maximum of 3 inches shall operate an interlock that removes the driving force from the tumbler and deenergizes the heat source within ten seconds."

The dryer shall be rated in volts and amperes or in watts (or kilowatts) when the overall power factor is 80 percent or more (as would be the usual case for an electric dryer).

Figure 21-10. Lint filter for dryer of laundry center is beneath dryer door opening. (Frigidaire Division, General Motors Corporation)

[5] Changes in the UL standard subsequent to July 24, 1968, are *not* part of the ANSI standard.

(b)

the cleaning agent (paragraph 214). (3) A warning shall be included in the marking on the dryer or in the operating instructions which shall consist of the following wording or its equivalent: "A. For a dryer which includes a "no-heat" setting in its controls. *Do not use heat when drying articles containing foam rubber or rubberlike materials.* B. For any other dryer. *Do not dry items containing foam rubber or rubberlike materials in this dryer.*"

(a)

Figure 21-11. Portable dryer is 24 inches wide, 33 inches high with casters and 32 inches high without casters, and 20 1/2 inches deep. (a) With casters removed, it can be placed on a counter or mounted on a wall. (Stacking rack shown in an accessory.) (b) Dryer may be installed next to washer. It uses a 115-volt appliance circuit. (Whirlpool Corporation)

New requirements relative to markings state: (1) If a dryer can be readily adapted for connection to a supply circuit of either of two different voltages, complete instructions for making the connections suitable for the different voltages shall be included in the permanent marking of the appliance (paragraph 213). (2) The dryer shall be plainly marked with a warning that it is intended for use only with fabrics which have been washed with water as

Figure 21-12. "Porta-dryer"—24 inches wide, 30 inches high (with casters), 15 inches deep, 77 pounds. Drying time selective up to 115 minutes. Used in "adequately wired circuit protected by 15-ampere fuse." Dryer may be mounted permanently. (Maytag Company)

Other requirements relative to markings specify: A dryer that provides means for collection of dust and lint shall be plainly marked to indicate the necessity for keeping the lint trap clean (paragraph 216). "Instructions describing the method of obtaining adequate venting for dryers that are designed for venting shall be permanently marked on the dryer or shall be included in the instruction book" (paragraph 223).

The requirements for family-type (Type 1) gas clothes dryers cover construction, performance, and definitions. Some construction requirements of special interest to consumers are given below.

The construction of all parts of a clothes dryer, whether specifically covered by the various provisions of the standard or not, shall be in accordance with reasonable concepts of safety, substantiality, and durability.

Dryers shall be provided with means for exhaust duct connection, or other means for disposal of moisture and lint.

A clothes dryer which requires more than the manual operation of a single control means external to the appliance to place it in service shall have clearly defined and complete instructions on the appliance for lighting and shutting down the appliance.

Clothes dryers shall be accompanied by printed instructions and diagrams adequate for their proper field installation and safe operation. On clothes dryers for installation in mobile homes or travel trailers, these instructions shall also specify that an exhaust duct to the outside must be installed and provisions made for the introduction of outside air into the dryer.

The rating plate shall give: manufacturer's name; the manufacturer's number of the appliance; the manufacturer's hourly Btu input rating; type(s) of gas for which tested—natural, manufactured, mixed, liquid petroleum, or liquid petroleum gas-air mixtures; type of gas for which equipped; the manifold pressure recommended by the manufacturer; and symbol of the organization making the tests for compliance with the standard.

Performance requirements of special interest include the following.

A clothes dryer shall produce no carbon monoxide. This provision is deemed met when a concentration not in excess of 0.04 percent is present in the air-free products of combustion when the dryer is tested in a room having approximately a normal oxygen supply.

Dials of the thermostats provided with temperature markings shall indicate a temperature within 20° F of the temperature of the air at the outlet of the drying chamber.

IRONS

Irons, historically known as electric flat irons, include dry irons; steam and dry irons; spray, steam, and dry irons; and travel irons. In the United States, the steam and dry iron and the spray, steam, and dry model are especially popular.

CONSTRUCTION

Steam and dry irons have a water chamber in which a small quantity of water is heated at a time to provide continuous generation of steam through holes in the

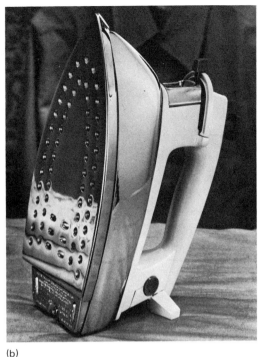

(a)   (b)

Figure 21-13. Iron controls and soleplate. (a) Iron controls, water-level indicator, and fabric-setting guide (under handle). Ahead of spray mist—shot-of-steam button is thumb-tip steam or dry switch and ahead of thermostat dial is thumb-tip temperature control. (b) Steam vents in soleplate and nameplate information. (Sunbeam Corporation)

soleplate. Spray, steam, and dry models also have at least one extra control to supply a wet spray or mist at the front of the iron. Some have an additional extra control to supply extra steam as a "shot" or a jet of steam (Figs. 21-13 and 21-14). The irons convert to steam or dry by operating the steam-dry control.

The soleplate is smooth and made of a rust-resistant material or has a rust-resistant finish. Aluminum alloys often are used—sometimes with a Teflon finish. Stainless steel and chrome-plated steel also are used.

An especially important characteristic for ease in use is the fit of the iron to the hand of the user and a prospective purchaser should lift and test the balance of an iron in her hand. Handles and controls are made of Bakelite or other heat-insulator-type materials. Most standard irons

have cords attached at the rear or on the right side at the rear. The right side place of attachment usually is preferred by household equipment students for ironing from right to left with the right hand. A left-handed person is likely to prefer that the cord be attached on the left side. A connection for left-side attachment of the cord is provided under the disk at the rear of the handle for the model shown in Figure 21-14.

Weight of iron in pounds is given on the specification sheet. The best weight to select is a matter of personal preference. The weight given usually is exclusive of the weight of added water for a steam and dry iron. One-half to three-fourths cup of water is a common capacity. These quantities of water are likely to suffice for 25 to 45 minutes of steam production at the lowest setting. Some models have ad-

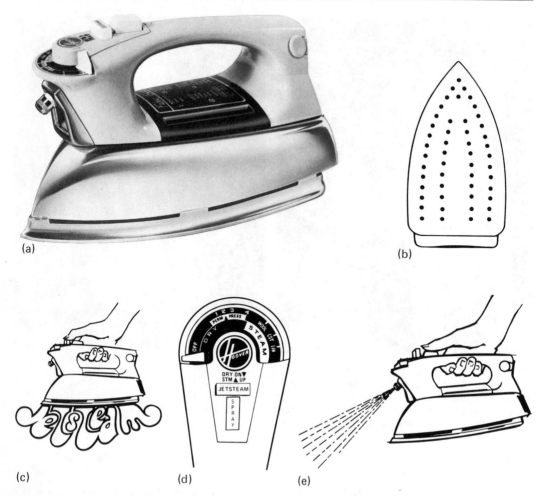

Figure 21-14. (a) Spray, steam, and dry iron with jetsteam. (b) Scratch-resistant stainless steel soleplate for smoother ironing. Bevel-edge helps iron under buttons with ease. (c) Pressing a button releases extra steam for loosening wrinkles. (d) Dial settings for different fabrics. (e) Push-button spray. (Hoover Company)

justable steam volumes according to the setting of the steam control (See Experiment 4, at the end of the chapter.) Irons are tilted to the rear on their heel rest to raise the soleplate off fabric or ironing board. A purchaser should check that the iron is stable in this position.

Irons have a single element mounted above or in the soleplate. This is either nichrome ribbon wrapped around mica sheets or a tubular-type element. Standard-size irons usually are rated at approximately 1,200 watts, 120 volts, alternating current only, and are thermostatically controlled to give different temperatures of the soleplate according to dial setting.

If thermostatically controlled irons designed only for alternating current are used on a direct current supply, the "points" of the thermostat switch that open and close the electric circuit are likely to fuse together and make the thermostat inoperative. Direct current is not common in the United States but occasionally it is used in hotels in large cities.

Travel Irons

Dry travel irons and spray, steam, and dry travel irons are available (Figs. 21-15 and 21-16). Both models illustrated are described as weighing less than 2 pounds. The translucent water bulb is removable.

See-thru plastic bulb holds 20-mins. of steam ironing time.

Handle folds for easy packing in attache case or overnight bag.

Comes with quality zipper travel bag.

Lightweight—only 1¾ lbs.

Figure 21-15. Portable spray, steam, and dry travel iron for use with 120 volts AC. Plastic bulb is described as holding water for 20 minutes of spray and steam coverage. (General Electric Company)

The model shown in Figure 21-15 is rated at 650 watts, 120 volts, AC only. The model shown in Figure 21-16 is rated at 700 watts, 120 volts AC or DC, 230 volts AC only. The 120-volt DC rating permits the iron to be used in those occasional United States hotels referred to in the previous paragraph; the 230-volt AC rating makes the iron widely useful in countries where 220 volts is the common supply voltage for portable appliances.

USE AND CARE

Always unplug the iron after use. Allow it to cool before storing and store on the heel rest with the cord wound loosely around the iron.

Do not use an abrasive on the soleplate.

Although some manufacturers state that tap water can be used in steam and dry irons, the iron will not be useful for as long as it would be if only distilled or demineralized water is used. Mechanically softened water or water to which a packaged softener is added contains minerals and on that account is not recommended for use with steam irons.

Follow the manufacturer's instructions with regard to care of the spray nozzle. Also follow the manufacturer's instructions relative to use or nonuse of commercial cleaners for the water chamber of an iron.

Many loosely woven and lightweight washable fabrics and many woolens can be steam ironed successfully when dry. Fabrics such as polished cottons, silks,

For domestic or overseas use
on 120 Volts AC/DC and
230 Volts AC—adaptor
plugs included.

See-thru plastic bulb holds
20 mins. steam ironing time.

Handle folds for easy packing
in attache case or overnight
bag.

Comes with quality zipper
travel bag.

Figure 21-16. Portable spray, steam, and dry travel iron for use with 120 volts, AC or DC, or with 230 volts AC. Note adapter plugs for use in different types of receptacle outlets. (General Electric Company)

or rayons that water spot should not be steam ironed when dry. Also do not steam iron dry materials which are more successfully ironed when dampened; these would include materials laundered with a vegetable starch, most linens, heavy cottons, and some medium and heavyweight rayons. When instructions are not given on finished garments or on yard goods for special fabrics, experiment with a small patch to determine whether steam or dry ironing is better.

To save time, iron first the articles made of fabrics that require the lower heat settings. Iron each part of an article dry before proceeding to another part, to avoid creases and wrinkles. Do not rearrange an article unnecessarily. As far as possible, utilize the full length of the ironing board —moving the iron smoothly with the grain

of the fabric. Press, that is, move iron up and down lightly, articles such as knits that might stretch.

Ironing is facilitated by use of a flat, well-padded, stable ironing board. The surface of the board should be slightly above elbow height. (Homemakers might well find ironing less fatiguing if they experimented with the height.) Padding should be fluffed occasionally in a no-heat cycle of the dryer. The cover should be smooth and tight on the board.

IRON STANDARD

Construction and performance of irons is covered by the following standard: UL 141. *Electric Flatirons and Ironing Machines,* April 1954. Reprinted with revisions included January 1971.

Special requirements covering the steam feature are not given in this standard. With one exception, the general safety requirements on soleplate temperature for an iron as received at Underwriters' Laboratories *and* after specified abuse (dropped in specified ways five times) is that the maximum soleplate temperature shall not be more than 350° C (662° F) during an initial interval of operation consisting of the first "on" period plus the first five minutes following the first thermostat cut-off.

Effective February 1, 1971, the length of the cord shall be no less than 8 feet. Effective January 14, *1974*, special safety features are specified for an iron that may be folded for storage or for traveling convenience.

EXPERIMENTS

Experiment 1. Operating Characteristics of Dryers

1. Drum temperature of gas and electric dryers for different fabric cycles: Measure temperature with the dryer empty and the control set for different heat or automatic control settings. Use a thermocouple connected to an indicating instrument. Temperature measurements should be taken at intervals of one minute until maximum temperature is reached. The thermocouple junction is placed in the drum in a way to respond to air temperature only and the door of the dryer is closed. An indicating instrument such as a Temco pyrometer is suitable.

2. Electric energy, cycling times, and efficiency of electric dryer for a load of terry towels:

    a. Connect a 240-volt electric dryer in a circuit with two ammeters so that current can be measured separately in each "leg" of the three-wire supply. Measure voltage of each leg separately or measure voltage of "240-volt" supply as convenient. Connect a 120-volt dryer in a circuit with one ammeter and a 150-volt voltmeter.

    b. Use *wet* load of terry towels that has a 10-pound dry weight and a known weight of water. It is convenient to use a load that has been put through a complete washer cycle. For the efficiency determination assume that the initial temperature of the water in the wet load is the temperature of the rinse water of the washer.

    c. Set the dryer control for automatic dry or for a time that is long enough for complete drying.

    d. Record amperes and volts during the dry cycle. Measure with a stop watch or an electric timer, the length of time the heating element is "on" in order to be able to calculate total watt-hours. *Note:* Calculation of watt-hours from measurements of amperes and volts is reasonable for electric dryers since the power factor is nearly one.

e. Remove load from dryer and weigh to determine pounds of water removed.

f. To calculate efficiency, assume that the water evaporated at the maximum temperature reached in the dryer for setting used. (This temperature has to be determined in a separate experiment.)

$$\text{eff} = \frac{\text{wt of water removed } (970 + \text{av final temp} - \text{initial temp}) \, 100}{3.411 \times \text{watt-hours used by dryer}}$$

where eff is efficiency in percent

wt of water removed is in pounds

970 is the approximate value for heat of vaporization of water in British thermal units

final temp is average final air temperature in °F in drum

initial temp is temperature in °F of rinse water used to wet load

3.411 is the number of British thermal units equivalent to one watt-hour

3. Do step 2, electric dryer, with a 5-pound load of pillowcases that have been put through an easy-care cycle in the washer.

4. Measure cubic feet of gas and, if appropriate, cycling characteristics for a gas dryer with a load of terry towels. Calculate efficiency from the equation given for an electric dryer *except* use Btu input in the denominator: cu ft of gas times Btu per cu ft.

5. Do step 4, gas dryer, with a load of pillowcases.

Experiment 2. Overdrying and/or Overloading

1. Observe wrinkles in overdried loads by placing the dried articles on wood horses in the laboratory and looking at them "straight on" and tangentially.

2. Observe wrinkles when an excessively large load is used.

3. Compare the appearance of the articles with that of similar ones properly dried and, where appropriate, hand smoothed on a flat surface.

Experiment 3. Special Uses of Dryers

Check the user's booklet for instructions on such special uses as freshening draperies, drying a blanket, drying feather and/or Dacron-filled pillows, drying knits of cotton and man-made fabrics, drying permanent-press shirts. Try these special uses with articles brought from home. Assess results.

Experiment 4. Operating and Use Characteristics of Irons

1. Measure temperature at center of soleplate for three different thermostat settings with a surface-temperature measuring instrument such as an Alnor pyrometer. Use a low setting first, then a

medium setting, and finally a high setting. Take measurements at intervals of one-half minute for a long enough time so that you can plot the pattern of temperature versus time for each of the three thermostat settings.

2. Operate the iron as a dry iron and make a scorch pattern at the highest setting as follows: place a blotter on a thick piece of asbestos paper, then place the iron with soleplate down on the blotter and operate the iron on high setting for three minutes.

3. Add distilled water to iron and heat the iron. Check whether steam output is increased as the control is moved to settings higher than the lowest steam setting. Iron fabrics that are temperature-sensitive at different steam settings to determine whether scorching will occur while steam is present.

4. Evaluate ease of use for ironing flat work with long strokes and transfer of iron from right to left hand at end of stroke. As nearly as possible use the full length of the ironing board in this operation.

5. Evaluate ease of use and satisfactoriness of results for *pressing* knits and double thicknesses of fabric.

BUYING GUIDE

DRYERS

Before getting information on various models, a careful purchaser might well consider whether: (1) she or he wants the dryer to "match" (take the load of) a washer of moderate or large capacity and whether the dwelling in which it will be used has the necessary heating source—a 120/240-volt individual circuit or a gas supply—and suitable space; (2) she wants a portable and/or a space-saver model.

General

1. Does it have the Underwriters' Laboratories seal for an electric dryer or the American Gas Association and Underwriters' Laboratories seal for a gas dryer?

2. What is the manufacturer's name and address? What is the model or catalogue number of the dryer?

3. Dimensions: What is the width, depth, height in inches? Depth with the door open 90 degrees?

4. Installation requirements: Is the dryer designed for flush-to-wall installation? Gas dryers generally will be vented. If the local code permits installation without venting and if you do not propose to vent, ascertain the AGA recommendation on inches of clearance behind an exhaust deflector.

5. Does it have leveling feet or glides?
6. What material(s) are used on the top, sides, and drum?
7. What is the wattage(s) at rated voltage or Btu/hr rating? Horse-power of motor? (A 1/6-horsepower motor is used with many compacts and a 1/4- or 1/3-horsepower motor with standard and large-size models.)
8. What is the cubic foot capacity of drum—8, 6, 5, other?
9. What is the cubic foot per minute rating of the fan or blower—200, 175, 135, other?
10. What arrangements are available for servicing?
11. What does the warranty promise?

Operating Characteristics

1. What is the pattern of air movement inside the appliance. Is a negative pressure system used? That is, is the blower so located that the pressure in the drying chamber is lower than that out-side? (This is stated to be an aid for keeping lint out of the room.)
2. If a gas dryer, what mechanism is used for lighting the main burner: constant-burning pilot, electric ignition, or another?
3. What special cycles are provided and what are the characteristics of each? For example, what temperature(s) are used for drying in a permanent-press setting? What is the time allowed for cool-down (drum rotating but no heat)? What spin speed is used for each cycle?
4. Are timed *and* automatic drying provided? If automatic drying is provided, what mechanism is used for controlling degree of dry-ness? Is any special provision made for small loads?
5. Are superspeed or other similar settings provided? If so, how do they differ from the regular setting?
6. Is a sanitizing cycle or (sanitizing) modification of a regular cycle provided? What is involved?

Convenience and Other Special Features

1. Is there a cycle-end signal?
2. Full-width console light?
3. Interior light?
4. Adjustable drying rack?
5. Others?

ELECTRIC IRONS

1. What are voltage and wattage or voltage and current ratings? Note particularly whether alternating current only is specified on the nameplate. Is a model or catalogue number specified? Are manufacturer's name and address given? Does the iron carry the Underwriters' Laboratories seal of approval?

2. Observe overall construction features.
   a. Is the handle easy to grasp?
   b. Is the thermostat dial operated easily? Is the dial located so that the user will not touch the hot iron when turning the dial?
   c. Lift the iron by the handle. Does it balance well in the hand or does it tip forward or backward?
   d. Is it convenient to tilt the iron?
   e. How many pounds does the iron weigh?
   f. Is the cord attached at a convenient location for your ironing habits? Is a protector provided at the place of attachment of the cord? Is the cord plug large enough to grasp easily when the iron is connected or disconnected from the power supply?
   g. If the iron is a steam-dry type, how is water added? If possible, add water to determine how easy or difficult it is to do this. Notice the pattern of the steam openings in the soleplate. What is the water capacity? What is the claimed minimum steaming time between fills? Does the user's manual give instructions for adjustable steam?
   h. Does the iron have a spray control? An extra-shot-of-steam control? Does the user's booklet give information on use of these controls?
3. What kind of warranty is provided?

SUMMARY

The satisfied consumer has informed himself (herself) of what is available and to the extent he is able, how well various kinds and models of equipment will perform for his family's living pattern. The concerned consumer may work to increase the information manufacturers routinely make available. Further he will use equipment to conserve the world's energy supply and add the minimum pollution to the world's environment.

INDEX

November 1980

3